Leading the Change:

Johns Hopkins Medicine from 2012 to 2022

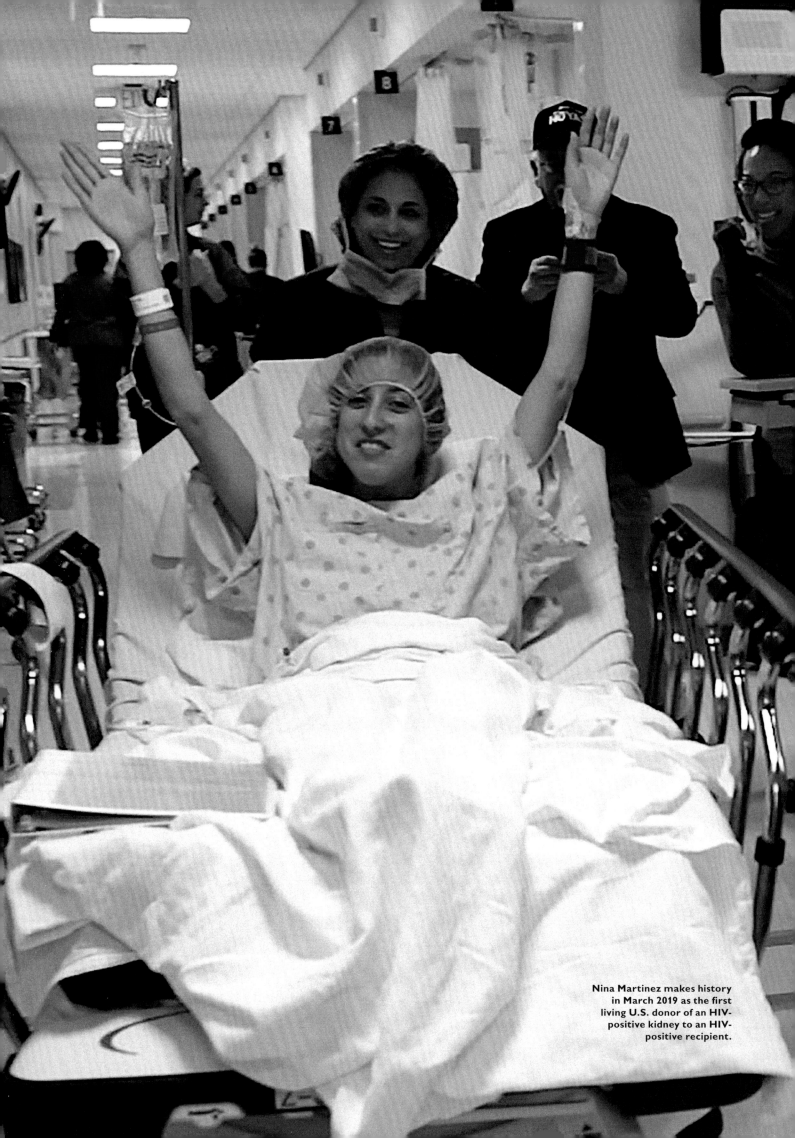

Nina Martinez makes history in March 2019 as the first living U.S. donor of an HIV-positive kidney to an HIV-positive recipient.

Leading the Change:
Johns Hopkins Medicine from 2012 to 2022

KAREN NITKIN

Johns Hopkins Medicine in association with
Johns Hopkins University Press
Baltimore, Maryland

9 8 7 6 5 4 3 2 1

www.hopkinsmedicine.org

Printed in the United States by Schmitz Press

Johns Hopkins Medicine
Marketing and Communications
901 South Bond Street
Suite 550
Baltimore, Maryland 21231-3359

Distributed by Johns Hopkins University Press
2715 North Charles Street
Baltimore, Maryland 21218
www.press.jhu.edu

Library of Congress Control Number: 2023932375

ISBN 978-1-4214-4801-5 (hardcover)

Composed in Adobe Garamond Pro (designed by Robert Slimbach), based
on the roman types of Claude Garamond (c. 1480–1561) and the italic types
of Robert Granjon (1513–1589), and Gill Sans MT Standard, based on the
famous set of humanist sans serif fonts originally designed by Eric Gill and
the Monotype Type Drawing Office, first appearing in 1928.

Editor: Sue De Pasquale

Director of Communications and Design: Michael Keating

Production Manager: Maria Patterson

Copy Editors: Justin Kovalsky, Victoria Banks, Karen Blum

Cover Photo: Keith Weller

Book Design: Max Boam

Dedication

To **Paul B. Rothman**, dean of the school of medicine and chief executive officer of Johns Hopkins Medicine from 2012 to 2022; **Theodore DeWeese**, interim dean/CEO; and **Kevin Sowers**, president of the Johns Hopkins Health System and executive vice president of Johns Hopkins Medicine. Together, they provide the strong leadership, moral compass and strategic vision that make Johns Hopkins Medicine extraordinary.

This book is also dedicated to all the physicians, researchers, nurses, residents, interns, fellows, employees, volunteers and trustees of Johns Hopkins Medicine, as well as to the patients, family members and residents of the surrounding communities.

The Johns Hopkins Hospital was along the flight route for a flyover in May 2020 that paid tribute to those on the front lines of the COVID-19 response.

Table of Contents

Paul B. Rothman

Introduction

A Time of Growth and Change

HEN PAUL B. ROTHMAN became dean of the medical faculty and CEO of Johns Hopkins Medicine on July 1, 2012, he took the reins of a storied institution in the midst of tremendous growth and change.

The health system had recently added three community hospitals: Sibley Memorial Hospital, in Washington, D.C.; Suburban Hospital, in Bethesda, Maryland; and All Children's Hospital in St. Petersburg, Florida.

Two massive clinical towers were about to open on the East Baltimore campus, the Sheikh Zayed Cardiovascular and Critical Care Tower and The Charlotte R. Bloomberg Children's Center.

The Genes to Society medical school curriculum, launched in 2009, taught future clinicians to recognize patients as individuals whose health is impacted by their genetic makeup and accumulated experiences.

At the same time, the health system was beginning the complex process of moving to Epic, the electronic medical records system, creating the infrastructure for a big data revolution that was just beginning, one that piggybacks on the Genes to Society idea by helping clinicians understand the complexities of each patient's health and respond accordingly.

As just the second dean/CEO of Johns Hopkins Medicine (JHM), Rothman ushered in a generational change, following the leaders who had brought the health system and medical school together to create the governing structure known as JHM in 1996.

Kevin W. Sowers would soon become the second president of the health system and executive vice president of Johns Hopkins Medicine, Robert Kasdin would join JHM as its first senior vice president and chief operating officer, and legal matters would be helmed by G. Daniel Shealer Jr., followed by Annemarie Martin-Boyan.

On March 10, 2022, Rothman announced he would retire on June 30 — 10 years, to the day, after becoming dean and CEO.

Theodore DeWeese, vice dean for clinical affairs and president of the Johns Hopkins Clinical Practice Association, began serving as interim dean and CEO on July 1, 2022.

"I have long envisioned myself as a 10-year dean/CEO," said Rothman, who was 64 at the time of his announcement. "A decade felt like the right time horizon to help advance the mission of JHM," he continued in a letter to JHM colleagues.

> *"That vision was crystallized by the COVID-19 pandemic, which demanded so much of our institution and our community. Two years later, I believe that we have navigated the worst of the pandemic, and it is time for a new leader to guide us forward."*

> *"The pandemic took a devastating toll on patients and their caregivers, the community, and our colleagues — both professionally and personally. It tested our strength and resilience. Lives were lost. Lives were changed. The challenges continue. However, we are optimistic and hopeful about what lies ahead."*

Throughout the pandemic, JHM responded with both expected strength and surprising nimbleness, setting itself apart as a beacon of light through a time of confusion and darkness. Under the leadership of Rothman and Sowers, Johns Hopkins provided caring, evidence-based treatment and clear-eyed information.

In the first days of the pandemic, Johns Hopkins Medicine set up a Unified Incident Command Center with experts and leaders from across the health system and university, who would navigate rapidly changing conditions and information to craft compassionate, science-based decisions and policies.

To limit the spread of the virus, leaders expanded the use of telemedicine and at-home support for patients, virtual learning for students, and remote work for employees.

Speed was of the essence in the scramble to learn about the new virus and how to contain it.

Clinical microbiologists Karen Carroll and Heba Mostafa created an in-house COVID-19 test that gave results in hours instead of days. Facilities teams quickly converted regular patient rooms and floors to ones with negative air pressure to contain the highly infectious coronavirus.

Researchers went into high gear to learn all they could about the disease, focusing on how to predict disease trajectory and tailor treatments to individuals.

Rothman and Sowers teamed with community partners to bring COVID-19 testing, information and vaccines to senior citizens, people with disabilities and other populations with barriers to health care.

Rothman said two principles guided his COVID-19 leadership: follow the science, and do everything possible to keep patients and staff safe.

"The decisions we made were hard, but I think we did the right thing," he said. "If you're always guided by the values that we're guided by — honesty, integrity, making sure we're improving the health of our community — ... If you follow that North Star, you're never really going to go wrong."

A Legacy of Leadership

Rothman arrived at Johns Hopkins Medicine from the University of Iowa and led the institution through a time of immense innovation and integration.

As Rothman shepherded the institution into a more integrated era, one of his first actions was to bring in about 150 stakeholders to participate in the development of JHM's first comprehensive five-year strategic plan, using a "bottom-up" approach to ensure buy-in across the sprawling organization.

The resulting plan, launched in June 2013, had six priority areas:

- **People:** Attract, engage, develop and retain the world's best people.
- **Biomedical Discovery:** Become the exemplary model for biomedical research by advancing and integrating discovery, innovation, translation and dissemination.
- **Patient- and Family-Centered Care:** Be the national leader in the safety, science, teaching and provision of patient- and family-centered care.
- **Education:** Lead the world in the education and training of physicians and biomedical scientists.
- **Integration:** Become the model for an academically-based, integrated health care delivery and financing system.
- **Performance:** Create sustainable financial success and implement continuous performance improvement.

Note that people are first.

"This institution is so mission-driven, and people here have such a great set of values that it allows us to attract people who come here to make an impact on the broader community, rather than just themselves," Rothman said in 2022. "I think that's the key to our success."

Rothman led creation of the Office of Diversity and Inclusion and the Office of Well-Being to foster an environment in which all clinicians, researchers, nurses and staff can thrive personally and professionally. He guided the successful appointments of 30 stellar interdisciplinary faculty through Johns Hopkins University's Bloomberg Distinguished Professorships Program.

He helped to create a more diverse class of medical students through an initiative, launched in 2022, that aims to replace education loans with scholarships.

The precision medicine initiatives Rothman championed are transforming medicine by using data from Epic and other sources to better understand how diseases vary by individual and how treatments can be tailored to specific patients.

Johns Hopkins Medicine now supports dozens of precision medicine centers of excellence, which are breaking new ground in the diagnosis and treatment of diseases, including Alzheimer's, cystic fibrosis, prostate cancer and COVID-19, and in ensuring that all patients get the treatment that is right for them.

He supported creation of the Catalyst and Discovery Awards programs to help early-career faculty and cross-divisional teams, and championed the Johns Hopkins Medicine Clinical Awards for Physicians and Care Teams to support excellent clinicians.

Theodore DeWeese: From Precarious Childhood to the Highest Perch

Theodore DeWeese is as surprised as anyone else at the turns his life has taken.

Having grown up in public housing in the early years of his life, he is now at the pinnacle of one of the world's leading academic medical centers.

"I'm the luckiest person in this room, I can guarantee you," DeWeese said during a March 2022 Town Hall, shortly after he was named interim dean of the Johns Hopkins University School of Medicine and CEO of Johns Hopkins Medicine.

DeWeese, the Sidney Kimmel Professor of Radiation Oncology and Molecular Radiation Sciences, joined the medical faculty in 1995 and was most recently vice dean for clinical affairs and president of the Johns Hopkins Clinical Practice Association.

It was a long way from a childhood he described as precarious. When DeWeese was three, his father died, leaving "my mom, two sisters and me with no house, no money, no job and no food," he said.

The family moved into public housing in Denver, where they were one of the few families of European descent living there.

DeWeese got to know neighbors who had little, but shared what they did have. In particular, a mother of six who lived next door, an immigrant from Mexico, would bring DeWeese's family plates of delicious beans and tortillas, he said.

Nobody in his family had gone to college, and DeWeese didn't plan to, either. After high school, he became a mechanic and auto parts person at a Ford dealership in Denver, and raced motorcycles. With encouragement from his mother and others, he went to what is now Metropolitan State University of Denver, where he took a chemistry class that ignited his interest in science.

DeWeese had mentors who believed in him, and he said he couldn't help but notice that his Black and brown friends didn't get the same support. "I didn't fully understand at the time, but I understand now that the lack of this same support was, in fact, a form of structural racism," he said during the Town Hall. "It was a valuable lesson to me — the importance of how you give a hand up, and to whom."

After medical school at the University of Colorado, De-Weese came to Johns Hopkins for a residency in radiation oncology and then pursued a laboratory research fellowship in cancer biology before joining the faculty.

When the school of medicine created a new Department of Radiation Oncology in 2003, DeWeese became founding director and professor. In 2019, he stepped down as director of the department and as radiation oncologist-in-chief of The Johns Hopkins Health System to become the vice dean of clinical affairs.

Recognized as an international expert in the health management of men with prostate cancer, DeWeese has conducted multiple clinical trials that have sought to improve the quality of the lives of men with the disease. This work has included several "first-in-man" clinical translations of novel therapies.

"If we keep our patient right at the center of what we do, [then] our research, our care of patients, and our education of trainees will always be focused in the correct way — for Baltimore, our region and the world," he said.

To foster innovation, Rothman helped marshal philanthropic support to launch several important institutes, including the Bloomberg-Kimmel Institute for Cancer Immunotherapy, the Skip Viragh Center for Pancreatic Cancer Clinical Research and Patient Care, the Armstrong Institute for Patient Safety and Quality's Center for Diagnostic Excellence, the Cochlear Center for Hearing and Public Health, and the Center for Psychedelic and Consciousness Research.

Johns Hopkins Technology Ventures, which started in 2014, represented a new era for Johns Hopkins inventors — and the many people who have benefited from the technologies, products and services these researchers have been able to bring to market.

In its first seven years, JHTV fostered more than 170 commercial ventures based on the discoveries of Johns Hopkins faculty. One of those discoveries was CancerSEEK, a noninvasive blood test developed by Ludwig Center co-directors Bert Vogelstein and Ken Kinzler, which screens for eight common cancer types.

During Rothman's tenure, doctors performed The Johns Hopkins Hospital's first bilateral arm transplant and a world-first penis and scrotum transplant. Genetic medicine professor Gregg Semenza won the 2019 Nobel Prize in Physiology or Medicine.

Looking Ahead

As Johns Hopkins Medicine positions itself for the future, it is following its "Clinical Road Map" that is expanding outpatient offerings and strengthening referrals to specialized care. Green Spring Station Pavilion III, which opened in 2019, is an important part of that plan, as is the current development of Belward Farm, in Montgomery County, Maryland.

In 2019, the health system and university unveiled an ambitious project to renovate the former Johns Hopkins Hospital Children's Medical and Surgical Center (CMSC) and construct a 12-story tower on the site of the former Brady Building.

The project has been strongly supported by Sowers, who succeeded Ronald R. Peterson, now president emeritus of The Johns Hopkins Health System.

Rothman, self-effacing by nature, said the successes of Johns Hopkins Medicine over the past 10 years are due to its talented and caring people, "the very best that mankind has to offer," as Rothman said in the March 10 letter to colleagues.

Bill Conway, chairman of the Johns Hopkins Medicine Board of Trustees, and Ronald J. Daniels, president of The Johns Hopkins University, said in a March 10, 2022, letter to the Johns Hopkins community:

> *"Paul has led Johns Hopkins Medicine through a remarkable 10 years as it continued its preeminence in education, discovery and patient care and served as the nation's — and indeed, the world's — most trusted resource during the COVID-19 pandemic.*
>
> *"It is Paul's humanity, humility and willingness to do 'whatever it takes' that has kept our patient care and research missions moving forward. This was never more evident than in his steadfast and humane leadership during the pandemic. Whether briefing policy makers or leading the quick and safe reopening of Johns Hopkins clinical services, Paul's commitment to caring for others, rooted in the best science, helped us all navigate the rapid and ever-evolving virus and its variants, from alpha to omicron.*
>
> *"And we will never forget the time in the wake of Hurricane Sandy when Paul donned waders, rolled up his sleeves, and worked cheek by jowl with colleagues to move research materials out of the flooded basement of our research building in East Baltimore, salvaging countless hours of work and ensuring that potential discoveries were not lost to the elements. In moments like these, and so many others, Paul demonstrated the essential ethos of persistence, collaboration and care that define Johns Hopkins Medicine."*

Paul Rothman (center) and wife Frances Jane Meyer in February 2012, talking with (left to right) Johns Hopkins University President Ron Daniels, Johns Hopkins Medicine CEO Ed Miller, and Ron Peterson, executive vice president of JHM and president of the Johns Hopkins Hospital and Health System. Rothman visited Johns Hopkins several times before becoming dean/CEO on July 1, 2012.

Chapter 1

New Leader, New Strategy — The Paul Rothman Era Begins

PAUL B. ROTHMAN WASN'T LOOKING FOR A NEW JOB when a recruiter telephoned him seemingly out of the blue on a summer day in 2011.

He had been dean of the Carver College of Medicine at the University of Iowa and leader of that university's clinical practice plan for three years. He liked it there.

What's more, he and wife Frances Jane Meyer, a clinical gastroenterologist on the Carver College faculty, had promised their three children they wouldn't move while any of them were in high school.

But Rothman was intrigued when he learned that the job in question was one of the most prestigious and important in academic medicine: dean of the Johns Hopkins University School of Medicine and chief executive officer of Johns Hopkins Medicine.

"This job is unique in several ways that were very attractive to me," Rothman said in a 2021 interview. "One is that it's Johns Hopkins, the leading academic medical center in the world. Iowa has great research and is a great medical center, and we loved it there. But Hopkins is just bigger, with a broader scope and more impact.

"Another difference was that in Iowa the dean's role was limited to being able to impact the school of medicine, there called the college of medicine, and not the academic health center. This job had this role of not only being the dean of the school of medicine but also being able to strategically help the entire institution through the CEO role."[1]

Rothman was interested. And he realized he could keep his promise to his children. "We have three kids. That year, my middle son was graduating high school and going to college and my youngest son was entering high school, so technically none of them were in high school for a two-month period the summer of 2012. I said, 'Well, I won't break my promise to my kids,' and I agreed to look at the position."

In the fall of 2011, Rothman traveled to Baltimore several times for interviews with leaders of The Johns Hopkins University and Johns Hopkins Medicine. "The key thing was I got to spend time with [Johns Hopkins University President] Ron Daniels," Rothman said in 2021. "We were both in New York City for

Paul B. Rothman

Paul B. Rothman, a native New Yorker who grew up in Bayside, Queens, was a 1980 Phi Beta Kappa graduate of the Massachusetts Institute of Technology, where he was also captain of the varsity crew team.

He earned his medical degree from Yale University in 1984, then returned to New York to train at New York-Presbyterian/Columbia University Irving Medical Center in internal medicine and rheumatology and accepted a postdoctoral fellowship at Columbia University prior to joining its medical school faculty.

As a researcher at Columbia, Rothman investigated the signaling pathways required for allergic immune responses and cancer, a focus he maintained while at Iowa. His research focused on immune system molecules known as cytokines, and the role these molecules play in the normal development of blood cells, as well as the abnormal development of these blood cells that leads to leukemia.

He served from 2004 to 2008 as professor and head of the Department of Internal Medicine at Iowa, with secondary appointments in the Department of Microbiology and the Department of Physiology and Biophysics. In 2008, he became dean of the Carver College of Medicine at the University of Iowa and leader of that university's clinical practice plan.

His professional memberships included the American Association for the Advancement of Science, the American College of Physicians, the Council of the Association of American Physicians, the American Academy of Allergy, Asthma and Immunology Grant Review Committee, and the American Thoracic Society Asthma Immunology and Inflammation Program Committee.*

Though Rothman had not studied or trained at Johns Hopkins, he had a connection through Myron "Mike" Weisfeldt, a member of the search committee and the William Osler Professor and director of the Department of Medicine at Johns Hopkins. Weisfeldt was chair of Medicine at Columbia University early in Rothman's faculty tenure there and was impressed with his work as a physician, researcher and educator.

"I saw young, talented people flock to his laboratory, publish and progress in their careers," Weisfeldt said in December 2011. "I also saw his clinical commitment to excellence and heard reports of a flair for teaching. I knew he had it all, including leadership and administration."†

Weisfeldt had appointed Rothman to a divisional leadership post at Columbia and later nominated him for the Department of Medicine chairmanship he held at Iowa before stepping up to the Dean's Office there.

"His becoming dean at Iowa three years ago clearly occurred because of his excellence as chair," Weisfeldt said. "His characteristic traits throughout have been a commitment to excellence, charisma, hard work and collegiality."

Also on the search committee was Carol Greider, the Daniel Nathans Professor and director of the Department of Molecular Biology and Genetics at Johns Hopkins, and a winner of the Nobel Prize in medicine in 2009.

"Paul's work in the late '80s laid a foundation for our understanding of cell-to-cell communication in the immune system and its role in allergic diseases," Greider said in 2011.

Greider noted that Rothman had been named a Pew Scholar in 1992, recognition given to "the very top young biomedical investigators in the country." In that role, she said, he "developed molecular methods for dissecting the pathways that become dysfunctional in asthma and other inflammatory diseases.

"As a clinician," Greider added, "Paul has been able to translate his fundamental studies of cell signaling to models for modulation of the immune response to treat patients with severe allergic reactions. Paul personifies the goal of translational medicine; he has a true, deep understanding of the molecules and their subtle interactions that govern the immune system, while at the same time maintaining an overview of how these interactions play out in diseases such as asthma."‡

* O'Shea, Dennis, "A Brief Bio of Paul B. Rothman, M.D.," Johns Hopkins University press release, Dec. 19, 2011, https://releases.jhu.edu/2011/12/19/dean-ceo-bio/.

† "A New Leader for Johns Hopkins Medicine," *The JHU Gazette*, Dec. 19, 2011, https://gazette.jhu.edu/2011/12/19/rothman-calls-johns-hopkins-position-his-dream-job/.

‡ O'Shea press release, Dec. 19, 2011.

Thanksgiving, so we spent several hours together at a cafe. We had a lot of time to talk, and that was important for both of us."[2]

Rothman accepted the job in December 2011, and started July 1, 2012.

The formal announcement, made on Dec. 19, 2011, in the expansive 780-seat Turner Auditorium on the Johns Hopkins Medical Campus, was a reminder to Rothman that he wasn't in Iowa anymore. "The auditorium was packed, and cameras and the press were there," he said. "In Iowa, a press conference is when two guys from the *Iowa City Press-Citizen* show up, so this was a little different."[3]

A Time of Growth and Change

Rothman would be just the second CEO of Johns Hopkins Medicine. The first was Edward D. Miller, who was retiring after 15 years as both the School of Medicine's 13th dean and the first CEO of Johns Hopkins Medicine, an organizational structure that began in 1996 to ensure the school of medicine and the Johns Hopkins Health System were aligned in priorities, strategies and mission.

Rothman was taking the helm of a school of medicine with more than 2,500 full-time faculty members, nearly 1,300 part-time faculty, and more than 1,400 medical and graduate students, as well as more federal research support than any other U.S. medical school.

As CEO, he would also govern — in partnership with Ronald R. Peterson, then executive vice president of JHM and president of the Johns Hopkins Hospital and Health System — a system of six hospitals in Maryland; Washington, D.C.; and Florida. Johns Hopkins Medicine also comprises several suburban health care and surgery centers; more than 30 outpatient primary health care sites; a full-service home care provider; managed care plans; and hospital management, consulting and clinical education services in the Americas, Europe, the Middle East and Asia.

The East Baltimore campus was undergoing a massive expansion and transformation, with the $1.1 billion, 1.6-million-square-foot Sheikh Zayed bin Sultan Al Nahyan Cardiovascular and Critical Care Tower and The Charlotte R. Bloomberg Children's Center set to open in April 2012.

And the health system had expanded by three hospitals in as many years. Suburban Hospital in Bethesda, Maryland, had become a wholly owned subsidiary of the Johns Hopkins Health System on June 30, 2009. A year later, in May 2010, the system acquired Sibley Memorial Hospital in Washington, D.C.; and on April 1, 2011, it added All Children's Hospital, in St. Petersburg, Florida.

This gave more people access to a health system that for 95 years had been represented by a single hospital in a single city: The Johns Hopkins Hospital in Baltimore. The only previous sustained mergers had been Baltimore City Hospitals (later named the Johns Hopkins Bayview Medical Center) in 1984, and nearby Howard County General Hospital in 1998.[4]

But it also created new challenges for managing and integrating the growing health care footprint.

Miller described the seismic changes in a September 2011 column that appeared in *Dome*, a publication that provides news for faculty and staff members at all Johns Hopkins Medicine member organizations.

Construction of the Sheikh Zayed bin Sultan Al Nahyan Cardiovascular and Critical Care Tower and The Charlotte R. Bloomberg Children's Center. Bottom photo: Project executives Donna Brinkmeyer, Chris Smith and Richard Brill, and Facilities Senior Director Howard Reel watch a worker cut steel wall supports.

Farewell to Edward Miller

Rothman was just the second person to hold the CEO title at Johns Hopkins Medicine, the powerful administrative and governance structure created in 1996.

The first was Edward D. Miller, a star anesthesiologist who had been recruited from Columbia University in 1994 to revitalize the Department of Anesthesiology and Critical Care Medicine at Johns Hopkins. As a 2012 *Hopkins Medicine* magazine article explained, he wound up doing much more:

> The early 1990s was a time of upheaval on the East Baltimore campus. Those running the School of Medicine and the Hospital and Health System clashed bitterly over Hopkins' direction in a rapidly changing health care world. Faculty tensions ran high.
>
> When neither Dean Michael Johns nor Hospital President James Block could bridge their differences, both resigned, and a new governing structure — Johns Hopkins Medicine — was put in place. Henceforth, those historically separate entities would be run by the same boss.
>
> Miller, a neutral figure in that dispute, was asked to fill in as interim dean on March 1, 1996, while the trustees conducted a national search for Hopkins Medicine's first Dean/CEO. [Ronald R. Peterson was appointed interim president of the health system and hospital in September 1996.] Almost immediately, the easygoing anesthesiologist proved remarkably adept at healing wounds and resolving previously intractable problems.
>
> The trustees took notice. In January 1997, they announced that Ed Miller was exactly what they wanted: a unifier and consensus builder. Thus began the Miller era, a 15-year span marked by new buildings that transformed the campus, global outreach, regional expansions, an innovative medical curriculum, new departmental leadership, cross-disciplinary institutes, growing student and faculty diversity, and continued preeminence in research, teaching, and patient care.*

---

*   "View from the Top," *Hopkins Medicine* magazine, May 14, 2012, https://www.hopkinsmedicine.org/news/publications/hopkins_medicine_magazine/archives/springsummer_2012/view_from_the_top.

In pretty rapid succession, our school of medicine has grown in its research and educational impact, and our system has added three more hospitals: Suburban and Sibley Memorial in the D.C. area and All Children's Hospital in St. Petersburg, Fla. Johns Hopkins HealthCare has become a billion-dollar managed care business. Home Care has expanded and now also plays an important role in extending patient care and managing patient populations.

We have the largest primary care practice in Maryland with Johns Hopkins Community Physicians, and we've tripled our ambulatory care locations. Johns Hopkins Medicine International has expanded its reach globally. In short, we've outgrown ourselves.[5]

A reorganization known as JHM 3.0 would make the new, larger Johns Hopkins Medicine "more efficient and responsive," he wrote.

A key element of the new structure was an Operating Committee with representatives from across the enterprise, that would make decisions on strategic issues, such as the pending launch of Epic, an electronic medical records system.

The massive multiyear undertaking would unify the enterprise, improving patient care and eventually providing the information that would fuel big data and precision medicine research.

JHM 3.0 also created the Office of Johns Hopkins Physicians to ensure that as the JHM family grew, Johns Hopkins' faculty remained at its core and continued to thrive.[6]

It was a challenging time financially as well. Federal grant money for research had been trending downward, and the Affordable Care Act, enacted March 23, 2010, meant reduced reimbursements from Medicare and Medicaid, even as millions more Americans gained insurance. (The law went fully into effect in 2014.)

Changes in the hospital reimbursement rates from the state Health Services Cost Review Commission created additional uncertainty, as did costs associated with construction and maintenance of the new clinical towers.

Peterson enumerated some of those challenges in an April 11, 2012, column in *Dome*.

> *We are hearing from reliable sources on the federal level that reductions in the Medicare budget are likely. Since Medicare represents a significant percentage of our patient population, this is a concern for both our hospitals and for our faculty clinicians and other physicians who practice throughout our care delivery system.*
>
> *Then there is Medicaid, which jointly is funded by federal and state dollars, and there is pressure on both of those levels to make cuts. Already we are hearing that for the upcoming year, Maryland must approve additional Medicaid cuts. This would affect not only our hospitals — because Medicaid patients make up one of our sizeable patient populations — but also our managed care organization, Priority Partners, one of the largest in the state.*
>
> *We are being told that there will be either reductions or flattening of NIH funding, possibly furthering financial pressures on trying to cover the costs of research projects.[7]*

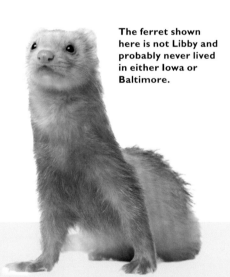

The ferret shown here is not Libby and probably never lived in either Iowa or Baltimore.

Libby, the World's First Cloned Ferret

Paul Rothman and his wife Frances Jane Meyer, a clinical gastroenterologist, arrived in Baltimore with their youngest son, a high school freshman; Arwen, a rescue dog that was part black Lab, part Australian sheep dog; and Libby, the world's first cloned ferret. Their older children were in college.

Rothman explained that when he left Columbia University for the University of Iowa in 2004, his children were 11, 10 and 6, and weren't happy about leaving New York. "I made a lot of promises at the time; one was that they could get a dog, a rabbit and a fish," he said in 2021.

"The dog was easy and we never got the fish." When they went to a local pet store to look at rabbits, they left with a ferret they named Milo. "He was the cutest, sweetest thing," said Rothman. "He was playful, and he used kitty litter." But ferrets are social and this one seemed lonely.

As it happens, ferrets are good models for human lung disease, and John Engelhardt, chairman of the Department of Anatomy and Cell Biology at the Carver College of Medicine, had created the first cloned ferret and then cloned one that he used for cystic fibrosis research.

"Engelhardt didn't need the first ferret, so he asked me if I wanted her," said Rothman. "We thought she would be a good friend for Milo. She outlived Milo, who died before we moved. Libby lived for a year in Baltimore and then passed."*

---

* Rothman interview, Oct. 13, 2021.

This April 2012 photo shows the transformed East Baltimore campus.

Campus Transformation

Between January 2012 and his official start in July, Rothman traveled to Baltimore every other week, to spend two days getting to know the people and the issues of the health system and school of medicine. "I had six months to get to know a lot about the issues so I could hit the ground running when I started July 1," he said.

Several months before his official start date, Rothman attended an event that symbolized the increasing size and sophistication of the institution he would soon lead.

On April 12, 2012, more than 1,000 people attended a lavish dedication ceremony for Johns Hopkins Medicine's new Sheikh Zayed bin Sultan Al Nahyan Cardiovascular and Critical Care Tower and The Charlotte R. Bloomberg Children's Center, one of the largest and most ambitious construction projects in Hopkins history.

Musicians and dancers entertained the audience between speeches from government officials, hospital leaders, and the towers' major benefactors: then New York City Mayor and Johns Hopkins University alumnus Michael R. Bloomberg and United Arab Emirates President Sheikh Khalifa bin Zayed Al Nahyan.

The two 12-story light-filled towers were constructed with patients and families, clinicians, medical students, and future generations in mind. They are, Peterson said that night, "the best of who we are, as a hospital, as a university as a community."[8]

George Dover

The moment was the result of years of planning and work that began with the 2002 update of the East Baltimore campus master plan. As Peterson noted during the 2006 groundbreaking: "Hopkins has been able to maintain its world-class reputation, despite the challenges associated with its aging infrastructure, but it's abundantly clear that we must now bring our buildings and supporting facilities up to modern standards."[9]

The $1.2 billion, 1.6-million-square-foot building represented the largest hospital construction project in the nation at the time. Combined, the towers have 560 all-private patient rooms, 33 expansive new operating rooms, state-of-the-art imaging equipment and new emergency departments for adults and children among other features.[10]

The Zayed Tower became the home of the Johns Hopkins Heart and Vascular Institute, with a full range of cardiovascular services; advanced neurological and neurosurgical services; transplant surgery; trauma care, orthopaedics; general surgery; and labor and delivery.

It has 355 private patient rooms, including 224 for acute care, 96 for intensive care, and 35 for labor and delivery. The rooftop of the Sheikh Zayed Tower has a heliport for patients who arrive by helicopter.

The Charlotte R. Bloomberg Children's Center is named in honor of Bloomberg's late mother. The state-of-the-art facility has 10 surgical suites designated exclusively for pediatric patients and 205 all-private patient rooms, including a 45-bed neonatal intensive care unit and a 40-bed pediatric intensive care unit.

"Every detail of our children's hospital is designed to enhance the care and comfort of our young patients and make it convenient for parents to stay at their child's bedside," said George Dover, who was director of Johns Hopkins Children's Center. "We have sleeping accommodations for parents in all of the patient rooms, as well as kitchen and laundry facilities on pediatric floors. Our goal is to make it easier and more comfortable for parents to be involved in their children's care."[11]

With healing gardens, soaring lobbies, a hand-picked art collection and cheerful, light-filled patient rooms, the new towers were designed to provide a welcoming and caring environment to advance the healing process.

Each patient room has a private bathroom and sleeping accommodations for family members. Sound-absorbing features in patient care corridors, ranging from acoustical ceiling tiles to a quiet nurse call system, contribute to the tranquil environment.

Oncology, pediatric, obstetric and medicine patients could now order meals on demand, and all patients could choose from a wide range of dining options customized to meet their nutritional needs and preferences. Also new were Balducci's Cafe, located in the Zayed Tower, and additional dining options in the Cobblestone Café.[12]

"For more than a century, patients have come to Johns Hopkins from throughout Baltimore, the nation and the world for the best possible, evidence-based, patient-centered care," said Miller in December 2011, as construction neared completion. "Our new facilities will enable us to provide that excellent care with

The Whimsical Sculptures of the Bloomberg Children's Center

The art and architecture of the new clinical towers were designed to soothe, heal and delight. The exteriors are enveloped in shimmering walls of fritted glass created by Brooklyn artist Spencer Finch, with colors, also chosen by Finch, inspired by the impressionist works of French painter Claude Monet.

The interior features some 500 works of carefully selected and custom-designed art from more than 70 artists from across the United States, the result of an unusual collaboration with artists, architects, an art curator, Bloomberg Philanthropies, patients, community members and the leadership and staff of Johns Hopkins.

The artwork includes 11 super-sized sculptures created by set designer Robert Israel, pictorial window shades, 300 paintings inspired by popular children's books, and more than 200 works inspired by nature.

"Through these exceptional artists and architects, we have created a unique space that incorporates art and design thoughtfully and with attention to detail," said Bloomberg.*

The whimsical sculptures include a 22-foot-tall, brightly colored ostrich that appears to float in the atrium, a family of pufferfish that seems to swim above patients and visitors, a cow with a nine-foot wingspan that flies above the main information desk, and a big rhinoceros with a little rhinoceros on its back that stands outside the entrance to Children's Center Emergency Department. †

* "Unveiling a New Standard of Care: The New Johns Hopkins Hospital Building," Johns Hopkins Medicine press release, April 12, 2012, https://www.hopkinsmedicine.org/news/media/releases/unveiling_a_new_standard_of_care_the_new_johns_hopkins_hospital_building.

† "Sculptures and Furniture," Johns Hopkins Hospital web page, https://www.hopkinsmedicine.org/the_johns_hopkins_hospital/about/enhanced_facilities/art_architecture/sculptures-furniture.html.

Moving Days

On Halsted 7, a kinetic sense of purpose filled the air, as caregivers on the cardiovascular surgical intensive care unit prepared to move an acutely ill mock patient to the unit's new home in the Sheikh Zayed Tower. His belongings were packed, his medications and IV fluids were ready, and his identification and "Ticket to Ride" were confirmed for transport.

Because the patient was sedated and on a ventilator, his move had been assigned to a clinician transport team from the Department of Anesthesiology and Critical Care Medicine. The team arrived at 11:20 a.m. and departed at 11:30 a.m. with the patient in his bed, along with a welter of medical devices and a small phalanx of clinicians to monitor his condition en route.

Everything was fine until the transport team reached the Halsted elevators, where it took a disconcerting minute to maneuver the patient, his ventilator, oxygen tank, monitor, IV pole and medical entourage into the deep but narrow space.

Emerging from the elevator, the transport team zigzagged through corridors to a connector between the old and new buildings. Without further incident, the group arrived at Zayed 5, where a receiving care team settled the patient into a sunny, yellow room.

Another "mock patient" move was complete.

Opening a new health care facility that encompassed the broad scope of the Sheikh Zayed Tower and The Charlotte R. Bloomberg Children's Center doesn't happen with a mere flip of a switch. Johns Hopkins Hospital administrators, clinicians, support staff and an outside consultant spent more than two years preparing for the move.*

The planning paid off on April 29 and 30, 2012, when 271 patients were transported to the new towers without incident.

"From my perspective, the move was flawless," said Karen Haller, then vice president for nursing and patient care services at the hospital. "When patients' families arrived on the units, they

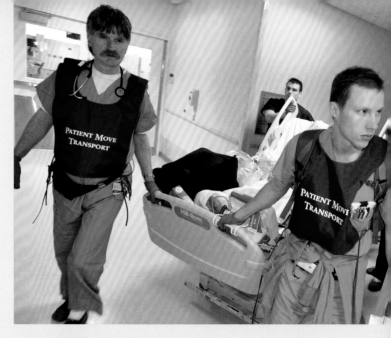

In April 2012, 271 patients were moved from The Johns Hopkins Hospital's historic buildings to The Charlotte R. Bloomberg Children's Center and the Sheikh Zayed Tower.

exuded confidence in the teams. And the patients waved and cheered with us. It was fun."

Many clinicians offered tender farewells to the crowded units, dark corridors and out-of-fashion color schemes they were leaving behind. While combing through old log books, photos and other memorabilia, they thought of lives saved and lost.

Finally, after years of meticulous planning and colossal effort, the day had come to occupy the blue- and green-hued towers.†

---

* Shapiro, Stephanie, "The Home Stretch," *Dome*, April 2012, https://www.hopkinsmedicine.org/news/publications/dome/april_2012/the_home_stretch_.

† "A Job Well Done, *Dome*, May 2012, https://www.hopkinsmedicine.org/news/publications/dome/dome_may_2012/a_job_well_done.

Sally MacConnell

greater comfort and privacy for our patients and their families in a state-of-the-art environment."[13]

Sally MacConnell, who has been with Johns Hopkins since 1983, led the planning, design and construction of the building as senior vice president of facilities. She said in a 2021 interview that the project was the largest and most complex of her career. As with any major hospital construction, it required the ability to peer into the future and prepare for inevitable and rapid changes in health care delivery and technology.

"We knew that people in hospitals were going to be sicker and sicker and that air handling was going to be all important," she said. As a result, the building was fully equipped with a HEPA (high efficiency particulate air) filtration system, which provided essential infection control during the COVID-19 pandemic.

A key part of the planning process was gathering insights and suggestions from groups who would use the building, including doctors, nurses, security and nutrition staff members, and patients and families. A decade after it opened, MacConnell said in 2021, "the building is working quite well."[14]

More Clinical Care Options for Patients

Parallel to the ambitious expansion of The Johns Hopkins Hospital, the institution was increasing its community physician offerings, giving patients more places to receive outpatient care and to get referrals, if needed, for specialty care in the Johns Hopkins health system.

In 2013 alone, Johns Hopkins Medicine opened community-based practices in Fulton, Rockville, and Chevy Chase, Maryland; and in Washington, D.C. The ones in Fulton and Washington were for internal medicine and family care, while the Rockville and Chevy Chase locations became part of Johns Hopkins Community Physicians Heart Care, a group offering a full roster of services for cardiovascular diseases and conditions such as heart failure and high blood pressure.

The Fulton location opened in May 2013 on Maple Lawn Boulevard in the fast-growing community southwest of Baltimore in Howard County.

"Now, patients will not have to travel far to experience outstanding care from one of the most trusted names in medicine," said Johns Hopkins Community Physicians (JHCP) President Steven Kravet when the office opened. "JHCP Fulton

Hurricane Sandy and the Great Bucket Brigade Rescue

Hurricane Sandy, which made landfall in New Jersey on Oct. 29, 2012, was one of the most destructive storms in U.S. history, causing an estimated $74 billion in damage and claiming at least 233 lives.

Though Maryland was not as hard hit as nearby New York and New Jersey, hundreds of thousands of residents lost power as fierce winds and torrential rain walloped the state.* On the East Baltimore campus of Johns Hopkins, the David H. Koch Cancer Research Building lost power, and stormwater from clogged city drains rose through cracks in the floor.

At risk were tissue samples that had been meticulously collected and curated over decades and were stored in special freezers; as well as lab mice and rats that, without power, lacked water

Johns Hopkins University's Homewood campus the day after Hurricane Sandy.

and ventilation.

About 200 lab workers, faculty members, facility managers, students and Johns Hopkins Medicine leaders rolled up their sleeves to rescue years of research from the storm-paralyzed facility. These included Paul Rothman, dean of the medical faculty and CEO of Johns Hopkins Medicine; Rich Grossi, JHM chief financial officer; and Daniel Ford, vice dean for clinical investigation.

Human chains formed spontaneously to pass research materials from the depths of the building up three flights of stairs. Standing at the summit of this evacuation route was Landon King, vice dean for research, and others who helped to manage the relocation of lab animals and biospecimens.

By day's end, thousands of rodents had found temporary lodgings in well-ventilated areas, while frozen blood, tissue and stool samples were relocated safely to other containers or

protected by dry ice.

"Although in no way was Tuesday a good day, it was an extraordinary day because it encapsulated an awful lot of what makes Johns Hopkins such a special place," King said. †

Years later, as Rothman reflected on his 10 years as dean and CEO, he remembered Terry Langbaum, then administrative officer of the Cancer Center, calling his office at 6:30 that morning to say that help was needed, and he fondly recalled that experience, of wading in sewage and working together, as a highlight of his tenure. "It really showed the power of this place," he said.‡

---

* Calvert, Scott, and the Baltimore Sun, "Maryland 'Dodged a Bullet' with Powerful Storm," *Baltimore Sun*, Oct. 31, 2012, https://www.baltimoresun.com/weather/bs-md-sandy-overview-20121030-story.html.

† Smith, Linell, "Research and Rescue Mission," *Dome*, Jan. 4, 2013, https://www.hopkinsmedicine.org/news/publications/dome/dome_january_2013/research_and_rescue_mission.

‡ Rothman quoted at JHM Town Hall, March 31, 2022, https://webcast.jhu.edu/Mediasite/Play/6dde844b259b49e485dca6438b06a65f1d

will offer the most advanced and comprehensive health care for patients in all stages of life."[15]

At 6,458 square feet, JHCP Fulton was built to grow, and that's what it did. The practice started with two doctors providing family medicine care, and by 2021 was staffed with 11 physicians, including internal and family medicine doctors, and Gyn/Obs.

In July 2013, Johns Hopkins Medicine opened a community location in Washington, D.C., at the McPherson Square metro station, 1400 I Street NW and immediately began accepting patients for internal medicine care.[16]

In addition, Johns Hopkins Community Physicians (JHCP) Heart Care opened two offices, one in Rockville in late 2013 and one in Chevy Chase in early 2014, adding to existing locations in Bethesda and Silver Spring. This brought the total number of JHCP Heart Care cardiologists to 20.[17]

These symbols represent the six priorities of the institution's first five-year strategic plan:

• **People:** Attract, engage, develop and retain the world's best people.

• **Biomedical Discovery:** Become the exemplary model for biomedical research by advancing and integrating discovery, innovation, translation and dissemination.

• **Patient- and Family-Centered Care:** Be the national leader in the safety, science, teaching and provision of patient- and family-centered care.

• **Education:** Lead the world in the education and training of physicians and biomedical scientists.

• **Integration:** Become the model for an academically-based, integrated health care delivery and financing system.

• **Performance:** Create sustainable financial success and implement continuous performance improvement.

The Strategic Plan

One of Rothman's early priorities was creating the first-ever comprehensive five-year strategic plan for Johns Hopkins Medicine. "The first thing I wanted to focus on was to get people into place and to think about how we were going to strategically think about the institution moving forward," Rothman said in 2021. "And that's what we did that first year."[18]

The 18-month process involved more than 150 faculty members, administrators and staff members throughout the enterprise.[19]

"We had an excellent process that let us hear from a lot of people and have a plan bubble up from the bottom up and then top down and then bottom up again," Rothman said in 2021. "There is a great deal of value just in the strategic planning process. You hear from a lot of constituencies you might not otherwise hear from, and often that informs the direction of the institution."[20]

The resulting plan was launched in June 2013.

Rothman was determined to create a "living, breathing" plan that would not sit on a bookshelf but instead would serve as the backbone for important decisions about programs and resources.

He chose a five-year time frame because health care and medicine are always changing, he said. "The problem with three years is by the time you start implementing it you have to redo it, so we thought five had a balance between giving you some of the ability to execute in that type of time frame but not be so out of date that the world has changed around you and the plan you developed has no salience to what you are doing on a day-to-day basis," he said in 2021.[21]

A second strategic plan, known as Innovation 2023, would subsequently guide the institution from 2019 through 2023.

Rothman's original strategic plan, referred to as "Leading the Change," guided the institution in carrying out improvements that included:

- increasing the racial and gender diversity of JHM executive leadership
- creating precision medicine centers of excellence
- launching an accountable care organization
- improving employee health through workplace wellness programs
- increasing patient access to treatment with dedicated phone lines and expanded hours
- creating clinical communities to standardize high-quality care across the enterprise
- introducing a primary care leadership track for students in the school of medicine
- establishing the Office of Telemedicine.[22]

Epic Rollout

One key manifestation of the strategic plan began with the rollout of Epic, the state-of-the-art electronic medical record system.

Like many health care organizations, Johns Hopkins Medicine had, over the years, developed a hodgepodge system of electronic medical record use, with different hospitals and specialties using their own systems.

The patchwork system also created barriers to research because data about patients and diseases were stored in different ways and in different places.

In 2011, Johns Hopkins trustees set aside $100 million for the transition to a single electronic medical record system. Johns Hopkins leaders chose Epic, a private health care software firm based in Verona, Wisconsin, as its vendor. Among its selling points, Epic's systems were already in wide use, with a proven track record; and they would deliver seamless and integrated patient records encompassing inpatient and outpatient care across the Johns Hopkins Health System.[23]

The decision came at a time of tremendous growth and change for Johns Hopkins Medicine, which had recently acquired three hospitals — Sibley Memorial, in Washington, D.C.; Suburban Hospital, in Bethesda, Maryland; and All Children's Hospital, in St. Petersburg, Florida.

At the same time, a new federal mandate for health care providers was looming as part of the 2009 Health Information Technology for Economic and Clinical Health (HITECH) Act, calling for "meaningful use" of electronic medical records by Jan. 1, 2014.

The move to Epic began in 2013, led by then Chief Information Officer Stephanie Reel and Chief Medical Information Officer Peter Greene. It would allow Johns Hopkins to integrate operations and move toward its goal of acting as a single organization. It would improve care by ensuring that clinicians at different Johns Hopkins locations had the same patient information and were operating with the same protocols and best practices.

And it would enhance patient-provider communication through the MyChart patient portal.

"No longer are lab results sent by fax and scanned into the patient record. Referrals and paper notes that once were jotted by doctors on sheets of paper are

Eppie, a character developed by graphic designers at Johns Hopkins Medicine, appeared in internal communications throughout the Epic rollout.

today a part of Epic and a more complete portrait of patients' histories," noted JHM's 2014 Biennial Report.

"Doctors can also 'share' patients — when caring for the same ones — since all information becomes part of a single digital chart. Patient files in Epic include information as far back as April 2003. This enhances patient safety, now that doctors have ready access to complete information, including prescriptions and prior test results."[24]

Epic also dramatically accelerated the progress of "big data" research that was already helping clinicians tailor treatments to individuals. A few years after Epic rolled out across the health system, Johns Hopkins would create Precision Medicine Centers of Excellence (PMCOEs), which used data from Epic and other sources to provide entirely new insights about diseases and treatments.

The multiyear transition began April 4, 2013, when the first wave of outpatient practices migrated to Epic, including Johns Hopkins Community Physicians practices; School of Medicine providers who practiced in JHCP suites; Johns Hopkins at Odenton; and Wilmer at Frederick, Bel Air, Bethesda and Wyman Park.

It ended July 1, 2016, when The Johns Hopkins Hospital and Kennedy Krieger Institute went live — though new features and shortcuts have been continually added via regularly offered Epic updates and the ongoing process of tailoring the system to the needs of Johns Hopkins clinicians.[25]

"It's been a huge investment, but it's paid off in spades," Rothman said in a 2021 interview. "When we look back, I think we'll see this is one of the most important investments we made."

The Transition

In that same interview, Rothman said he had learned from his previous experience bringing Epic to the University of Iowa health system. "We 'Big Banged' the entire place in one day," he recalled. Although the new system worked fine, it took a few months for the people using it every day to get used to it, he said. At Iowa, he said, "I don't think we invested in training the way we needed to." [26]

At Johns Hopkins, the system was rolled out gradually. And all Johns Hopkins employees who used Epic had to first learn about the system through a combination of online and in-person instruction, tailored to specific roles and updated, based on the comments of past participants.[27]

In the four weeks immediately following each "go-live," a 24-hour help desk took on the most pressing concerns in a process called "break-fix." Meanwhile, a battalion of specially trained "super users" helped Epic users adjust to the system by providing on-site and remote help. Then came the fine-tuning. More than 50 "physician champions" were trained to serve as liaisons between the software experts configuring Epic and the clinicians using it. Through this process, for example, physician champion Michael Boland, an ophthalmologist, was able to help his fellow clinicians at Wilmer Eye Institute streamline "some fairly complicated workflows" related to documenting in-office procedures, such as ocular imaging.

"The goal," he explained, "is to actually train clinicians to help."[28]

Epic's Rewards

One immediate and tangible benefit of Epic was that it allowed creation of the MyChart patient portal system, which gave patients and their caregivers a secure conduit to see test results, make appointments and pay bills. (Starting in 2016, providers could add an OpenNotes feature, making their exam notes visible to patients — a move widely praised for giving patients and their caregivers more control and information.)[29]

A "care everywhere" function simplified record-sharing across thousands of Epic-using hospitals and clinics, and decision support tools gave clinicians up-to-date decision trees that saved time and promoted best practices.

An Epic update in 2015 added features, including a keyword search function that clinicians could use to find specific information in patient charts; Fast Pass, which let patients on waiting lists book earlier appointments when they became available; and Quick List, which let clinicians select from targeted sets of orders for frequent complaints, such as chest pain, and also create their own Quick List order sets for the issues they encountered most.

Epic also allowed integrated rollout of tools such as Dragon Medical One voice recognition in 2018, which saved clinicians significant time by transcribing their spoken notes directly into patient records.[30]

SlicerDicer, an Epic tool that was part of the 2015 update, let users analyze data and develop preliminary research ideas. "Let's say you think patients on one statin have a different frequency of abnormal liver enzymes compared to another," said internist Howard Levy, an Epic physician champion and co-chair of the Epic design team focusing on patient- and family-centered care. "With SlicerDicer, you can ask those questions."[31]

Patient and Family Advisory Councils

Patient and Family Advisory Councils, known as PFACs, support the Johns Hopkins Medicine mission of providing patient- and family- centered care. Patients, family members, and caregivers work closely with hospital staff to generate ideas for improving patient care, address specific concerns, and generally enhance communication between providers and those receiving care.

A 2019 photo showing members of the Adult Patient and Family Advisory Council at The Johns Hopkins Hospital.

Ask any patient and family advisor why they joined a PFAC and the answer rarely varies. They want to make the path easier for the next patient. "I literally got involved because the entrance to the MRI during a snowstorm was not as accessible," said Stacy Kaye, explaining why she became a member of the PFAC at Sibley Memorial Hospital.

"I think PFAC is the best example of how to make a difference, whether it's something you realize that could be better, or it's a small grievance that you may not have thought was important; if it's important to you, it's going to be important to someone else," she said.[32]

Nationwide, the first PFACs were formed in children's hospitals in the 1990s. By 2017, about 40 percent of hospitals had at least one PFAC, though they varied widely in structure and mission, according to an AARP Public Policy Institute report.[33]

Johns Hopkins launched its first PFAC in 2007, at its Children's Center. It started as an eight-member steering committee and, by 2013, grew into a council with more than 50 parent and staff members. The council's successes included creation of a full-time parent adviser position in pediatrics in 2009, which has given parents a voice at the highest levels in the Children's Center and greater influence on issues ranging from patient safety and infection control to hospital amenities and navigating the buildings.[34]

The Johns Hopkins Hospital launched an Adult PFAC in 2011, and two years later, the concept expanded to the Johns Hopkins Bayview Medical Center. Within a few years, every Johns Hopkins hospital had at least one such council, with most having several. Across Johns Hopkins Medicine's six hospitals, Home Care Group, and Johns Hopkins Community Physicians practices, there are more than 20 active PFACs, with more than 500 members, usually meeting monthly.[35]

When Johns Hopkins hospitals began treating patients with COVID-19 in March 2020, PFACs shifted to online meetings, while also taking on some of the most difficult questions posed by the pandemic, particularly related to visitor restrictions.

Care within the Community

J-CHiP and TAP

The patient had diabetes, and her blood sugar levels remained stubbornly high, despite regular doctor visits and insulin prescriptions.

Finding the problem took the shared insights of her Johns Hopkins Community Health Partnership (J-CHiP) team, which included a community health worker, pharmacist, home care nurse and Michael Fingerhood, an internist and addiction medicine specialist.

And it took their combined efforts to fix it.

"She couldn't see well enough to draw up the insulin dose and was embarrassed to admit this," Fingerhood said. "When we put the pieces together, she was prescribed insulin pens and encouraged to see an eye specialist." With the team's help, she followed through. Three months later, her blood sugar levels had improved significantly.[36]

Michael Fingerhood

Patricia M.C. Brown, former president of Johns Hopkins HealthCare LLC, led the launch of J-CHiP in June 2012 and became deputy project director for the program, which provided team-based coordinated care for underserved, high-risk people in East Baltimore neighborhoods, as well as Medicare and Medicaid adult inpatients being discharged from The Johns Hopkins Hospital and Johns Hopkins Bayview Medical Center.

Brown and Edward D. Miller, then dean/CEO of Johns Hopkins Medicine, as well as Johns Hopkins physicians, faculty, administrators, medical directors, behavioral health experts, nurses and other staff, had worked for months to refine and improve the J-CHiP program while awaiting a funding decision.

It was a potent example of how Johns Hopkins provides care that is tailored to the specific needs of individual patients, while serving the residents of the Baltimore neighborhoods that surround its hospitals. And it fulfilled three components of the strategic plan by providing patient- and family-centered care, integrating services and improving financial performance.

Many East Baltimore residents faced multiple challenges to their health and well-being, including limited access to nutritious food or difficulty traveling to medical appointments. As a result, their life expectancies could be 20 years shorter than those who live in more affluent communities nearby.[37]

With J-CHiP, patients who were going home from the hospital received a coordinated plan to keep them from returning for the same problem. Post-discharge phone calls from nurses, patient education materials, and visits from J-CHiP staffers provided patients and their families or caregivers with encouragement and instructions for follow-up care.

"I make sure that patients are able to get to appointments with their primary care doctors and specialists, in some cases by arranging transportation for them," explained Aaron Rash, a patient transition guide and nurse. "If patients are doing poorly, I can get them in to see the doctor more quickly or arrange home health services. When patients don't have a primary care doctor, I help them find one, along with any other community resources they may need."[38]

J-CHiP, which ended in 2016, was funded with a three-year, $19.9 million grant from the Centers for Medicare and Medicaid Services (CMS), part of CMS's $1 billion Health Care Innovation Awards, a competitive initiative supporting programs that improve population health, reduce per capita costs, cut hospital admissions and readmissions, and improve care experiences for individuals.[39]"

The program, which built on existing efforts, such as the Johns Hopkins Health System Readmissions Task Force, eventually enrolled more than 80,000 residents,

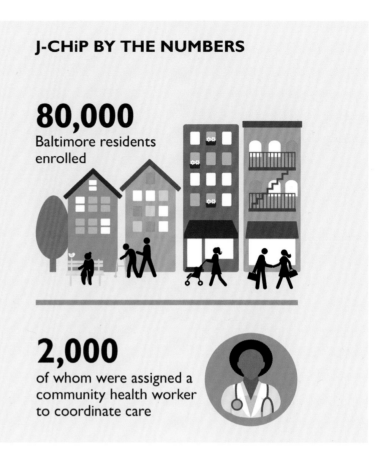

J-CHiP BY THE NUMBERS

80,000
Baltimore residents enrolled

2,000
of whom were assigned a community health worker to coordinate care

The J-CHiP program enrolled 80,000 Baltimore residents.

Patricia M.C. Brown

most with complex health care needs, including about 2,000 local community members who were assigned a community health worker to coordinate care.

Multidisciplinary teams assigned to high-risk Medicare and Medicaid beneficiaries included physicians, care managers, health behavioral specialists and community health workers. Also key to success was J-CHiP's collaboration with community-based organizations, particularly Sisters Together and Reaching (STAR) and the Men and Families Center. The nonprofits supplied additional staff members and expertise to support neighborhood residents at high risk for hospitalization.[40]

A 2014 article in *USA TODAY*, headlined "Programs Focus on High-Risk Patients to Reduce Spending," described how J-CHiP and specifically case worker Donnie Missouri helped a patient who was not adequately managing her hepatitis C, HIV and depression, and relied on emergency room visits when she needed care. That changed when she connected with J-CHiP, as the article details:

> *In between doctors' appointments, [the patient] receives a call at least once a week to make sure she's taking her medications, to see whether she needs help with food or housing, or just to see whether she needs to talk. Missouri works with a care coordinator, social worker, nurse practitioner and primary care physician to keep tabs on hundreds of low-income Medicare and Medicaid recipients throughout Baltimore.*
>
> *"I think they care," [the patient] said. "I trust them."*
>
> *They may care, but they are also motivated by the belief that by helping the super-utilizers with coordinated care that includes housing, food, bus tokens and other needs, they can cut millions of dollars in health spending.[41]*

A 2018 study authored by Johns Hopkins physicians, researchers, nurses, administrators and health policy experts, published in the *Journal of the American Medical Association (JAMA) Network Open,* found that J-CHiP's inpatient care coordination plan contributed to a savings of $29.2 million for Medicare and nearly $60 million for Medicaid over its four years. The community high-risk population intervention contributed to a savings of $24.4 million for Medicaid. [42]

In the same 2012 to 2016 time period, the inpatient care coordination plan for Medicaid beneficiaries saw a dramatic decrease in emergency department visits that occurred within 90 days after hospitalization, reducing that number by 133 per 1,000 patients, or nearly 1,900 emergency visits.[43]

Likewise, the community intervention for Medicaid beneficiaries saw reductions of hospitalizations, emergency department visits and 30-day readmissions.[44]

"We set out to assemble a care coordination model that could help keep people healthy and out of the hospital and to keep those already hospitalized from being readmitted," said cardiologist Scott Berkowitz, who was senior medical director of accountable care for the Office of Johns Hopkins Physicians, and lead author of the *JAMA Network Open* article. Berkowitz now serves as the chief population health officer and vice president of population health for Johns Hopkins Medicine.

"We believed we could also do it efficiently enough to improve care and at the same time save payers money," he said. "Our study suggests that this model of care

"Doing Something for Someone"

The 5-year-old girl from Guatemala needed cardiac surgery. First, however, she had to see an oral surgeon because her teeth had decay that could cause a heart infection. Moreover, her unstable hip required treatment. Although she and her family members lived close to The Johns Hopkins Hospital, they lacked the knowledge and the resources to obtain its services.

Fortunately, they had Reina Gonzalez to guide them.

Gonzalez, a bilingual medical office assistant at Johns Hopkins Community Physicians – East Baltimore Medical Center (EBMC), is a trusted medical compass for uninsured Spanish-speaking patients in East Baltimore.

Few of the people she helps speak English. Fearful of the health care system, they don't know what to expect. They worry that they won't understand what they are told, and that the medical staff will reject them because they do not have insurance.

Gonzalez makes appointments for as many as 116 clients a month, and reminds them, often, about those appointments.

The diminutive woman has become a towering figure within the organization's world of charitable care. An admiring co-worker fashioned a poster that praised her compassion, gave her a crown and saluted her as "Queen Reina."

"What I like most is my communication with the patients … feeling like I'm doing something for someone," she said. *

*   Smith, Linell, "The Access Partnership at Johns Hopkins Celebrates a Decade of Providing Low-Cost Services — and Hope," *Dome*, June 6, 2019, https://www. hopkinsmedicine.org/news/articles/johns-hopkins-offers-access-to-healthcare-for-underinsured-and-uninsured-patients-in-east-baltimore.

Reina Gonzalez is a patient navigator with The Access Partnership (TAP), which serves uninsured and underinsured people in the communities surrounding The Johns Hopkins Hospital and Johns Hopkins Bayview Medical Center.

coordination was associated with improved health outcomes and substantial cost reductions in an urban, academic health system."[45]

The Access Partnership (TAP)

J-CHiP was not the only JHM program reaching medically complex and underserved Baltimoreans.

Through another Johns Hopkins program, The Access Partnership (TAP), launched in May 2009, specialists donate their services to Baltimore residents who are eligible to receive primary care through state or federal insurance but can't afford to pay for the next level of tests and treatment that their physicians recommend. Patients who participate in TAP pay an advance fee of $20 every three months, and receive no additional bills for services received.

On TAP's 10-year anniversary in 2019, the program had treated 7,411 patients and provided 15,636 referrals.

"It represents some of the best of what we aspire to be here at Hopkins," said John Colmers, senior vice president for health care transformation and strategic planning. "We can be enormously proud of being able to provide these types of services to patients who are in the most need."[46]

John Zampella, who learned on Match Day that he would do his residency at Johns Hopkins, was in the first class to complete all four years of the Genes to Society curriculum.

Medical Education, from Classroom to Community

Genes to Society

On May 23, 2013, the first school of medicine students to complete the new Genes to Society curriculum received their M.D.s in a convocation ceremony at the Joseph Meyerhoff Symphony Hall in Baltimore.

Their medical education had covered topics such as ethics, population health, and the structure of health care systems. They had learned about health care disparities, pain management, patient safety and palliative care. And they had started interacting with patients and learning clinical interviewing skills in their first week of medical school.

"The old model, in which students learned in the classroom for the first two years and then began seeing patients in the third year, is a thing of the past," said Roy Ziegelstein, vice dean for education. Learning is more interactive and team-based now, he noted, and the very first course of medical school addresses disparities and inequities in health and health care.

"Today's students are different from students 25 years ago. They learn in different ways," said Ziegelstein. "So, in Genes to Society, we are constantly exploring new ways to teach our students."[47]

That included expanded use of online learning (efforts that accelerated during COVID-19), hands-on time in the simulation center, and individualized learning paths made possible through "precision education," which uses big data to provide more personalized instruction and performance assessment.

The goal of Genes to Society (GTS) was nothing less than "reframing the context of health and illness more broadly, to encourage students to explore the biologic properties of a patient's health within a larger, integrated system, including social, cultural, psychological, and environmental variables," noted a 2010 article in *Academic Medicine,* written by some of the curriculum's architects, including Charlie Wiener, Patricia A. Thomas, Elizabeth Goodspeed, David Valle and David G. Nichols. [48]

One catalyst for the change was *Genetic Medicine—A Logic of Disease*, a 1999 book by pediatric faculty member Barton Childs. It proposed a new framework for teaching and practicing medicine, based on recent advances in knowledge about human DNA, and a new understanding of how genes and environment contributed to disease.[49]

"It was called Genes to Society because we recognized that individual variability had two large sources," said Valle, director of the Department of Genetic Medicine, in a 2013 Johns Hopkins Medicine video. "One was the genetic makeup of each person, and the second was the accumulated environmental experiences of each person. So to do one without the other made no sense."[50]

Charles Wiener

The Human Genome Project, completed in 2003, and ongoing related advances in genetics and genomics, exposed for the first time the true extent of genetic variation from one individual to the next. It was time for medical education to incorporate and build on this new level of understanding of human individuality.

The new curriculum was the result of years of work that involved the entire school of medicine community. The project began in 2002, and included repeated invitations to faculty for critiques and ideas, frequent briefings to trustees and alumni, and feature articles in student newsletters and other Johns Hopkins publications.

A phased rollout, with additional opportunities to refine the curriculum, took place from May 2006 to May 2009, with the official launch in October 2009. "We learned that changing a medical curriculum is really a big deal. It's like turning a gigantic oil tanker," said Valle.

Meghana Desale completed the GTS curriculum and earned her Johns Hopkins medical degree in 2013. "I think it was a great way to learn medicine," she said. "Every time you go through each of the different systems you're not just learning, 'Here's the anatomy, here's the physiology and the pathophysiology.' You're learning everything from how [patients'] diets affect their care to how certain policies affect it. So you're really thinking about the big picture."[51]

Medicine for the Greater Good

Panagis Galiatsatos roamed the cavernous community room of Poe Homes, a public housing complex in West Baltimore. A couple of months earlier, he had been in the same room with advice about heart health. Now, he was explaining how to treat sprains and concussions.

His audience, again, was about 25 children and teens, the drummers and dancers of the Christian Warriors Marching Band. Again, they sat on folding chairs arranged in a circle around Galiatsatos, then a pulmonary and critical care fellow in the Johns Hopkins University School of Medicine. Galiatsatos had about 20 minutes before the marchers resumed their twice-weekly practice.

"Do you remember me?" he asked.

"Yeeesss," came the drawn-out chorus of a reply.

"Do you remember my name?"

"Nooooo."

Galiatsatos chuckled and told his young audience to call him Dr. G. He knows his name is hard to remember. He also knows he's forming relationships, even as the marchers shift and whisper in their seats. Maybe some of the young people in this room will remember his advice for treating a sprain. Or maybe they'll remember, next time they visit a doctor, that this one made them smile.[52]

Galiatsatos is a co-founder and director of Medicine for the Greater Good (MGG), which helps residents and other health care providers recognize social

Colleen Christmas

Nursing student Kathleen Littleton and pulmonary and critical care fellow Panagis Galiatsatos visit Poe Homes in 2016 to discuss concussions and sprains. (Galiatsatos later became an assistant professor in pulmonary and critical care medicine.)

factors that contribute to health, develop advocacy skills, and develop community programs and partnerships that help people overcome barriers to optimal health.

MGG began at the Johns Hopkins Bayview Medical Center in 2011 as a series of workshops that encouraged trainees and experts to discuss nonclinical topics, like health policy, behavioral counseling and social determinants of health. In 2013, it expanded to include a requirement that all internal medicine residents at Bayview complete at least one project that benefits the community. Many do more.[53]

To date, more than 100 residents and students have been trained by nearly two dozen faculty to engage in projects that have reached more than 7,000 Baltimore City residents.[54]

Projects include the Lay Health Educators Program, which teaches community members how to set up health initiatives such as blood pressure screenings in their communities; and Caregiver Cafes, where caregivers can get together over coffee to learn about resources and discuss their experiences. Members of MGG also give talks in schools, churches and public housing about topics such as sports injuries or managing diabetes.

The program was created by Colleen Christmas, an associate professor in geriatric medicine and gerontology who was the Internal Medicine residency

program director at Johns Hopkins Bayview during its formation; Galiatsatos, who was an assistant chief of service during the creation and is now an assistant professor in pulmonary and critical care medicine; and Ariel Green, who was one of the residents who initially proposed such training in 2011 and is now an associate professor of geriatric medicine and gerontology. The directors in 2021 were Christmas and Galiatsatos, as well as Thomas Cudjoe, assistant professor in geriatric medicine and gerontology; and Erica Johnson, assistant professor in the division of infectious diseases and current internal medicine residency program director at Johns Hopkins Bayview.

"More than 50 percent of what we do when we're taking care of patients is not strictly medical," said Green. "It's confronting other issues, like poverty. That's why it's important for doctors to get beyond the walls of the hospital."[55]

Thomas Cudjoe

Erica Johnson

Centro SOL

A drive to provide culturally sensitive community-based care led a passionate group of Johns Hopkins doctors and other Johns Hopkins clinicians and volunteers in Baltimore to create the Center of Excellence for Latino Health — better known as Centro SOL, an acronym for Center for Salud/Health and Opportunities for Latinos.

Centro SOL, which opened in the summer of 2013, took a family-centered, multidisciplinary approach, bringing together experts in medicine, pediatrics, gynecology and obstetrics, and psychiatry to improve health and opportunities for the predominantly Spanish-speaking people living in the neighborhoods surrounding the Johns Hopkins Bayview Medical Center.

For example, one program, called Testimonios, launched in May 2014 and provided a physically and emotionally safe location for recent immigrants to discuss their stresses and concerns, guided by bilingual faculty and staff members with backgrounds in social work, psychology and psychiatry.[56]

In 2020 and 2021, Centro SOL was hugely instrumental in bringing COVID-19 testing, information and vaccines to Latino residents, largely through partnerships with the Sacred Heart church and other locations that hosted pop-up testing and vaccination clinics staffed by bilingual clinicians.[57]

Tina Cheng, formerly director of pediatrics at Johns Hopkins, was the center's first director. "Addressing the health care needs of Latino families in a comprehensive way will be our number one priority," she said when it opened.

Also leading the effort were Sarah Polk, a pediatrician in the Children's Medical Practice at Johns Hopkins Bayview; and Kathleen Page and Adriana Andrade, infectious diseases specialists at The Johns Hopkins Hospital.[58]

Centro SOL built on efforts started by Page and Andrade a few years earlier, when they noticed an uptick in Latinos with HIV who faced significant language barriers. In 2010, they created the Hopkins Organization for Latino Awareness, or HOLA, a network of 15 Latino faculty members and others working to improve care for Latinos at Johns Hopkins through scholarship, education and policy leadership.

Their efforts included cultural competency courses for medical students and

Myron "Mike" Weisfeldt

a free clinic for immigrants, supervised by HOLA faculty. Ongoing community outreach by the center's faculty includes a bimonthly health column published in a Latino periodical; a Spanish language program on El Zol radio station; the largest Latino HIV outreach program in Baltimore; and pro bono provision of care at Esperanza Center, the resource center for immigrants run by Catholic Charities of Baltimore.[59]

Centro SOL was made possible by a gift from the Aaron and Lillie Straus Foundation and matching funds provided by leaders of the Johns Hopkins University School of Medicine, the Johns Hopkins Health System and Johns Hopkins Bayview Medical Center.

Urban Health Programs

In 2010, Department of Medicine Director Myron "Mike" Weisfeldt decided it was time for his department to do more to improve the health of people living in the Baltimore communities surrounding Johns Hopkins hospitals. The result was the Urban Health Programs, launched that July, which prepare physicians to care for disadvantaged and underserved patients.

It is divided into two tracks: internal medicine (UHIM), a three-year, adult-focused track designed for 12 residents that started in 2011; and combined internal medicine-pediatrics (UHMP), a four-year residency that started in 2010 and trains 16 residents per year to care for the entire family. The UHMP residents typically switch between those departments every two to three months and end up with board certification in both disciplines.

The first residents of the new programs graduated in 2014.

"As a resident in the Internal Medicine-Pediatrics Urban Health Residency Program, I was given membership in a group that led marches against gun violence, advocated for undocumented families, distributed naloxone to inpatients, improved HIV care in underserved Latinx communities and created curricula on trauma," said Justin Berk, a 2019 UHMP graduate.[60]

The program's urban health curriculum pairs traditional inpatient rotations with community-based continuity clinic practice at the East Baltimore Medical Center and community-based rotations with federally qualified health center partners like Chase Brexton, Health Care for the Homeless, and Baltimore Medical Systems. Other rotations, including addiction medicine and psychiatry, are held on-site at The Johns Hopkins Hospital and Johns Hopkins Bayview Medical Center. The program has community partners in such places as the House of Ruth, a domestic violence center; and the Helping Up Mission, an addiction treatment program.

Since graduating its first residents, the Urban Health programs have successfully placed greater than 80 percent of graduates in urban health primary care roles, with an ever-growing list of graduates in leadership positions, said program

Medical student Ramy El-Diwany, shown in 2012 with Baltimore residents Sallie Lovelist, Leona Johnson and Stephon Hand, helped found Charm City Care Connection, which connects Baltimore residents to health resources. El-Diwany went on to become a general surgery resident at Johns Hopkins.

The Biomedical Careers Initiative: More Options for Ph.D.s

Students earning Ph.D.s in medical subjects like neuroscience or immunology often find themselves competing for a shrinking number of the tenure-track faculty research posts that were once considered the sole logical next step for people with their advanced credentials.

As trainee career interests have become more diverse, a program called Biomedical Careers Initiative (BCI), launched in 2013, helps them learn about other options.

In a 2015 survey of the roughly 800 biomedical graduate students in the school of medicine, research faculty jobs remained the top choice for just 27 percent, with the remaining 73 percent saying they'd prefer jobs in government, industry, K-12 schools or nonprofit organizations.*

For such career paths, graduate students can turn to BCI for internships, networking events and educational seminars. BCI's website provides information such as recommended coursework across Johns Hopkins schools, volunteering ideas and "day in the life" vignettes from alums in the field.

Since its inception in 2013 with support from the Ph.D. Innovation Initiative, BCI has helped

more than 85 students secure internships ranging from industry research and development to science writing to consulting. †

* Nitkin, Karen, "Charting a Course Outside the Academy," *Dome*, Aug. 26, 2015, https://www.hopkinsmedicine.org/news/articles/charting-a-course-outside-the-academy.

† "Biomedical Careers Initiative," Johns Hopkins School of Medicine webpage, https://bci.jhu.edu/

Director Leonard Feldman. Wherever Urban Health graduates practice, "they are shaping the future of primary care in urban, underserved settings on local, regional, and national levels — and they're doing it with deep empathy, humility, and commitment to equity and diversity," he said.[61]

Center of Excellence

The renewed focus on community-based care came while Johns Hopkins Medicine was also expanding its national and global reach.

In October 2013, Johns Hopkins HealthCare, the managed care arm of Johns Hopkins Medicine, scored a major win when Johns Hopkins Bayview became the only hospital on the entire East Coast chosen as part of a newly formed National Employers Centers of Excellence Network, which gives employees of Walmart, Lowe's and other large companies and their dependents better care at lower cost.

Bayview was named a Center of Excellence for Joint Replacements within the network, meaning that some 1.5 million employees and their dependents of those companies could have knee and hip replacement surgeries at Johns Hopkins Bayview without incurring any costs. Consultations and care would be covered at 100 percent, and the companies would cover travel and lodging expenses of each patient and a companion to travel to Baltimore for the procedures.

"We are very pleased to provide excellent, patient-centered care for employees covered under this agreement who are seeking hip and knee replacements," said Patricia M.C. Brown.

"The joint replacement team at Johns Hopkins Bayview Medical Center provides the highest quality of care for primary, complex and revision surgeries using the latest technology and best practices, which should enable employees to recover and return to work as soon as possible."[62]

Johns Hopkins HealthCare, which dates to 1994, is a joint venture of the university and the health system that develops and administers contractual health care relationships with managed care organizations, government programs, employers and health care providers.

Its lines of business include four health insurance companies: Johns Hopkins Employer Health Programs (EHP), Priority Partners, Johns Hopkins US Family Health Plan, and Johns Hopkins Advantage MD, as well as Johns Hopkins HealthCare Solutions, which uses Johns Hopkins science to develop and offer commercial ventures locally and globally.

East Baltimore Development Inc.

By 2000, the East Baltimore community adjacent to The Johns Hopkins Hospital and medical campus had been in decline for decades. Residents reported worse-than-average outcomes on measures related to employment, health and educational achievement.

Against a rising tide of home vacancies, a community-based partnership called the Historic East Baltimore Community Action Coalition Inc. (HEBCAC) was working to restore homes, and eventually turned to city leaders for help. The city in turn commissioned an analysis that envisioned a solution in the form of a research park and mixed-income housing for new and current residents.[63]

In 2002, the City of Baltimore, under then Mayor Martin O'Malley's leadership, worked with a coalition of public and private institutions to create East Baltimore Development Inc. (EBDI), the group that would acquire and revitalize 88 acres in a neighborhood known as Middle East.

In addition to the City of Baltimore, the coalition included The Johns Hopkins University, Johns Hopkins Medicine, the Annie E. Casey Foundation, the State of Maryland, and community groups including HEBCAC, the Middle East Community Organization (MECO), and the Save Middle East Action Committee (SMEAC).

Eventually, the 88-acre project area would feature for-sale and for-rent homes, community green spaces, a science and technology park including tech incubators, and a generous mix of retail stores and restaurants, including an Atwater's, Starbucks and Walgreens.

"What this project has done is present us a unique opportunity to demonstrate to the community our commitment to be good neighbors," Ronald R. Peterson said in January 2013.

"The vibrancy of the surrounding area is certainly important to us as we recruit staff, faculty, and students, and for the patients and families who come to our hospital. There is no denying we have an important institutional interest in this area," he said. "But having a more livable, vibrant community is good for the residents of this area. We are eager to demonstrate to young and old alike our commitment to this project."[64]

However, the transformation had its difficulties, including a poorly timed interruption during the financial crisis that started in 2007. The pause came after some 742 families had been relocated but before major construction had begun,

Mural Depicts 'Charm and Resilience' of East Baltimore

A colorful mural showcasing the residents and rowhouses of East Baltimore, unveiled in August 2021, was the first project of EBDI's Public Arts and Placemaking workgroup.

The group includes Eager Park residents and community partners, as well as representatives from EBDI, the Baltimore Office of Promotion and the Arts, and Johns Hopkins.

Local artist LaToya D. Peoples

met several times with community members before developing her vision for the mural on the north face of the Henderson-Hopkins School.

"In each meeting, there was keen focus on the unique story of the neighborhood, the ways that the people who have been here the longest contribute to the beauty that it holds, and what it means to bridge the gap between the old and the new," she said. "I

attempted to create a visual experience of that charm, resilience and complex narrative."*

The cheery mural, featuring bright colored flowers and panels of purple and blue, shows architectural features of a typical rowhouse and gives a nod to the evolving yet historic makeup of the community.

The flowers represent the vegetables and blooms often found in backyard gardens in historic East Baltimore. The intricate patterns painted onto the blue and purple panels represent the circle of life and aim to memorialize Henrietta Lacks, a patient at Johns Hopkins Hospital in 1951 who became the source of the HeLa cells that are still impacting modern medicine today.

Featured in the mural are Henderson-Hopkins students; an

image of Randolph Scott, a lifelong Middle East/Eager Park resident and retired Army veteran; and an image of Delegate Hattie Harrison, a lifelong resident of East Baltimore who represented the 45th District in the Maryland General Assembly from 1973 until she died in 2013.

"This mural showcases all that is the best of East Baltimore — the diversity, the culture, the history — in one piece of art," said Baltimore Mayor Brandon M. Scott. "It will serve as an inspiration for other public arts projects to come about in this neighborhood and continue to promote intersectional conversations and relationship-building in our community — something that is so important for our city to grow again and become the best version of itself."†

---

* "EBDI Debuts Public Mural in Eager Park," by EBDI, *Community Health Stories*, Aug. 19, 2021, https://www.hopkinsmedicine.org/news/articles/ebdi-debuts-public-mural-in-eager-park.

† Watson, Keyarah, "New Mural in East Baltimore Celebrates, Depicts Notable Residents, WMAR2 News, Aug. 14, 2021, https://www.wmar2news.com/news/local-news/new-mural-in-east-baltimore-celebrates-depicts-notable-residents.

making it seem as though disruptions in the neighborhood had been for naught.

EBDI fulfilled its commitment to establish an Opportunity to Return program that would help relocated homeowners and renters return to Eager Park should they choose, and post-relocation surveys of impacted residents indicated that more than 80 percent of residents considered themselves better off.[65]

Henderson-Hopkins

A turning point came on June 11, 2012, when more than 300 East Baltimore students joined Johns Hopkins University President Ronald J. Daniels, then Baltimore Mayor Stephanie Rawlings-Blake and other city and state officials to break ground for the first new school in East Baltimore in 25 years, Elmer A. Henderson: A Johns Hopkins Partnership School.

"The Henderson-Hopkins school represents the most important investment to date towards East Baltimore Development Inc.'s goal of creating a thriving, vital and inclusive community," Daniels said. "We envision this world-class new campus to become the center of the community, open to everyone — every day, evenings and weekends."[66]

The $43 million facility, which opened on Jan. 2, 2014, is a community school for nearly 600 students in kindergarten through grade 8, operated by the Johns Hopkins University School of Education in partnership with Morgan State

Students tour the Henderson-Hopkins school, which opened Jan. 2, 2014.

Sisters Shannon and Sheena Seopaul at the EBDI housing lottery event on Sept. 10, 2016. Shannon purchased a two-bedroom townhouse using a one-day $36,000 grant from Live Near Your Work, a Johns Hopkins program that helps employees buy homes in Baltimore.

The Rangos Building

University's School of Education and Urban Studies.

A few months after Henderson-Hopkins opened, it was lauded by the *New York Times* as the centerpiece of a "grand urban experiment" that "aspires to be a campus for the whole area — with a community center, library, auditorium and gym — as well as a hub for economic renewal."

As the *Times* noted: "Putting a cutting-edge school run by a great university in a historically troubled neighborhood flips the usual urban imbalance of power and changes the status quo of public education steeped in blight."[67]

In 2021, the school ranked No. 7 among Baltimore's elementary schools and No. 14 among the city's middle schools in a new ranking of public elementary and middle schools from *U.S. News and World Report*.[68]

In September 2014, the school got a new neighbor: the Harry and Jeanette Weinberg Early Childhood Center, a licensed childcare and early learning center for children from 6 weeks to 5 years old. A Johns Hopkins Hub article announcing the opening noted the center would "offer the latest research-based approaches to early education" and explained it was open to children in federally subsidized Early Head Start and Head Start programs, as well as to those whose parents worked at Johns Hopkins or nearby.[69]

Another leap forward for EBDI came on Sept. 10, 2016, when 47 university and health system employees reserved not-yet-built or under-construction townhouses in a highly touted lottery. The buyers were lured by one-day-only grants of $36,000 from Live Near Your Work, a Johns Hopkins program that helps employees buy homes in Baltimore. Additional incentives for the one-car garage townhouses, which started at $270,000, came from Baltimore City and Ryan Homes lender NVR Mortgage.

One recipient was Shannon Seopaul, at the time a program coordinator in the school of nursing, who moved from an Upper Fells Point rental. "When they sent out the emails about the lottery and how much the incentive was, I figured why not," she said in a 2021 phone interview.

Since then, she left Johns Hopkins to work for the National Institutes of Health but returned in June 2021 to join the Bloomberg School of Public Health faculty as a senior research associate for global disease epidemiology and control.

She said the townhouse, with two bedrooms, two-and-a-half baths, a deck and a garage, is in walking distance to the main campus and about a 10-minute drive to her Inner Harbor office. "This morning I walked to get my flu shot," she said. "Living close to the hospital is super helpful and was one reason I came back to Hopkins."[70]

Science + Technology at EBDI

EBDI's first big ribbon-cutting was for the John G. Rangos Sr. Building in April 2008. The 278,000-square-foot research facility at 855 N. Wolfe St. was the first of several buildings that would make up the Science + Technology Park at Johns Hopkins.

Johns Hopkins was the anchor tenant, leasing one-third of the space for the school of medicine's Institute for Basic Biomedical Sciences. "To my knowledge,

Pederson Brain Science Institute

Founded in 2007, the Brain Science Institute, located in the Rangos Building, brings together experts from across the university to study brain development, function and disease and use the findings to support brain health and healing.

"It was thoroughly designed around the concept of translation, of turning faculty research into drug discovery," said neurology and neuroscience professor Jeffrey Rothstein, the John W. Griffin director of the institute, as well as director of the Robert Packard Center for ALS Research.

The catalyst was one of the largest cash gifts to Johns Hopkins — a pledge of $100 million from George and Marilyn Pederson to promote brain research and health. George Pederson in 1968 had co-founded ManTech International Corp., a publicly traded company that provides cybersecurity and data services to federal agencies.

"Mrs. Pederson was interested in brain pathways and the arts," said Rothstein. "The Pedersons approached us and said, 'How do we invest in brain science?' Our idea was to integrate. There are more than 450 neuroscientists at Johns Hopkins University. We're all doing great research, but how do we translate that work so it goes back to patients?" Miller, the former dean/CEO, helped secure the funding.

Janice Clements   **Jack Griffin**   **Richard L. Huganir**   **Jeffrey Rothstein**

The institute's founding director was University Distinguished Professor of Neurology Jack Griffin, along with co-directors Rothstein and Richard L. Huganir. Janice Clements was executive director.

When Griffin died in 2011, Rothstein took the helm. The ambitious institute is divided into several parts:

- **Synapses, Circuits and Cognitive Disorders** (SCCD), a research core launched in 2010 and run by Huganir, co-director of the BSi and professor and director of the Solomon H. Snyder Department of Neuroscience. Its faculty, with co-appointments in the neurosciences and the BSi, study autism spectrum disorders, schizophrenia and neuroplasticity, as well as touch perception, motor control, social behaviors, reward, mood and decision-making.

- **The NeuroTranslational Drug Discovery Program**, formed in 2010, began when the institute hired a team from the pharmaceutical industry to work with Johns Hopkins scientists, engineers and clinicians to bring therapies to market. In 2015, the group formed a formal division within the BSi, called Johns Hopkins Drug Discovery and led by neurology professor Barbara Slusher. To date, it has spun out five biotechnology companies.

- **The Neurodegeneration Program**, launched in 2012, brings together Johns Hopkins researchers, neurologists and neuroscientists to learn about neurodegenerative disorders, including Alzheimer's disease, amyotrophic lateral sclerosis (ALS), Huntington's disease

and Parkinson's disease.

- **The International Arts and Mind Lab** (IAM Lab), with Susan Magsamen as executive director, launched in 2016 with a mission to study the intersection of brain sciences and the arts and to bring research to practice. Inspired by Marilyn Pederson, BSi hosted the first Science of the Arts speaker series in October 2010, bringing together representatives of the arts, art historians, educators, psychologists and brain researchers to discuss the emerging field of neuroaesthetics — often described as "your brain on the arts." It was so successful that the International Arts + Mind Lab was formed. In 2019, it evolved into a full-fledged center.*

---

*   "International Arts + Mind Lab," webpage, Pederson Brain Science Institute, Johns Hopkins Medicine, https://www.hopkinsmedicine.org/pedersen-brain-science-institute/international-arts-mind-lab.html.

this is the first time Johns Hopkins has put a significant chunk of basic scientists from the clinical departments [together] with true basic scientists, who are working on the most recent discoveries," said Weisfeldt, in an interview before the dedication.[71]

Lieber Institute for Brain Development

November 2011 saw the launch of the Lieber Institute for Brain Development, which established a permanent research facility at the Science + Technology Park in order to work closely with scientists of the Pederson Brain Science Institute (BSi).[72]

Lieber is a neuroscience research institute dedicated to developing novel treatments, diagnostic tests and insights into disorders arising from abnormalities in brain development.

It has the world's most extensive curated collection of human brain specimens — more than 1,300, spanning the normal human lifespan from fetal life to late life and including hundreds of brain samples from individuals who had developmental brain disorders during life. This rare collection of brain samples offers an opportunity to study how genes and the environment construct a brain, and to potentially identify new and effective treatments.[73]

It is named for Steven A. Lieber; his wife, Connie; and their son and daughter-in-law, who committed $100 million to establish the institute. An additional $20 million was donated by the Milton and Tamar Maltz Family Foundation in Cleveland to fund what they described as "a Manhattan Project" to tackle neurodevelopmental health issues.[74]

Kavli Neuroscience Discovery Institute

In 2016, Johns Hopkins launched the Kavli Neuroscience Discovery Institute, which integrates neuroscience, engineering and data science to understand the relationship between the brain and behavior.

The institute, led by Huganir, with biomedical engineering professor Michael I. Miller as co-director, was created with a shared $20 million commitment from Johns Hopkins and the Kavli Foundation, a California-based nonprofit established by entrepreneur Fred Kavli in 2000 to advance science for the benefit of humanity.

Dwight Bergles, professor of neuroscience, took over for Huganir as director in 2020; in 2021, The Johns Hopkins University and the Kavli Foundation committed another $20 million to increase the institute's endowment to $40 million.

The Kavli Institute at Johns Hopkins became part of an international network of seven Kavli Institutes carrying out fundamental research in neuroscience and a broader network of 20 such institutes dedicated to astrophysics, nanoscience, neuroscience and theoretical physics.[75]

The 45 initial members of the Kavli Neuroscience Discovery Institute were drawn from 14 departments in the Johns Hopkins schools of medicine, engineering, arts and sciences, and public health, and the Applied Physics Laboratory.

"Neuroscience is inherently interdisciplinary," Huganir said. "You can study the biochemistry of the brain, but how does that relate to circuits and behavior? It's tough to answer that in a single laboratory. It necessitates interaction and collaboration, and with the Kavli Neuroscience Discovery Institute, we're trying to take that to a new level to understand the brain."[76]

FastForward 1812

Another momentous development for EBDI was the grand opening of the FastForward 1812 innovation hub on April 26, 2017.

The bright and modern facility at 1812 Ashland Ave. became the flagship site for Johns Hopkins Technology Ventures, launched in 2014 and led by Christy Wyskiel, to bring Johns Hopkins innovations to market. It boasted an open-concept floor plan with office and communal workspaces that sits above 15,000 square feet of lab space.

The FastForward 1812 innovation hub opened on April 26, 2017.

"When you think about the innovation and the technology and the biotech companies that can grow right here … it doesn't get much better than that," then Baltimore Mayor Catherine Pugh said at the dedication ceremony.

Landon S. King, executive vice dean of the Johns Hopkins University School of Medicine, echoed that aspiration of nurturing products that "get started in Baltimore … while the businesses stay in Baltimore." The new space replaced FastForward's interim hub in the Rangos Building, increasing available lab space sevenfold for the startup community in East Baltimore, he said.[77]

Protein Library

A shoestring budget and nearly 9,000 miles between members of a 72-person research team couldn't prevent Akhilesh Pandey from completing a draft of the human proteome, a library of 30,057 proteins in the human body.

Scientists around the world had been trying for years to complete this daunting project. Working from his lab at Johns Hopkins, Pandey, then a professor at the McKusick-Nathans Institute of Genetic Medicine and of biological chemistry, pathology and oncology, led his team — much of it based in India — to the finish line first and catalogued 84 percent of all human proteins. The team also identified about 200 previously unknown proteins.

The findings were published in the May 29, 2014, issue of *Nature*.*

While genes determine many of the characteristics of an organism, they do so by providing instructions for making proteins, the building blocks and workhorses of cells, and therefore of tissues and organs. For this reason, many investigators consider a catalog of human proteins — and their location within the body — to be even more instructive and useful than the catalog of genes in the human genome.†

"By generating a comprehensive human protein data set, we have made it easier for other researchers to identify the proteins in their experiments," said Pandey, who is now at the Mayo Clinic. "We believe our data will become the gold standard in the field, especially because they were generated using uniform methods and analysis, and state-of-the-art machines.‡"

*   Kim, Min-Sik, et. al., "A Draft Map of the Human Proteome," *Nature*, May 29, 2014, https://www.ncbi.nlm.nih.gov/pmc/articles/PMC4403737/.
†   "Extensive Cataloging of Human Proteins Uncovers 193 Never Known to Exist," Johns Hopkins Medicine press release, May 28, 2014, https://www.hopkinsmedicine.org/news/media/releases/extensive_cataloging_of_human_proteins_uncovers_193_never_known_to_exist.
‡   "2014 Year in Review," *Dome*, Jan. 7, 2015, https://www.hopkinsmedicine.org/news/stories/2014_year_in_review.html.

Johns Hopkins researcher Akhilesh Pandey led a 72-person research team that completed a draft of the human proteome (a library of 30,057 proteins in the human body), which was published in 2014.

Double Arm Transplant

On April 2009, Brendan M. Marrocco lost both arms and both legs in a roadside bomb attack in Iraq.

On December 18, 2012, a surgical team led by Johns Hopkins doctors Jaimie Shores and Gerald Brandacher gave the 26-year-old infantryman arms from a deceased donor. The hospital's first bilateral arm transplant was the most extensive and complicated limb transplant procedure to date in the United States.

Marrocco's new right arm was attached above the elbow and the left was connected below the elbow.

During the 13-hour procedure, the team connected bones, blood vessels, muscles, tendons, nerves and skin on both arms in a procedure known as vascularized composite allotransplantation

Brendan Marrocco catches a ball with his new arms.

(VCA), a form of transplant that uses multiple tissue types.*

VCA "is a new and we think exciting field of transplant," said surgeon W. P. Andrew Lee in a 2018 video about Marrocco's arm transplants.

The successful operation earned international coverage in publications including the *Washington Post, USA Today* and Israel National News, and was featured

in numerous radio and television reports.†

During a Jan. 23, 2013, press conference at The Johns Hopkins Hospital, Marrocco used his new left arm to push his hair off his forehead. "It's given me a lot of hope for the future," he said. "I hated not having arms. I was all right with not having legs. Not having arms takes so much away from you, even your personality."‡

A July 2, 2014, *Washington Post* video shows Marrocco gripping and stacking wooden blocks, propelling himself in a wheelchair and doing pull-ups. "The last year and a half has been incredible," he says. "It's a whirlwind of new things and getting back to doing old things and just kind of being able to enjoy life again."§

---

*   "Johns Hopkins Surgeons Lead Double Arm Transplant," Johns Hopkins Medicine webpage, https://www.hopkinsmedicine.org/news/stories/double_arm_transplant.html.

†   "VCA Transplants Restore Hope for Quadruple Amputee, Health Resources and Services Administration," YouTube video on HRSAtube channel, July 12, 2018, https://www.youtube.com/watch?v=JOVJmMSU18Y.

‡   "Johns Hopkins Double Arm Transplant Press Conference," YouTube, Johns Hopkins Medicine channel, Jan. 29, 2013, https://www.youtube.com/watch?v=jFIAHAORs44&t=1s.

§   "Brendan Marrocco's New Arms," YouTube, *Washington Post* channel, July 2, 2014, https://www.youtube.com/watch?v=5VJeng7PH-s.

In 2021, EBDI was planning to develop a sister building to 1812 Ashland Ave. right next door at 1867 E. Eager St. EBDI intends for this building to offer additional commercial lab space to provide space for Baltimore's burgeoning tech innovation scene.

Nikita Levy: An Early Leadership Test

An early leadership challenge for Rothman came in February 2013, when an employee filed an internal complaint about Nikita Levy, a Johns Hopkins gynecologist working at the East Baltimore Medical Center since 1988.

The complaint alleged that Levy was wearing a device which looked like a writing pen around his neck while examining patients, and the employee believed it was a camera.[78]

"I remember the call telling me that a physician had been videotaping patients," said Rothman in a 2021 interview. "On one level, you hear that, your jaw drops because it's beyond the realm of anything you would have considered that you would be hearing, so at some level you just have to get over the disbelief that somebody would do that." [79]

The next day, Feb. 5, Levy acknowledged his actions under questioning and was escorted off the grounds by Johns Hopkins security. His upcoming patient appointments were cancelled, and Baltimore City Police began an investigation. Johns Hopkins sent letters to his patients, saying Levy was no longer affiliated with the institution; it also offered to help them find new providers.

On Feb. 18, Levy died by apparent suicide. [80]

Rothman credited the Johns Hopkins Medicine Marketing and Communications team with helping the institution navigate a situation that attracted substantial media attention, and he credited Peterson with providing transparency and leadership.

"Throughout the episode, we always thought: We're going to do the right thing," Rothman said in 2021. "We would do the right thing to keep people safe from this individual, and we would figure out what happened." [81]

Investigations by the FBI and police concluded without criminal charges, but a class action suit involving more than 7,000 of Levy's patients resulted in a $190 million settlement in July 2014 for what the *New York Times* characterized as "one of the largest medical malpractice cases of its kind." [82]

"This settlement, which has been formalized by the plaintiffs' attorneys and the Health System and given preliminary approval by the judge, will not in any way compromise the ability of the Health System to serve its patients, staff and community," noted a statement issued by Johns Hopkins Medicine on July 21, 2014.

"We assure you that one individual does not define Johns Hopkins. Johns Hopkins is defined by the tens of thousands of employees who come to work determined to provide world-class care for our patients and their families." [83]

Shortly after the situation came to light, Johns Hopkins Medicine launched a Speak Up and Speak Out campaign, reminding employees that they could anonymously report concerns about patient care or privacy over a dedicated hotline.

"With the Speak Up and Speak Out campaign, we made it easy for anyone who believed that something concerning was going on would feel very comfortable reporting it up," Rothman said in 2021." [84]

Rothman addressed the situation in the annual State of Johns Hopkins town hall address on Dec. 2, 2014. [85]

"Johns Hopkins has faced serious challenges in the past, and whenever an issue has occurred, we have done our very best to take lessons learned and use them to improve our patient care and the practice of medicine," he said. "One thing we have been reminded of throughout this difficult situation is how important it is that we foster an environment that encourages us all to speak up. Because an employee was willing to step forward, we were able to put a stop to Dr. Levy's behavior as soon as we learned of it." [86]

Johns Hopkins Medicine International

Feb. 1, 2014, saw the launch of Johns Hopkins Aramco Healthcare (JHAH), a first of its kind joint venture between Saudi Aramco, the world's largest oil company, and Johns Hopkins Medicine International (JHMI), the organization initially formed as Johns Hopkins International on Jan. 6, 1999 to bring together the various international efforts within Johns Hopkins Medicine.

JHMI History

From its earliest days, The Johns Hopkins Hospital has welcomed patients from outside of the United States, who traveled to Baltimore for care.

Yet for every person who traveled to Baltimore "there are hundreds and thousands of people who could not get the best of what we can do," said Mohan Chellappa, the transplant surgeon who became a founder of JHMI, and later served as interim president, executive vice president and president of global ventures.

JHMI began with an idea, Chellappa said in a 2021 interview: "We wanted to share beyond our institution what we have learned about delivering the best of what medicine can provide. We decided to try to establish good relationships with partners who had the same mission."[87]

The first such collaboration was with the Singapore government in 1998. It created two entities that launched in 2000: Johns Hopkins Singapore Clinical Services, which provided oncology care; and Johns Hopkins Singapore Private Ltd., which delivered basic science education, physician training and clinical research. The clinical services agreement ended on November 2017 when the 1,200-bed Tan Tock Seng Hospital, home of the oncology service since 2005, took control of operations.[88]

Steven J. Thompson

Steven J. Thompson, at the time vice dean of administration for Johns Hopkins Medicine, was tapped by Miller to travel to Singapore to develop the programs. "True to Hopkins, it was hard for us to pull apart the missions of clinical care, research and education," he recalled in a 2021 interview. "Over a period of time as I was developing this, it became clear it needed to be all three."[89]

When Thompson got back to Baltimore, he became the first CEO of Johns Hopkins International.

"It was a good first experience to showcase what we can bring to the world," said Chellappa of the Singapore ventures. "As an institution that puts a lot of focus on research, it was good that our initial foray included that." By the time the partnership ended, it had been part of 65 clinical trials, including many that focused on diseases with high prevalence in the region, such as dengue or cancer of the nasopharynx.[90]

Pamela Paulk

Agreements in Abu Dhabi, in the United Arab Emirates; Santiago, Chile; Beirut, Lebanon; and many other countries followed, establishing Johns Hopkins as a partner that could help governments and large organizations around the world improve patient care, conduct research, and train a new generation of clinicians.

"We are able to deliver to wealthy countries as well as countries with limited resources," said Chellappa. "In all health care, you have to think globally and you have to deliver locally."[91]

At an event in January 2014, representatives of Johns Hopkins Medicine and Saudi Aramco unveiled Johns Hopkins Aramco Healthcare (JHAH), which combined Saudi Aramco's long-established health care delivery system with Johns Hopkins' clinical, education and research expertise.

Front row seated: Paul Rothman and Abdulaziz Al-Khayyal, senior vice president, industrial relations, Saudi Aramco.

Behind them are members of the JHAH board of directors. Left to right: Emad Al-Dugaither, Motassim Al-Ma'ashouq, Steven Thompson, Ron Peterson, Huda Al-Ghoson, Christopher Kersey, Mohan Chellappa, and Yasser Mufti.

The agreements were mostly in the Middle East, with some in Asia and South America, "places where the economies were growing and they were ready to invest in education and health care," said Pamela Paulk, JHMI president from 2015 to 2018, in a 2021 interview."[92]

Perdana University

One ambitious project lasted just a few years.

In 2010, Johns Hopkins partnered with the government of Malaysia to help create Perdana University Graduate School of Medicine, the country's first four-year graduate medical program, as well as a 600-bed hospital and a research institute.[93]

The agreement launched with considerable fanfare. On Nov. 2, 2010, then U.S. Secretary of State Hillary Rodham Clinton and Deputy Prime Minister of Malaysia Tan Sri Dato' Haji Muhyiddin bin Mohd Yassin observed the signing ceremony in Kuala Lumpur.[94]

The medical school, with Charlie Wiener as dean and CEO, opened in October 2011. The school used Johns Hopkins' Genes to Society curriculum, and maintained rigorous acceptance standards, selecting 25 students out of 70 applicants for its first class.[95]

Yet in 2014, financial disputes brought an early end to the agreement, before construction of the hospital and research institute.[96] Wiener noted in 2021 that Johns Hopkins lived up to its end of the contract. "We enrolled three classes. They all graduated and they're all doing exceptionally well," he said.[97]

Johns Hopkins Aramco Healthcare

Johns Hopkins Aramco Healthcare (JHAH) was larger than previous projects. Under the agreement, JHAH provided health care to some 175,000 eligible medical recipients by combining Saudi Aramco's well-regarded and long-established health care delivery system with Johns Hopkins' clinical, education and research expertise.

Saudi Aramco, which dates to 1933 and is owned by the government of Saudi Arabia, had always provided medical care to its employees, retirees and their family members, including parents. In 2011, it selected Johns Hopkins Medicine as its partner for a joint venture in health care, and three years later JHAH was inaugurated, with Johns Hopkins Medicine owning 20 percent.[98]

The shared goal was to enhance a health care organization serving the immediate needs of Saudi Aramco employees, retirees and their families, while moving toward a health system with enhanced specialty and subspecialty care.

Johns Hopkins contributions to JHAH include:

- Clinical program development, education and training
- Direct clinical care by Johns Hopkins faculty rotating at JHAH
- Guidance for improving patient safety and continuous quality improvement
- Operations development, integration and support

Patients are seen at the 330-bed Dhahran Medical Center, a hospital offering a high level of specialty care; a 55-bed hospital in Al-Hasa that opened in 2018; three satellite outpatient Johns Hopkins Aramco Healthcare facilities; and third-party medical designated facilities in the Eastern Province.[99]

Sophie Lanzkron

"Together, we will be greater than the sum of our parts, because this joint venture combines Saudi Aramco's existing health system with the transformative science, clinical care and education that Johns Hopkins is known for," Rothman said during the signing ceremony for the agreement.

"Johns Hopkins Aramco Healthcare will become an incubator for clinical and scientific progress and will address some of the region's most pressing health challenges, including cardiovascular disease, diabetes and other chronic conditions, which are on the rise worldwide."[100]

Leading the venture from Johns Hopkins Medicine were Keith Vander Kolk as chief executive officer; Linda Gilligan as chief operating officer; Zeina Khouri-Stevens, chief nursing officer; and Sharon Myers, chief quality and safety officer.

"Aramco, a company whose principal pursuit is not health care, has already built a great foundation in health care," said Vander Kolk. "To couple that with the intellect of Johns Hopkins and the expertise we can lend to them to build a great health care system is a very unique opportunity."[101]

Tangible benefits were almost immediate.

In 2015, JHAH was seeing up to 150 adult sickle cell cases each month, often in its emergency department. The high patient volumes stemmed, in part, from the 25 percent prevalence of the genetic trait for sickle cell disease found in the Eastern Province of Saudi Arabia, the site of JHAH's main hospital.

In response, Sophie Lanzkron, director of JHM's Sickle Cell Center for Adults, led a group of JHAH and JHM clinicians to assemble a care team and open a dedicated infusion center with tailored treatment plans.

The results were dramatic: Between 2015 and 2017, both emergency department visits and hospital admissions dropped by 85 percent. JHAH directed the majority of patients to the new infusion clinic, where visits lasted less than 12 hours — compared with an average five-day hospital stay in 2015 — and where the average time to receive pain medication decreased by nearly half.[102]

In addition, Johns Hopkins faculty:

- Conducted clinical rotations in specialties such as bariatric surgery, colorectal surgery, radiology, ophthalmology, vascular surgery and cardiology
- Developed programs and services not previously offered at JHAH, including robotic surgery, cardiac surgery, cardiac rehabilitation, complex electrophysiology, palliative care, endovascular therapies and noninvasive imaging
- Introduced inpatient and outpatient palliative care
- Developed a center of excellence in bariatrics and weight management
- Established the Office of Emergency Management for Disaster/Emergency Preparedness
- Performed the first robotic surgery in the Kingdom of Saudi Arabia, using the da Vinci Xi Surgical System
- Performed Johns Hopkins Aramco Healthcare's first Descemet Membrane Endothelial Keratoplasty procedure to correct the innermost layer of the cornea, the endothelium[103]

JHMI Leadership

Thompson was CEO of Johns Hopkins Medicine International from 1999 to 2007, and again from 2011 to 2014. Thompson, who had joined Johns Hopkins in 1994 as administrator of the Department of Anesthesiology and Critical Care Medicine, had left the JHMI leadership post in 2007 to become senior vice president of Johns Hopkins Medicine.

Harris Benny was CEO from 2007 to 2011.

Chellappa served as interim president after Thompson left Johns Hopkins in 2015. Chellappa, a fellow of the Royal College of Surgeons of Edinburgh and the American College of Surgeons, also holds a business degree. He received special training in surgical gastroenterology at the University of Leeds and Hammersmith Hospital in England in 1982, and training in liver and kidney transplantation at the University of Louvain in Belgium in 1986.

Chellappa pioneered laparoscopic gastrointestinal surgery in Asia and subsequently provided training in laparoscopic techniques to many surgeons from the Middle East and the Asia-Pacific region.[104]

On March 1, 2015, Paulk became president of JHMI. Paulk, who holds a master's degree in social work from Florida State University and a master's degree in business administration from The Johns Hopkins University, was vice president of human resources for the Johns Hopkins Health System from 2000 to 2010, and senior vice president for human resources for Johns Hopkins Medicine from 2010 to 2015.

A few years after Johns Hopkins International was formed, she added the role of vice president of global services at JHMI to her responsibilities. Her work helping the organization identify potential collaborations was instrumental in the growth of JHMI.

"Pamela possesses a truly unique combination of skills — a stellar operational track record and outstanding leadership skills — that will be necessary to take Johns Hopkins Medicine International to the next level," said Christopher W. Kersey, chairman of the board of Johns Hopkins Medicine International, at the time.[105]

Paulk described her tenure with JHMI as "a joy," particularly her experiences traveling nearly half of every year and getting to know each country's approach to health and medicine. "It was such a steep learning curve because every country was different," she said.

"The thing that was consistent was that every place we went, the mutual goal was to provide better health care for their people. There was this commonality that made it very welcoming. The esteem that the Johns Hopkins name holds around the world always humbles me."[106]

Wiener has been JHMI president since January 1, 2019. He earned his bachelor's degree from Duke University and his medical degree from the University of Miami. He was a fellow in pulmonary and critical care medicine at Johns Hopkins from 1987 to 1991.

At Johns Hopkins, he served as chair of the committee that created the school of medicine's "Genes to Society" curriculum, and as director emeritus of the Osler Internal Medicine Training Program, the first U.S. medical residency program. From 2015 to 2019, he was the director of the medicine, science and the humanities undergraduate major at the Johns Hopkins University Krieger School of Arts and Sciences.[107]

Wiener, in 2021, noted that the COVID-19 pandemic limited the travel that is essential for JHMI partnerships, but also reminded the world that Johns Hopkins is a world leader in clinical care, research and education. "We want to keep increasing our footprint internationally," he said. "The goal is to look for places where our mission aligns with the mission of others."[108]

Happy 125th Birthday to The Johns Hopkins Hospital

On May 7, 2014, The Johns Hopkins Hospital marked a major milestone as it celebrated 125 years since its grand opening.

It was a moment for reflecting on how far the hospital had come, while remaining true to the vision articulated by its eponymous founder, to treat both rich and poor, and to unite a hospital and a medical school in a previously unimagined way.

"On May 7, 1889, a large crowd gathered in the rotunda beneath the great dome of what is now known as the Billings Administration Building to celebrate the opening of The Johns Hopkins Hospital," wrote Ronald R. Peterson in a May 2014 *Dome* column.

Then the president of The Johns Hopkins Hospital and Health System and executive vice president of Johns Hopkins Medicine, he continued:

- They listened as Johns Hopkins' first president, Daniel Coit Gilman, and the hospital's designer, John Shaw Billings, spoke of their aspirations for

Karen B. Haller Endowed Scholarship

For more than three decades, Karen Haller has been a leader in patient-centered care, most recently as vice president for nursing and clinical affairs for Johns Hopkins Medicine International.*

In 2015, The Johns Hopkins Hospital's Department of Nursing and the Johns Hopkins University School of Nursing honored her work by creating the Karen B. Haller Endowed Scholarship.

Through the scholarship, an award is made each year to a Johns Hopkins University School of Nursing student in the Master of Science in Nursing: Entry into Nursing program.

Before joining Johns Hopkins Medicine International in 2015, Haller was chief nursing officer for The Johns Hopkins Hospital.

Haller, born in Detroit, earned a bachelor's degree, two master's degrees and a Ph.D., all from The University of Michigan. She completed a fellowship at The University of Michigan before becoming an associate professor of maternal-child health and a research scientist at Loyola University in Chicago.

She joined The Johns Hopkins Hospital in 1988 as director of nursing for research and education, and became associate professor and associate dean for clinical affairs at the Johns Hopkins University School of Nursing in 1989. In 1993, she was promoted to director of nursing for medicine.

Haller has published more than 100 articles and editorials on subjects ranging from leadership and management in patient care delivery systems to end-of-life decision-making and patient safety. For 10 years, she was editor of the *Journal of Obstetric, Gynecologic & Neonatal Nursing*. In 2001, she received the National Caring Award from the Association of Women's Health, Obstetric and Neonatal Nurses for her outstanding contributions to nursing and nursing research.†

Karen Haller

---

* Hub staff report, "JHH, JHU School of Nursing Honor Karen Haller with Scholarship," *Hub*, Dec. 22, 2015, https://hub.jhu.edu/at-work/2015/12/22/karen-haller-scholarship/.

† "Karen Haller," Johns Hopkins Medicine webpage, Portrait Collection: The Johns Hopkins Medical Institutions, https://portraitcollection.jhmi.edu/portraits/haller-karen.

---

what was then believed to be the largest hospital in the country.

- A century to the day later, on May 7, 1989, a similarly large group — including 125 collateral descendants of Johns Hopkins (a bachelor who had no children) — also gathered under the soaring dome to celebrate the hospital's 100th birthday.

- Now, as we mark the hospital's 125th anniversary, we can look back on an astounding quarter-century of achievements since our last celebration. These have had a profound impact on health care — not only in Baltimore and Maryland, but across the nation and around the world.[109]

The hospital marked the event with commemorative medallions that showed the hospital as it looked in 1889 on one side, and the new clinical towers on the other.

ENDNOTES

1 Karen Nitkin Zoom interview with Paul Rothman, Oct. 13, 2021.

2 Rothman interview, Oct. 13, 2021.

3 Rothman interview, Oct. 13, 2021.

4 "Johns Hopkins Medicine at 25," Johns Hopkins Medicine webpage, https://www.hopkinsmedicine.org/25-anniversary/milestones.html.

5 Miller, Edward D., "The Future of Johns Hopkins Medicine, *Dome*, Sept. 16, 2011.

6 Miller, Edward D., "The Future of Johns Hopkins Medicine, *Dome*, Sept. 16, 2011.

7 Peterson, Ronald R., "A Time for Belt Tightening," *Dome*, April 11, 2012.

8 "The Johns Hopkins Hospital Dedication Ceremony, April 12, 2012," YouTube video, Johns Hopkins Medicine, https://www.youtube.com/watch?v=HKlYsWl6cZM.

9 "Hopkins Breaks Ground for New East Baltimore Campus," June 2, 2016, press release, Johns Hopkins Medicine, https://www.hopkinsmedicine.org/Press_releases/2006/06_02_06.html.

10 "Johns Hopkins Contractors Turn Over the Keys to New Patient Care Facility," Johns Hopkins Medicine press release, Nov. 11, 2011, https://www.hopkinsmedicine.org/news/media/releases/johns_hopkins_contractors_turn_over_the_keys_to_new_patient_care_facility.

11 "It's Open! The Johns Hopkins Hospital's New $1.1-billion Building Welcomes First Patients, Johns Hopkins Medicine press release, May 1, 2012, https://www.hopkinsmedicine.org/news/media/releases/its_open_the_johns_hopkins_hospitals_new_11_billion_building_welcomes_first_patients.

12 "Facilities Designed with You in Mind," web page for The Johns Hopkins Hospital, https://www.hopkinsmedicine.org/the_johns_hopkins_hospital/about/enhanced_facilities/.

13 "New Era Approaches for The Johns Hopkins Hospital," Johns Hopkins Medicine press release, Dec. 30, 2011, https://www.hopkinsmedicine.org/news/media/releases/new_era_approaches_for_the_johns_hopkins_hospital

14 Karen Nitkin phone interview with Sally MacConnell, Oct. 15, 2021

15 "Johns Hopkins Community Physicians to Open New Location in Fulton," Johns Hopkins Medicine press release, May 3, 2013, https://www.hopkinsmedicine.org/news/media/releases/johns_hopkins_community_physicians_to_open_new_location_in_fulton.

16 "Johns Hopkins Community Physicians Locations: Washington D.C., I Street," Johns Hopkins Medicine webpage, https://www.hopkinsmedicine.org/community_physicians/locations/washington_dc/IStreet.html.

17 "JHCP Heart Care Expands Services," *New Directions*, Winter 2013, https://www.hopkinsmedicine.org/news/publications/johns_hopkins_medicine_suburban/new_directions_winter_2013/jhcp_heart_care_expands_services.

18 Rothman interview, Oct. 13, 2021.

19 "25 Years of Progress," Johns Hopkins Medicine webpage, https://www.hopkinsmedicine.org/25-anniversary/events-achievements.html#8.

20 Rothman interview, Oct. 13, 2021.

21 Rothman interview, Oct. 13, 2021.

22 "25 Years of Progress," Johns Hopkins Medicine webpage, https://www.hopkinsmedicine.org/25-anniversary/events-achievements.html#15.

23 Shapiro, Stephanie, "An Epic Journey," *Dome*, September 2011, https://www.hopkinsmedicine.org/news/publications/dome/september_2011/an_epic_journey.

24 2014 Biennial Report, Johns Hopkins Medicine.

25 Shapiro, Stephanie, "The Age of Epic," webpage, Epic at Johns Hopkins Medicine, https://www.hopkinsmedicine.org/epic/news/The%20Age%20of%20Epic.

26 Rothman interview, Oct. 13, 2021.

27 Nitkin, Karen, "Training for an Epic Assignment," *Dome*, March 2014, https://www.hopkinsmedicine.org/news/publications/dome/dome_march_2014/training_for_an_epic_assignment.

28 Nitkin, Karen, "Engineering a Smooth Path to Epic Improvements," *Dome*, May 2014, https://www.hopkinsmedicine.org/news/publications/dome/dome_may_2014/engineering_a_smooth_path_to_epic_improvements.

29 DuVernay, Christina, "The Case for Open Notes Access," *Dome*, November 2016, https://www.hopkinsmedicine.org/news/articles/the-case-for-open-notes-access.

30 Nitkin, Karen, "Dragon Medical One Brings Voice Recognition to Epic," *Hopkins Insider*, November 2018, https://www.hopkinsmedicine.org/news/articles/hed-dragon-medical-one-brings-voice-recognition-to-epic.

31 Nitkin, Karen, "Epic 2014 Ushers in New Features, Easier Navigation," *Dome*, April 2015, https://www.hopkinsmedicine.org/news/articles/epic-2014-ushers-in-new-features-easier-navigation.

32 "You're Invited! Join a Patient and Family Advisory Council," YouTube, Johns Hopkins Medicine, https://www.youtube.com/watch?v=BSRief4hXHY&t=180s.

33 47 Feinberg, Lynn Friss, "Patient and Family Advisory Councils in Hospitals: Building Partnerships to Improve Care, AARP Public Policy Institute, November 2017, https://www.aarp.org/content/dam/aarp/ppi/2017/10/patient-and-family-advisory-councils-in-hospitals-building-partnerships-to-improve-care.pdf.

34 Minkove, Judy F., with Gilbert, Patrick, "Valued Partners," *Dome*, May 2013, https://www.hopkinsmedicine.org/news/publications/dome/dome_may_2013/valued_partners_

35 "Patient and Family Advisory Councils, Johns Hopkins Medicine webpage, https://www.hopkinsmedicine.org/PFAC.

36 DuVernay, Christina, "A Health Care Delivery Model that Delivers," *Dome*, December 2014, https://www.hopkinsmedicine.org/news/publications/dome/dome_december_2014/a_health_care_delivery_model_that_delivers.

37 Berkowitz, Scott, et. al., "Association of a Care Coordination Model with Health Care Costs and Utilization," *JAMA Network Open*, Nov. 2, 2018, https://jamanetwork.com/journals/jamanetworkopen/fullarticle/2712183.

38 2014 Biennial Report, Johns Hopkins Medicine.

39 "Johns Hopkins Medicine Awarded $19.9M Innovation Grant from CMS for its J-CHiP Program," Johns Hopkins Medicine press release, June 18, 2012, https://www.hopkinsmedicine.org/news/media/releases/johns_hopkins_medicine_awarded_199m_innovation_grant_from_cms_for_its_j_chip_program.

40 Blum, Karen, "Improving Community Health, Reducing Health Disparities," *Dome*, March 2, 2017, https://www.hopkinsmedicine.org/news/articles/improving-community-health-reducing-health-disparities.

41 Kennedy, Kelly, "Focus on Super Utilizers Cuts Costs," *USA TODAY*, Sept. 16, 2014, https://www.usatoday.com/story/news/nation/2014/09/16/focus-on-super-utilizers-cuts-costs/14648069/.

42 Berkowitz, Scott, et. al., "Association of a Care Coordination Model with Health Care Costs and Utilization," *JAMA Network Open*, Nov. 2, 2018, https://jamanetwork.com/journals/jamanetworkopen/fullarticle/2712183.

43 Smith, Patrick, "Community Health Partnership Results: Better, Less-Expensive Health Care," *Dome*, Dec. 20, 2018, https://www.hopkinsmedicine.org/news/articles/community-health-partnership-results-better-less-expensive-health-care-1.

44 Smith, Patrick, "Community Health Partnership Results: Better, Less-Expensive Health Care," *Dome*, Dec. 20, 2018, https://www.hopkinsmedicine.org/news/articles/community-health-partnership-results-better-less-expensive-health-care-1.

45 Smith, Patrick, "By Removing Barriers, Community Health Workers Help Baltimoreans Focus on Getting Better," *Dome*, Dec. 6, 2019, https://www.hopkinsmedicine.org/news/articles/by-removing-barriers-community-health-workers-help-baltimoreans-focus-on-getting-better.

46 Smith, Linell, "The Access Partnership at Johns Hopkins Celebrates a Decade of Providing Low-Cost Services — and Hope," *Dome*, June 6, 2019, https://www.hopkinsmedicine.org/news/articles/johns-hopkins-offers-access-to-healthcare-for-underinsured-and-uninsured-patients-in-east-baltimore.

47 "First Full Genes to Society Class of Johns Hopkins Medical Students Graduates on May 23," Johns Hopkins Medicine press release, May 23, 2013, https://www.hopkinsmedicine.org/news/media/releases/first_full_genes_to_society_class_of_johns_hopkins_medical_students_graduates_on_may_23.

48 Wiener, Charlie, et. al., "'Genes to society'--the logic and process of the new curriculum for the Johns Hopkins University School of Medicine," *Academic Medicine*, March 2010, https://pubmed.ncbi.nlm.nih.gov/20182127/.

49 Childs, Barton, Genetic Medicine: *A Logic of Disease*, Johns Hopkins University Press, Sept. 15, 2003.

50 "Genes to Society: Overview," YouTube, Johns Hopkins Medicine channel, https://www.youtube.com/watch?v=ma5na-y4LpQ.

51 "Genes to Society: Overview," YouTube, Johns Hopkins Medicine channel, https://www.youtube.com/watch?v=ma5na-y4LpQ.

52 Nitkin, Karen, "Breaking Down Barriers," *Dome*, Aug. 2, 2016, https://www.hopkinsmedicine.org/news/articles/breaking-down-barriers.

53 Nitkin, Karen, "Breaking Down Barriers," *Dome*, Aug. 2, 2016, https://www.hopkinsmedicine.org/news/articles/breaking-down-barriers.

54 "Medicine for the Greater Good," webpage, Johns Hopkins Medicine, https://www.medicineforthegreatergood.org/mgg/.

55 Nitkin, Karen, "Breaking Down Barriers," *Dome*, Aug. 2, 2016, https://www.hopkinsmedicine.org/news/articles/breaking-down-barriers.

56 Nitkin, Karen, and Wallach, Rachel, "Sharing Stories in Spanish, *Dome*, May 2015, https://www.hopkinsmedicine.org/news/publications/_docs/dome_may_2015.pdf.

57 "Trabajando Juntos – Johns Hopkins Medicine Working to Vaccinate the Latino Community," Johns Hopkins Medicine webpage, Coronavirus (COVID-19) Information and Updates, May 12, 2021, https://www.hopkinsmedicine.org/coronavirus/equity/latino-community-vaccination.html.

58 "New Center Focuses on Latino Health Needs," *Johns Hopkins Bayview News*, Winter 2014, https://www.hopkinsmedicine.org/news/publications/jh_bayview_news/winter_2014/new_center_focuses_on_latino_health_needs.

59 Blum, Karen, "It's All About Familia," *Dome*, October 2013, https://www.hopkinsmedicine.org/news/publications/dome/dome_october_2013_vol_64/its_all_about_familia.

60 Minkove, Judy F., "Urban Health Internal Medicine Pediatrics Residency Bolsters Care for the Most Vulnerable Patients," *Dome*, Winter 2021, https://www.hopkinsmedicine.org/news/articles/urban-health-internal-medicine-pediatrics-residency-bolsters-care-for-the-most-vulnerable-patients.

61 Minkove, Judy F., "Urban Health Internal Medicine Pediatrics Residency Bolsters Care for the Most Vulnerable Patients," *Dome*, Winter 2021, https://www.hopkinsmedicine.org/news/articles/urban-health-internal-medicine-pediatrics-residency-bolsters-care-for-the-most-vulnerable-patients.

62 "Johns Hopkins Bayview Medical Center Named Center of Excellence for Joint Replacements by National Companies," Johns Hopkins Medicine press release, Oct 8, 2013, https://www.hopkinsmedicine.org/news/media/releases/_johns_hopkins_bayview_medical_center_named_center_of_excellence_for_joint_replacements_by_national_companies.

63 "Innovative Philanthropic Financing for Community Change: The East Baltimore Revitalization Initiative, prepared for the Annie E. Casey Foundation by Paul Brophy of Brophy & Reilly LLC and Tom Waldron of the Hatcher Group, 2014, https://assets.aecf.org/m/resourcedoc/aecf-InnovativePhilanthropicFinancingforCommunityChange-2014.pdf.

64 Rienzi, Greg, "The Changing Face of East Baltimore," *JHU Gazette*, January 2013, https://hub.jhu.edu/gazette/2013/january/east-baltimore-changes-development/.

65 "EBDI's Momentum," *Baltimore Sun* editorial, Sept. 12 2016, https://www.baltimoresun.com/opinion/editorial/bs-ed-ebdi-20160912-story.html.

66 "Ceremonial Groundbreaking of New Elementary School and Early Childhood Center in East Baltimore," Johns Hopkins University School of Education press release, June 11, 2012, https://releases.jhu.edu/2012/06/11/henderson-hopkins-school/.

67 Kimmelman, Michael, "Reading, Writing and Renewal (The Urban Kind)," *New York Times*, March 18, 2014, https://www.nytimes.com/2014/03/18/arts/design/reading-writing-and-renewal-the-urban-kind.html?hpw&rref=arts&_r=0.

68 *Hub* staff report, "Henderson Hopkins Named One of Baltimore's Top Schools by 'U.S. News,'" *The Hub*, Oct. 22, 2021, https://hub.jhu.edu/2021/10/22/henderson-hopkins-us-news-k-8-rankings-2022/.

69 Olesker, Michael, "Johns Hopkins-Run Early Childhood Center Opens in East Baltimore," *The Hub*, September-October 2014, https://hub.jhu.edu/gazette/2014/september-october/currents-east-baltimore-early-childhood-center/.

70 Karen Nitkin telephone interview with Shannon Seopaul, Sept. 22, 2021.

71 Rienzi, Greg, "New Era Begins in East Baltimore," *The JHU Gazette*, April 14, 2008, https://pages.jh.edu/gazette/2008/14apr08/14newera.html.

72 "BSi History," Webpage of the Pederson Brain Science Institute, Johns Hopkins Medicine, https://www.hopkinsmedicine.org/pedersen-brain-science-institute/history.html.

73 Hub staff report, JHU-Affiliated Lieber Institute Announces Brain Development Research Consortium, *The Hub*, April 3, 2014, https://hub.jhu.edu/2014/04/03/lieber-institute-partnership/.

74 "Gala Event Nov. 10 at Johns Hopkins to Mark Opening of Lieber Institute for Brain Development," Johns Hopkins Medicine media advisory, Nov. 10, 2011, https://www.hopkinsmedicine.org/news/media/releases/media_advisory__gala_event_nov_10_at_johns_hopkins_to_mark_opening_of_lieber_institute_for_brain_development.

75 "The Kavli Foundation and Johns Hopkins Announce Creation of Neuroscience Discovery Institute," Johns Hopkins Medicine press release, Oct. 1, 2015, https://www.hopkinsmedicine.org/news/media/releases/the_kavli_foundation_and_johns_hopkins_announce_creation_of_neuroscience_discovery_institute.

76 *Hub* staff report, "Johns Hopkins, Kavli Foundation Create New Institute Dedicated to Study of How the Brain Works," *The Hub*, Oct. 1, 2015, https://hub.jhu.edu/2015/10/01/kavli-neuroscience-discovery-institute/.

77 Pearce, Katie, "Startups Get Their Start at Johns Hopkins' New East Baltimore Incubation Space, *The Hub*, April 27, 2017, https://hub.jhu.edu/2017/04/27/fastforward-1812-grand-opening/.

78   "Nikita Levy: A Timeline of Events," Johns Hopkins Medicine webpage, Aug. 29, 2014, https://www.hopkinsmedicine.org/news/Nikita_Levy.html.

79 Rothman interview, Oct. 13, 2021.

80 Dance, Scott; Fenton, Justin; Anderson, Jessica; *Baltimore Sun*, "Accused Hopkins Gynecologist Suffocated Himself with Helium," *Baltimore Sun*, Feb. 20, 2013, https://www.baltimoresun.com/maryland/baltimore-county/bs-xpm-2013-02-20-bs-md-co-levy-death-20130220-story.html.

81 Rothman interview, Oct. 13, 2021.

82 Gabriel, Trip, "Hospital Agrees to Pay $190 Million Over Recording of Pelvic Exams," *New York Times*, July 21, 2014, https://www.nytimes.com/2014/07/22/us/johns-hopkins-settlement-190-million.html.

83 "Nikita Levy: A Timeline of Events," Johns Hopkins Medicine webpage, Aug. 29, 2014, https://www.hopkinsmedicine.org/news/Nikita_Levy.html.

84 Rothman interview, Oct. 13, 2021.

85 Power Point, The State of Johns Hopkins Medicine 2014.

86 Power Point, The State of Johns Hopkins Medicine 2014.

87 Karen Nitkin telephone interview with Mohan Chellappa, Oct. 21, 2021.

88 "Johns Hopkins Singapore," Johns Hopkins Medicine webpage, Johns Hopkins Medicine International, https://www.hopkinsmedicine. org/international/international_affiliations/asia_ pacific/singapore_international_medical_centre. html.

89 Karen Nitkin telephone interview with Steve Thompson, Oct. 29, 2021.

90 Chellappa interview, Oct. 21, 2021.

91 Chellappa interview, Oct. 21, 2021.

92 Karen Nitkin telephone interview with Pamela Paulk, Oct. 21, 2021.

93 Smith, Linell, "Exporting Innovative Education," *Dome*, Dec. 1, 2010, https:// www.hopkinsmedicine.org/news/publications/ dome/december_2010/exporting_innovative_ education.

94 "Johns Hopkins to Develop Medical School and Teaching Hospital," *JHU Gazette*, November 8, 2010, https://gazette.jhu.edu/2010/11/08/johns-hopkins-to-help-develop-medical-school-and-teaching-hospital/.

95 Smith, Linell, "An Amazing Feat," *Dome*, October 2011, https://www.hopkinsmedicine. org/news/publications/dome/october_2011/ an_amazing_feat_.

96 Sharma, Yojana, "US-Malaysia Medical School Collaboration Collapses," *University World News*, Aug. 19, 2014, https://www.universityworldnews. com/post.php?story=20140819111911158.

97 Wiener interview, Oct. 18, 2021.

98 Governs, Cymantha, "Global Reach: An International Partnership is Transforming Medical Care 6,700 Miles Away," *Dome*, March 30, 2015, https://www.hopkinsmedicine.org/ news/articles/global-reach.

99 "Johns Hopkins Aramco," Johns Hopkins Medicine webpage, Johns Hopkins Medicine International, https://www.hopkinsmedicine.org/ international/international_affiliations/middle_ east/johns_hopkins_aramco_healthcare.html.

100 "Saudi Aramco and Johns Hopkins Medicine Launch Health Care Joint Venture," Johns Hopkins Medicine press release, Jan. 28, 2014, https://www.hopkinsmedicine.org/news/media/ releases/saudi_aramco_and_johns_hopkins_ medicine_launch_health_care_joint_venture.

101 "Visionaries of Johns Hopkins Aramco Healthcare," YouTube, Johns Hopkins Medicine channel, https://www.youtube.com/ watch?v=QPNgrJAqIdo&t=96s.

102 "Johns Hopkins Medicine at 25," Johns Hopkins Medicine webpage, https://www. hopkinsmedicine.org/25-anniversary/milestones. html.

103 "Johns Hopkins Aramco," Johns Hopkins Medicine webpage, Johns Hopkins Medicine International, https://www.hopkinsmedicine.org/ international/international_affiliations/middle_ east/johns_hopkins_aramco_healthcare.html.

104 "Mohan Chellappa," *Dome*, Dec. 1, 2020, https://www.hopkinsmedicine.org/news/ publications/dome/december_2010/mohan_ chellappa .

105 "At the Helm: Meet Pamela Paulk, President of Johns Hopkins Medicine International," *Dome*, July 13, 2017, https://www.hopkinsmedicine.org/ news/media/releases/pamela_paulk_appointed_ president_of_johns_hopkins_medicine_ international.

106 Karen Nitkin telephone interview with Pamela Paulk, Oct. 21, 2021.

107 "Charles M. Wiener, M.D.," Johns Hopkins Medicine webpage, Johns Hopkins Medicine International Leadership Team, https://www. hopkinsmedicine.org/international/about/ leadership/charles_wiener.html.

108 Wiener interview, Oct. 21, 2021.

109 Peterson, Ronald R., "A Transformative Era," *Dome*, May 1, 2014, https://www. hopkinsmedicine.org/news/publications/dome/ dome_may_2014/a_transformative_era.

Chapter Two

Integration, Efficiency, Innovation (2014–2016)

HE YEARS AFTER THE LAUNCH of the Strategic Plan were a time of integration for Johns Hopkins Medicine. With the Epic electronic medical record system in place, attention turned to creating efficiencies and improving the performance of the sprawling six-hospital health system.

This involved several ambitious outward-facing initiatives, including the embrace of High-Value Health Care, which seeks ways to improve care while eliminating unnecessary tests and procedures; the launch of Johns Hopkins Technology Ventures to support the commercialization of Johns Hopkins innovations; and new research and clinical partnerships with other health systems.

Within Johns Hopkins, a move toward precision medicine would begin a dramatic shift toward using large amounts of data to provide care that is tailored to the individual patient. The One University initiative encouraged research collaborations across departments and divisions, and the first steps toward a supply chain consolidation would create efficiencies across the institution.

Recognizing that people are Johns Hopkins Medicine's most valuable asset, leaders also took steps to reward clinical excellence and promote the health and well-being of all JHM employees.

These years also saw several important hospital renovations and expansions.

Choosing Wisely and High-Value Care

In 2013, Johns Hopkins Medicine embraced Choosing Wisely, a nationwide challenge to clinicians to find concrete ways to increase health care value for patients by reducing tests and treatments that do not improve care but do increase cost.

It might sound counterintuitive, but tests and treatments can do more harm than good, particularly when they fail to give clinicians the information they need to make appropriate care decisions. The tests can be costly, invasive and time-consuming, and they often lead to yet more unnecessary procedures, exposing patients to additional stress, discomfort and, importantly, risk of harm.

Choosing Wisely — and other efforts at and led by Johns Hopkins — sought to change that.

> < Johns Hopkins Medicine innovations during this period included a move toward precision medicine, which uses large amounts of data to provide care that is tailored to the individual patient.

Michelle Sharp, a third-year resident in internal medicine at The Johns Hopkins Hospital, helped lead its Choosing Wisely initiative.

Roy Ziegelstein

Linda Mobula

"We've evolved a practice model in this country where it is fairly standard to do many tests and procedures as a reflex," said Roy Ziegelstein, vice dean for education at the Johns Hopkins University School of Medicine, in a 2014 *Dome* article.

"I try to teach students and residents that almost everything we do has the potential to do harm. A simple aspirin tablet has a potential to cause harm. Even something like a chest X-ray has radiation exposure."[1]

Choosing Wisely was introduced by the American Board of Internal Medicine (ABIM) Foundation in 2012. It was based on a 2010 *New England Journal of Medicine* article calling on medical specialty associations to identify five tests and treatments in their respective fields that were overused and did not provide meaningful benefit for patients.

"If we restrict ourselves to the most egregious causes of waste, we can demonstrate to a skeptical public that we are genuinely protecting patients' interests and not simply 'rationing' health care, regardless of the benefit, for cost-cutting purposes," argued author Howard Brody.[2]

Johns Hopkins joined about 80 national and state medical specialty societies, regional health collaboratives and consumer partners in taking on Choosing Wisely's challenge. Most physicians seemed to find value in the program, which put responsibility for cost-cutting on the clinicians who best know what works and what doesn't. At Johns Hopkins, one early success involved more judicious use of pulse oximeters to measure oxygen saturation in children. The device, which uses a small sensor on a patient's finger, can give erroneously low readings in young patients who move around a lot, leading to unnecessary tests and longer hospital stays.

In one national 2016 study, 92% of physicians in primary care, medical specialties, and surgical specialties agreed or somewhat agreed that Choosing Wisely was a legitimate source of guidance.[3]

High Value at Hopkins

Even before any formal involvement with Choosing Wisely, Johns Hopkins clinicians had been raising awareness about the need for high-value care.

Back in 2009, Ziegelstein helped form a group called Johns Hopkins Bayview Providers for Responsible Ordering (PRO), inspired by a presentation from internal medical resident Linda Mobula.[4]

Mobula told of a Johns Hopkins patient who had undergone many unnecessary and expensive tests to arrive at a diagnosis that was apparent without them.

The patient had been diagnosed with and was eventually treated for acute cholecystitis, or gallbladder inflammation, explained Mobula, now senior health specialist with the World Bank, a hospitalist physician with Johns Hopkins Community Physicians, and an assistant professor in the School of Medicine. But not before the diagnosis was confirmed with an ultrasound, a CAT scan, a nuclear medicine scan and magnetic resonance cholangiopancreatography (MRCP).

"This was wasteful care," said Ziegelstein.[5]

One of the members of PRO was cardiologist Jeffrey C. Trost, whose work reduced unnecessary diagnostic tests for patients with an acute coronary syndrome (ACS).

In August and September 2011, Trost and his colleagues led information sessions and created pop-up warnings (which could be easily overridden) that appeared on clinician computer screens when they ordered certain tests. These efforts, published in the *Journal of General Internal Medicine*, led to a 66% reduction in tests and savings of $1.25 million in the first year, with no decrease to diagnostic accuracy.[6]

A few years later, when Ziegelstein was vice dean for education, he approached Julia McMillan, associate dean for graduate medical education, with an idea. He wanted to challenge every Johns Hopkins residency and fellowship training program to identify one commonly used test or procedure that is unnecessary, and then design and implement an intervention to reduce or eliminate it.[7]

Julia McMillan   Pam Johnson

Ziegelstein made his pitch to the program directors during a graduate medical education committee meeting in the majestic Mary Elizabeth Garrett Boardroom.

"The response was not warm," Ziegelstein said in 2021. "They're busy people and to them I'm sure this just seemed like another thing to do."

But one residency director really embraced the idea: Pamela Johnson, in radiology. "She is a once-in-a-lifetime individual," said Ziegelstein. "She just took it and ran."

Johnson believed that the idea would only work with collaboration, and in 2015 she led the creation of the High Value Practice Alliance, a coalition of training program directors committed to advancing value-based quality improvement and engaging residents and fellows in the work.

At the same time, two hospitalists in the department of medicine, Lenny Feldman and Amit Pahwa, were embarking on their second year directing the Department of Medicine High Value Care Committee, which brought together representatives from Johns Hopkins' departments of medicine, pharmacy, infectious diseases, pathology, radiology and information technology to refine overuse of tests and treatments. They succeeded in reducing unnecessary use of labs (folate, *C. difficile*) and telemetry.

Recognizing that work to improve appropriate use of tests, procedures, treatments and other resources was being done by teams across the system, Ziegelstein and Johnson organized the first Johns Hopkins High Value Practice Research Symposium on Feb. 1, 2016. The conference featured more than 50 presentations, many from medical students or residents.

In March 2016, Redonda Miller, then vice president of medical affairs for The Johns Hopkins Hospital, and Renee Demski, vice president of quality, furthered this work by appointing Johnson, Feldman, Pahwa and Howard County General Hospital internist Trushar Dungarani to direct the Johns Hopkins Health System High Value Care Committee, designed to identify and scale effective performance improvement systemwide. Since inception, this team has reduced low-value lab and imaging tests to avoid millions of dollars in charges to patients and payers.

Allen Kachalia

Also on the heels of the Johns Hopkins high value research conference came the creation of the High Value Practice Academic Alliance (HVPAA), a national organization led by Johns Hopkins clinicians designed to broadly disseminate high value care quality improvement and education through a range of programs, including an annual conference that brought together hundreds of clinicians from dozens of health care systems and hospitals across the country.

HVPAA was founded by Johnson, who by then was vice chair of quality and safety in the Department of Radiology and Radiological Science and physician lead on the Johns Hopkins Department of Medicine High Value Care Committee; and Ziegelstein.

From its inception in 2017 to 2021, the conference facilitated sharing of more than 500 high value performance improvement initiatives from medical centers across the United States, Canada and other locations, including Japan and Norway. Additionally, the annual three-day meeting includes lectures, workshops, and panel discussions on high value health care.

In the 2019 conference's keynote speech, Allen Kachalia, senior vice president of patient safety and quality at Johns Hopkins Medicine, reviewed some of the barriers to providing high value care, including fear of litigation, a desire for answers even if test results won't change treatment, inertia (doing something the way it's always been done), and just being "safe" by ordering more tests.

"Clinical leadership to change how we do things is mandatory," said Kachalia, who is also director of the Armstrong Institute for Patient Safety and Quality, a role he assumed in 2018. "Be explicit about what you're trying to improve when you talk about improving value. It's about reducing waste and improving quality of care for patients."

Among the work presented at the 2019 conference:

- A study conducted by Reid Thompson, associate professor of pediatrics at the Johns Hopkins University School of Medicine, and graduate student Christine Kuryla, which reduced referrals to cardiologists for children with harmless heart murmurs. For the study, pediatricians used stethoscopes that record heart sounds, which were then uploaded through an app that used artificial intelligence to analyze whether the murmur required further evaluation.

- An adherence tool in the electronic medical record system, developed by Johns Hopkins pharmacists Alona Crowder and Jessica Merrey, used to check if patients were taking their medications or taking less than recommended. If cost was the reason, the pharmacists could recommend solutions including less expensive medications, discounts from manufacturers or copay cards.

- An intervention to decrease unnecessary imaging for hospitalized patients, led by Johnson and Johns Hopkins radiology fellow Nebiyu Adenaw. The radiologists developed an alert in Epic that let physicians know that an additional abdominal ultrasound within 72 hours of a previous CT (computed tomography) scan likely would not yield useful information. As a result, 16% of scan orders were canceled over a six-month period, with no sign that this compromised care.[8]

Updates to an Unusual Payment System

Unique among the states, Maryland has had a Medicare waiver since 1977. In most states, Medicare sets discounted rates for its patients, and hospitals set their own rates. That system incentivizes hospitals to charge more to private insurance companies.

In Maryland, a special commission, the Health Services Cost Review Commission (HSCRC), sets hospital rates for everyone.

Because of the waiver, hospital services delivered to Medicare patients in Maryland are paid by the federal government at a higher rate than they would without the waiver. To keep the waiver, Maryland must slow the rate at which total hospital costs are increasing. As health systems approach the global budget revenue limit, reimbursement rates drop to ensure it is not surpassed.[9]

On June 1, 2014, the state updated the system so that it rewarded efforts to lower patient care costs through population health measures and efficient treatments.

The implications for Johns Hopkins Medicine were clear.

"It means we will do our very best to provide preventive

The Office of Marketing and Communications created this graphic in 2015 to explain Maryland's reimbursement system.

Does the Medicare waiver change how The Johns Hopkins Hospital, Johns Hopkins Bayview Medical Center, Howard County General Hospital and Suburban Hospital are paid for care?

Yes. Previously, hospitals were paid based on admissions: More admissions equaled more revenue.

NOW, hospitals have a global revenue budget that they cannot exceed.

The new waiver rewards preventive care that keeps patients out of the hospital.

While admissions proceed as normal, we continue to be reimbursed at the approved rates.

However, as we approach the global revenue budget limit, and admissions grow ...

... the amount paid per admission drops to ensure that the global revenue budget is not surpassed.

Richard (Rich) Grossi   Ronald Werthman   Robert Kasdin   Daniel B. Smith

The Finance Folks

In 2015, Richard Grossi retired after 37 years at Johns Hopkins, including 18 as vice president and chief financial officer of Johns Hopkins Medicine.

Grossi had been a financial administrator at the medical school since 1978, and was promoted to the new job overseeing the financial health of both the health system and school of medicine when Johns Hopkins Medicine was formed in 1996.

Interviewed by reporter Sarah Gantz of the *Baltimore Business Journal* as he prepared to retire, Grossi said much had changed at Johns Hopkins in the decades since he joined, including a state regulation enacted in 2014 that rewarded hospitals for keeping people healthy, instead of paying them by the number of patients they admitted.

"We've gone from being a provider of health care to a more diversified player," he said. "We've gotten larger. Then we've been faced with a new reimbursement system in Maryland that requires [that] we're much more efficient."[*]

Robert Kasdin joined Johns Hopkins Medicine July 1, 2015, as its first senior vice president and chief operating officer, responsible for overall finances and operations, including strategic direction. Kasdin had come from Columbia University, where he had been senior executive vice president since 2002. In May 2022, Kasdin announced he would retire that July.

Ronald Werthman, who had been president and chief financial officer of the Johns Hopkins Health System and The Johns Hopkins Hospital since 1993, became senior vice president and chief financial officer for Johns Hopkins Medicine and the Johns Hopkins University School of Medicine.[†]

In July 2016, Daniel B. Smith became vice president of finance and chief financial officer for The Johns Hopkins Hospital and senior vice president of finance for the Johns Hopkins Health System.

Smith, like Grossi and Werthman, had been with Johns Hopkins for decades before becoming chief financial officer.

Smith had been part of the 1983 team that assessed the viability of acquiring what was then Baltimore City Hospitals, now Johns Hopkins Bayview Medical Center (JHBMC). He remained at JHBMC for 12 years, ultimately becoming the senior director of finance.

In 1995, he joined the Johns Hopkins Health System as the senior director of finance for budget development and financial analysis. During this time, he was instrumental in overseeing The Johns Hopkins Hospital budget process, financial analysis of Johns Hopkins Medicine business plans and implementation of a major technology-based accounting system. In addition to those duties, he has served as chief financial officer for the Johns Hopkins Home Care Group (JHHCG), becoming its president in 2005 after serving as the acting head of the organization for the previous year.[‡]

In a 2019 interview with Becker's Healthcare, Smith discussed some of the challenges of the job, such as quantifying the financial benefits of community health investments.

"The biggest philosophy influencing my leadership style would be the understanding that the CFO today is part of the strategic leadership of the organization," he said. "You have to be fully integrated with the leadership team and help drive the organization in understanding the financial implications of the strategic direction of the organization."[§]

*   Gantz, Sarah, "Best in Finance: Richard A. Grossi, Johns Hopkins Medicine," *Baltimore Business Journal*, April 17, 2015, https://www.bizjournals.com/baltimore/print-edition/2015/04/17/best-in-finance-richard-a-grossi-johns-hopkins.html.

†   Smith, Linell, "It All Adds Up," *Dome*, March 1, 2014, https://www.hopkinsmedicine.org/news/publications/dome/dome_march_2014/it_all_adds_up.

‡   "Daniel B. Smith Named President of Johns Hopkins Home Care Group," Johns Hopkins Medicine press release, April 4, 2005, https://www.hopkinsmedicine.org/Press_releases/2005/04_04_05.html.

§   Gooch, Kelly, "CFO Daniel Smith on the Value of EMRs," *Becker's Hospital CFO Report*, April 26, 2019, https://www.beckershospitalreview.com/finance/johns-hopkins-hospital-cfo-daniel-smith-on-the-value-of-ehrs.html.

care to keep patients as well as possible and out of the hospital for as long as possible," Ronald R. Peterson, president of The Johns Hopkins Hospital and Health System and executive vice president of Johns Hopkins Medicine, said at the time.

"When patients are admitted, it means that we make sure their course of care is done as efficiently as possible, and at the same time, we continue to provide a very good experience for patients."[10]

Scott Berkowitz

Because of the 2014 amendment, Johns Hopkins Medicine focused more than ever on treating patients in the most appropriate setting. The sickest patients and those with the most specialized needs might need care at The Johns Hopkins Hospital. Others could be treated in ambulatory settings such as those at White Marsh and Green Spring Station, which were expanded over time to provide more options, including outpatient surgery.

Also important was helping the general population stay as healthy as possible, to reduce admissions for preventable illnesses or emergencies.

In 2016, for instance, The Johns Hopkins Hospital reduced its readmission rate by 12.66 %. An after-care clinic at the Johns Hopkins Outpatient Center saved an estimated $1.4 million in avoided hospitalizations.[11]

At the time, Scott Berkowitz was executive director of the Johns Hopkins Medicine Alliance for Patients (JMAP), a partnership between the Office of Johns Hopkins Physicians and the Office of Managed Care and Population Health, aimed at reducing hospital readmissions by providing supported care plans to patients with multiple, complex health needs.

The new leadership team job, reporting directly to Kevin Sowers, president of the Johns Hopkins Health System and executive vice president of Johns Hopkins Medicine, calls for developing a population health strategy across the organization.[12]

Johns Hopkins Medicine Alliance for Patients (JMAP) – Better Care, at Lower Cost

In January 2014, Johns Hopkins Medicine formed the Johns Hopkins Medicine Alliance for Patients (JMAP), a Medicare Shared Savings Accountable Care Organization (ACO) that identifies high-risk patients and delivers tailored, coordinated care to both improve health outcomes and reduce costs.

The new network combined Johns Hopkins hospitals with primary and specialty care practices from within the health system as well as outside community practices, providing coordinated care to about 37,000 Medicare beneficiaries.

ACOs were a keystone of the Affordable Care Act, the sweeping health care reform signed into law in 2010. They encouraged providers and hospitals to form networks that coordinated patient care, keeping Americans healthier and reducing health costs at the same time.

When a Medicare Shared Savings ACO reached a savings threshold while attaining a number of quality targets, such as controlling high blood pressure or screening for fall risk, it shared in the savings generated. In 2015, only about a quarter of ACOs nationwide qualified for shared savings.

This map shows the many partner locations of the Johns Hopkins Medicine Alliance for Patients (JMAP).

JMAP achieved a 96.2% quality score in 2015 and reduced rates of hospital admissions and emergency department visits by 2% from 2014 to 2015.[13]

In 2014, JMAP also saved taxpayers more than $5 million, although this was not sufficient to achieve shared savings.[14]

William Baumgartner, senior vice president of the Office of Johns Hopkins Physicians (OJHP) and vice dean for Clinical Affairs for the school of medicine, and Patricia M.C. Brown, senior vice president of managed care and population health and president of Johns Hopkins HealthCare, directed the development of the new accountable care organization and co-chaired its board.

Berkowitz, a cardiologist and medical director of accountable care for Johns Hopkins Medicine, was the first executive director before becoming the chief population health officer and vice president of population health. In 2020 he became the Chief Population Health Officer and Vice President of Population Health for JHM. Matthew Poffenroth was the initial JMAP medical director, followed by Scott Feeser, both primary care physicians with Johns Hopkins Community Physicians.[15]

The JHM care coordination model provides a comprehensive and patient-centered approach to improving care for Medicare Fee-for-Service beneficiaries in the primary care setting, with nurses and licensed certified and clinical social workers working closely with patients to develop care plans that can include medication reviews, transportation to and from appointments, or improved access to specialists.

The care coordination model was developed with the following guiding principles in mind:

• Deliver evidence-based care coordination interventions
• Deliver care via a patient-centered and flexible care model
• Support seamless integration of care
• Maximize primary care practice presence of care team members
• Create a framework for structured coordination between providers and care team members
• Ensure accountability of outcomes among all care team members[16]

How JMAP Helped One Patient Stay Healthy

Marina, a patient at Johns Hopkins Community Physicians at Annapolis, was 82 years old, living alone and without a nearby support system. She had severe arthritis and was on a blood thinner. She needed some extra help.

Carole Campbell became her care manager in January 2015. Campbell helped get Marina a new bed with rails to prevent falls. She set her up for the Meals on Wheels program, and made sure her taxi vouchers were in order

so that she could get to her appointments.

Campbell has also made many phone calls for Marina, and has given her advice on how to stay healthy.

"All the advice that she is giving me is excellent," Marina said. "I follow the instructions that she gives me and I feel much better. And she helps me not just with my illness, but I feel like somebody cares — that when I talk, she listens to me. And that's what I love — that she listens to me."*

Carole Campbell

---

* "Patient Testimonials," Johns Hopkins Medicine webpage, Johns Hopkins Medicine Alliance for Patients, https://www. hopkinsmedicine.org/alliance-patients/patients-families/patient-stories.html.

"JMAP has been successful in bringing together talented team members from across Johns Hopkins Medicine — along with our primary care partners from Columbia Medical Practice and Potomac Physician Associates — in support of improved care for our patients," said Berkowitz in 2018. "Although there is important work still to be done, JMAP has provided a key foundation from which we can continue to optimize patient-centered care."[17]

"During our first 18 months, we enhanced care coordination services for more than 2,000 beneficiaries and launched a significant effort to improve urgent access to specialty care, provide population-based pharmacy services and promote quality improvement in care delivery, among other things," said Brown. "JMAP is serving as a catalyst for broader transformation within Johns Hopkins Medicine."[18]

In January 2017, JMAP announced that it had been approved by the Centers for Medicare & Medicaid Services to renew its participation in the Medicare Shared Savings Program for the next three years. The Shared Savings Program offers financial incentives to encourage ACOs to improve coordination, communication and overall care for Medicare fee-for-service beneficiaries while also reducing health care costs.

On January 1, 2019, JMAP entered into an agreement with the State of Maryland and Centers for Medicare & Medicaid Services (CMS) to participate as a Care Transformation Organization (CTO) in the Maryland Primary Care Program (MDPCP). The CTO is a new type of entity that hires and manages an interdisciplinary care management team to furnish an array of care coordination services.

The CTO uses JMAP's existing infrastructure to offer customizable care delivery, information technology, and quality improvement solutions to partner practices to realize value in care transformation.[19] The CTO Governing Committee is chaired by Jonathan Efron, who became the senior vice president of OJHP in 2017.

Individualized, Precision Medicine

By installing Epic, Johns Hopkins Medicine could accelerate its Big Data revolution, and create a precision medicine approach to research and treatment, which became known as Johns Hopkins inHealth and led to creation of Johns Hopkins Medicine's Precision Medicine Centers of Excellence.

Dean/CEO Paul Rothman believes the move toward data-based medicine was one of JHM's most important recent accomplishments.

"I always thought if we don't leverage all the work that everyone does to document in Epic, then we're failing them and our patients," he said in 2022. "Precision medicine, all the advances we're making in digital medicine, was a way to harvest all that hard work and advance our research, clinical care and education."[20]

Precision medicine researchers gather and study vast amounts of information on thousands of patients and their families to understand the subgroups of diseases and how they might progress in a given individual. As a result, patients can get the treatment that is right for them, avoiding unnecessary tests and therapies that can be invasive, risky and costly.

"Really, the goal of precision medicine is to learn from every patient, to enable the physician and the patient to use information to inform their decisions every time a medical decision was to be made," said Antony Rosen, vice dean for research, in a 2021 interview.

"Previously, a clinician would sit with a patient and use his or her skills and memory about patients they've seen and the literature they've read to try and place that patient into a category," he said. "That process is not a high-fidelity process. Sometimes you haven't seen the right patients before, or you haven't read a relevant study. With precision medicine, we can bring all that information in front of the clinician, displayed in ways that help them understand the probability of outcomes.

"It really changes the way that therapeutic encounter is framed in the sense that clinicians now have information at their fingertips not only about the patient in front of them, but also all similar patients," said Rosen.[21]

Launched in July 2012, inHealth leveraged universitywide assets from the Johns Hopkins University Applied Physics Laboratory (APL) and the schools of medicine, engineering, nursing and public health to develop patient-level insights that could guide care.

"Brain scans, genetic codes, family histories, eating habits and medical claim records are just a few examples of the information being collected and analyzed," noted a June 2014 article in *Dome*. "This paradigm shift will replace a 'one size fits all' concept of health care with an individualized approach that delivers better results at lower cost."[22]

One challenge was creating a powerful computing system that allowed all of this data to work together. The system had to be secure to protect patient information, and it had to make it easy for clinicians to access the information that would help them predict disease trajectories and recommend treatment plans, often during an appointment with a patient.

Antony Rosen

Bal Carter

Ken Pienta

In 2014, inHealth faculty and staff members began working with scientists and analysts at the APL to develop the Precision Medicine Analytics Platform (PMAP).

Researchers across Johns Hopkins Medicine could access and work with data using the Secure Analytics Framework Environment (SAFE), a virtual desktop and storage network.

The system funnels data from Epic and other sources into a cloud-based platform to help scientists connect treatments and outcomes with patterns of patient-related data, such as demographic or genetic information, or even data from wearable devices such as step trackers.

InHealth's pilot projects used big data to improve cancer screenings, cystic fibrosis treatment, heart care decisions, autoimmune disease management, and diagnosis and treatment of age-related diseases.

Patients with prostate cancer were among the early beneficiaries of inHealth.

H. Ballentine Carter, director of the Division of Adult Urology in the Brady Urological Institute of The Johns Hopkins Hospital, working with Kenneth Pienta, director of research at the institute, used decades of patient data to show that some men with prostate cancer would not benefit from surgery to remove their prostates because they were at low risk of disease progression and likely to die of causes other than prostate cancer.

The researchers took into account factors including a patient's age, chronic illnesses, life expectancy, race, family history and results from previous prostate-specific antigen (PSA) tests.

They used that information to develop an ActiveCare software tool that creates prediction models to help patients with prostate cancer understand their risks in surveillance and help their physicians make a care plan for monitoring their condition.[23]

As a result of their work, the field shifted from invasive and side-effect-prone prostate-removal surgeries to active surveillance programs that closely monitor prostate health.

Oncospace

One of the earliest uses of big data at Johns Hopkins was a project called Oncospace, a "learning health system" aimed at helping radiation oncologists craft treatment plans for patients with cancer.

Radiation oncology physicist Todd McNutt developed Oncospace as a database of diagnostic 3-D images of tumors from thousands of past patients with head and neck, prostate, or pancreatic cancers. Those images were matched to data about those patients — anatomy, comorbidities, radiation dose distribution, treatment side effects, case outcomes and more.

With Oncospace, radiation oncologists could "test" treatment plans by studying what happened when similar treatment plans were used with similar patients.

Early tests in both head and neck and pancreatic cancers found that using Oncospace boosted the effectiveness and safety of radiation regimens, said a Hopkins Medicine magazine article in 2015.

"Todd's work on this is really one of the first demonstrations of how we can develop large data warehouses of patient information and use it to make individualized treatment decisions for new patients," said Theodore DeWeese, then chair of Radiation Oncology and Molecular Radiation Sciences at Johns Hopkins, in 2015.[24]

In 2018, McNutt co-launched Oncospace, Inc., using technologies and intellectual property licensed from Johns Hopkins and supported by Johns Hopkins Technology Ventures.

In September 2021, the company, with McNutt as chief scientist, introduced a Predictive Planning solution that uses big data to help radiation oncology clinicians choose optimal treatment plans for prostate cancer patients.[25]

"This journey began with an ambition to create a structured, curated database of treatment planning information and to use machine learning and other methods in developing tools for physicians to evaluate achievable, personalized treatment pathways," said McNutt in 2021. "It is exciting to be on the threshold of taking this academic and clinical research program to a new phase, making it available to the wider clinical community."[26]

Todd McNutt

'Generating Wonder' Through Precision Medicine

In 2017, Johns Hopkins established its first Precision Medicine Centers of Excellence (PMCOEs), creating tools and protocols that built on the concepts of inHealth.

Researchers and clinicians with big data research ideas that could lead to specific interventions can apply for the designation, which gives them resources for research, including opportunities to collaborate with software developers, data analyzers and other experts, including the Johns Hopkins Technology Innovation Center — the group launched in 2014 to help clinicians in the Johns Hopkins Health System use technology to improve patient care.

The first PMCOEs were for prostate cancer and multiple sclerosis. Seven more PMCOEs were added in 2018, including the pancreatic cancer and arrhythmogenic right ventricular dysplasia/cardiomyopathy (ARVD/C) centers, as well as one for myositis, with Johns Hopkins rheumatologist Christopher Mecoli as director of research operations.

Myositis is a rare autoimmune disease characterized by inflammation that causes muscle weakness and pain, with symptoms that can include fatigue, rashes, difficulty swallowing or trouble breathing.

Mecoli has studied the chronic illness for more than a decade. He is particularly interested in the intersection between myositis and cancer.

Patients are at highest risk of cancer at the time when they develop myositis symptoms, he said, but only a few of those people actually get cancer. "We can't predict very well which patients will get cancer, so we screen everyone, very aggressively," he said.

Using the PMAP tool developed by Johns Hopkins Medicine and the APL, Mecoli could quickly filter results for patients with various biomarkers, to show that myositis patients with a biomarker known as TIF1 were at greater risk compared to patients with other biomarkers.

"It used to take three to four years to get the same observation I found in 30 or 40 seconds," he explained.[27]

By 2021, more than 20 PMCOEs had been established for illnesses including asthma, COVID-19 and mood disorders. Interviewed in 2021, Antony Rosen said he believes precision medicine, and its potential to help clinicians and researchers understand disease and tailor treatments to individuals, is still in its earliest stages.

He described its components as wonder (the ability to identify relevant patient subgroups), insight (using that information to inform clinical care) and value (understanding the quality and cost implications of acting optimally).[28]

"Our goal is to generate wonder, insight and value at scale," he said. "Human perception of wonder is a huge driver of discovery. People see something unusual and they wonder why. If you can bring the data together and start allowing the computer to help a person see patterns, that is a major generator of wonder, and allows understanding of subgroups of disease."

He said the system can also "use data and patterns to bring together cost and outcome information to find out where we should focus our attention. There are

Christy Wyskiel

certain things we do that aren't useful and are expensive, and others that are useful and inexpensive. We don't learn that at scale as an enterprise and we need to do that more."[29]

The potential for big data and PMCOEs is just beginning to be realized, said Rothman in 2021. "It will be the lifeblood of all activity centers moving forward," he predicted.

Johns Hopkins Technology Ventures

The launch of Johns Hopkins Technology Ventures in 2014 ushered in a new era for Johns Hopkins inventors — and the many people who have benefited from the technologies, products and services these researchers have been able to bring to market.

In its first seven years, JHTV fostered more than 170 commercial ventures based on the discoveries of Johns Hopkins faculty, said Christy Wyskiel, JHTV executive director and senior adviser to the president of The Johns Hopkins University for Innovation and Entrepreneurship.

The university's investment has also positioned Johns Hopkins as a Top 10 U.S. university for patent issuance since 2014, according to the National Academy of Investors.[30] Johns Hopkins also ranks in the Top 10 for startup creation as of 2018.[31]

Prior to JHTV, inventors say, the university was not particularly inviting to entrepreneurs, even after the Bayh-Dole Act became federal law in 1980, allowing universities to own, patent and commercialize inventions developed under federally funded research programs within their organizations.

"If you were doing entrepreneurial work, it was considered nonacademic, something below what the academic pursuit was supposed to be," said Drew Pardoll, director of the Bloomberg-Kimmel Institute for Cancer Immunotherapy.

In 2013, he joined with Jennifer Elisseeff, a professor at the Wilmer Eye Institute and Department of Biomedical Engineering, to co-chair a committee formed by Johns Hopkins University President Ronald J. Daniels and Rothman to find ways to make Johns Hopkins more inviting to entrepreneurs and ensure that scientific discoveries were put into practice through commercial mechanisms.

Elisseeff had already founded Aegeria, a startup that focuses on soft tissue regeneration and wound healing. She remembers contending with a lack of institutional resources, as well as with other barriers when she established it in 2009. Pardoll, too, had experience with translational medicine, as the founder of Potenza Therapeutics, a company that develops novel antibodies for immuno-oncology

Tech Ventures interior

therapies (which was recently acquired by Astellas Pharma Inc.).

The resulting *Report of the Committee on the Innovation Ecosystem*, released in May 2014, called for investment in three areas of institutional infrastructure: space for entrepreneurs to work, translational funding for their projects, and resources such as mentorship and corporate partnerships.[32]

Those recommendations led to the creation of JHTV, with Wyskiel as the head of the new commercialization center. She brought personal knowledge of the challenges facing academic startups through her previous work as an institutional investor and co-founder of two Baltimore companies that leveraged university technology.[33]

Jennifer Elisseeff in 2017, in the lab with biomedical engineering master's student Alexis Parrillo.

"I feel strongly that this agenda of service, and the work we are doing, ultimately will not only make the world a better place but has already brought vibrancy to Baltimore and will continue to make Baltimore a strong place for business to grow, entrepreneurs to thrive and students to remain after graduation," she said in 2019.[34]

Support for Inventors

As the JHTV team implemented the recommendations of the 2014 faculty-led committee, the opening of incubator space was the most visible manifestation of progress.

The first FastForward incubator space opened in the Stieff Silver Building near the Homewood campus in 2013, with leadership support from the Johns Hopkins Whiting School of Engineering. Two years later, the concept came to East Baltimore with the 6,000-square-foot FastForward East in the John G. Rangos Sr. Building.

JHTV's home since April 2017 has been the 23,000-square-foot FastForward 1812 innovation hub on the first floor of 1812 Ashland Ave., the heart of the Science + Technology Park at Johns Hopkins on the north side of the medical campus.

In August 2017, the FastForward Homewood facility moved to R. House in Remington, a few blocks from the Homewood campus, where it shares space with the food hall of the same name.[35]

These incubators provide affordable space as well as support services from the FastForward professional staff, a set of mentors-in-residence, and sponsors who provide pro bono services. The spaces also provide a vibrant environment for networking and exchange of learning, which are critical components of a thriving innovation hub.

Altogether, in 2021, more than 70 JHTV professionals were helping university

Corrie Health

Nationwide, one in six cardiac patients is readmitted to the hospital within 30 days.

In 2014, Johns Hopkins University School of Medicine student Francoise Marvel decided to address the problem by developing an app that would give people information and guidance as they recovered from heart attacks.

With support from Johns Hopkins Technology Ventures (JHTV), she eventually started a company called Corrie Health ("*Cor*" is Latin for heart) to commercialize her app and make it widely available.

In 2016, Marvel, then a third-year internal medical resident, teamed with preventive cardiologist and mobile health expert Seth Martin, and with Whiting School of Engineering students Matthias Lee and Gavi Rawson.

The resulting Corrie app, developed in partnership with Apple, monitors a patient's heartbeat, blood pressure and exercise, creating a record for patients and their doctors. It also alerts users when it's time to take medications, provides exercise plans and education, and makes it easy to schedule appointments.

"It's like a digital health buddy to keep patients on track," said Marvel, now an assistant professor in the cardiology department.

The deal between Corrie Health and The Johns Hopkins University gives the university equity in Corrie Health; the university and Marvel, Lee and Martin receive royalty distributions. More importantly, a study of the technology found that it cut 30-day readmission rates in half, compared to a comparable group of patients who had suffered heart attacks but did not use the app.*

---

* "Study Shows Use of Smartphone App Associated with Lower Hospital Readmission Rates for Heart Attack Survivors, Johns Hopkins Medicine press release, Sept. 21, 2021, https://www.hopkinsmedicine.org/news/newsroom/news-releases/study-shows-use-of-smartphone-app-associated-with-lower-hospital-readmission-rates-for-heart-attack-survivors.

Francoise Marvel (second from the left) collaborated with a multidisciplinary team to create the Corrie Health app.

inventors to protect intellectual property, license technology, establish corporate research collaborations, and launch startup businesses. The team secures over 100 technology licenses each year and has nearly 40 corporate sponsored research partners.

The innovation hubs are home to more than 30 companies, cementing Baltimore as one of the nation's emerging innovation economies. The full portfolio of Johns Hopkins startups has raised $3 billion since 2014, at a recent clip of $400 million on average per year.

Wyskiel credits the university's investment in infrastructure and JHTV staff with meeting the needs of faculty and students as well as investors and the business community.

"I'm so gratified that the faculty are focused on the big looming problems," she said. "They're not afraid to take the moonshot approach to research and technology development that will directly impact patients. We're just here to facilitate their dreams and make them come true."[36]

Technology Innovation Center

Adjacent to JHTV, the Technology Innovation Center (TIC) also launched in late 2014 to focus on the development of digital health innovations. The professional software development group and collaboration hub brings together clinical leaders,

Adarga Pharmaceuticals

In June 2021, Bayer AG purchased Adarga Pharmaceuticals, a startup founded by Barbara Slusher, director of Johns Hopkins Drug Discovery. Slusher had developed a shielding agent to protect a patient's healthy tissue during targeted radiotherapeutic treatment for prostate cancer.

"The science story here is the value of keeping your eyes open, because if you do, failures can turn into successes," said Slusher.

Slusher is not a prostate cancer expert — she is a professor of neurology and vice director of the Pedersen Brain Science Institute. For more than three decades, she has studied glutamate carboxypeptidase II (GCPII), an enzyme in the brain that generates glutamate — a chemical that nerve cells use to send signals to other cells.

GCPII is also found elsewhere in the body, and is known as prostate-specific membrane antigen (PSMA). Elevated PSMA levels are signs of prostate cancer, and can be found and killed with targeted radiotherapy.

As Slusher and her lab synthesized and tested potential brain-penetrable GCPII inhibitors without success, her postdoctoral fellow at the time, Michael Nedelcovych, posed a question: "If the drugs are not going into the brain, then where are they ending up?"

Working with Rana Rais, an associate professor of neurology, they conducted tissue distribution experiments to localize the GCPII drug in the body.

"What we basically stumbled upon," said Slusher, "was that our failed drugs for the brain were accumulating in the kidney and salivary glands."

Slusher's drugs, when administered before radiotherapy, bind to the PSMA in the kidneys and salivary glands and block the radiotherapy from attacking the healthy cells.

JHTV successfully sought patent protection for the drugs and helped Slusher form the startup company that was purchased by Bayer.*

Barbara Slusher

Rana Rais

---

* "How a Johns Hopkins Neuroscientist Founded – and Sold – a Startup Connected to Treating Prostate Cancer," Johns Hopkins Technology Ventures webpage, Aug. 23, 2021, https://ventures.jhu.edu/news/barbara-slusher-kidney-prostate-cancer-adarga/.

researchers and technologists to build and deploy these solutions within — and often beyond — the Johns Hopkins Health System.

In 2018, TIC joined with JHTV, the Johns Hopkins Medicine offices of the general counsel, marketing and communications, and information technology to create a review board ensuring that apps branded with the Johns Hopkins name meet its standards. Among those criteria, the apps must have clinical benefit, preserve patient privacy and improve on existing options.

Collaborations for Better Care and Research

The CAPRES Cardiac Research Group at Howard County, 2013. Left to right: Clinical Research Coordinator Keith Jackson, Research Program Administrator Mehro Akhtar, Principal Investigator Keith Friedman, Principal Investigator William R. Herzog, Jr., Clinical Research Coordinator Tanya Burley, Physician Assistant Sub-Investigator Maria E. Gregory.

In the early 2010s, Johns Hopkins Medicine was also forging research and patient care collaborations with other health care systems.

The agreements improved patient access to Johns Hopkins care, expertise and clinical trials, added diversity to research by incorporating more sites, and gave Johns Hopkins and its new partners competitive advantages in the fast-changing health care landscape.

The collaborations began within the Johns Hopkins family. Daniel Ford, at that time vice dean for clinical investigation and director of the Johns Hopkins Institute for Clinical and Translational Research (ICTR), recognized that the Johns Hopkins Health System's recently acquired community hospitals could be part of multisite investigations, giving patients more opportunities to participate in clinical trials while improving research by increasing the number and diversity of its research participants.

The resulting network, known as CAPRES (for Capital Region Research), was created by Ford and Brian Gragnolati, senior vice president of the Johns Hopkins Health System, in March 2011, as part of the ICTR.

Now led by CAPRES director Jackie Lobien and Mark Sulkowski, senior associate dean and director of the Office of Clinical Trials and of CAPRES, the network promotes research across the two National Capital Region (NCR) hospitals (Sibley Memorial Hospital in Washington, D.C., and Suburban Hospital in Bethesda, Maryland), as well as Howard County General Hospital in Columbia, Maryland.

At each of the three Johns Hopkins community hospitals, CAPRES serves as a central resource for research oversight and administration, including compliance with Johns Hopkins Medicine's institutional review boards (which review and monitor research conducted with human subjects), as well as budgets, contracts, research training, recruitment, clinical operations, marketing and finance.

CAPRES also works with the Johns Hopkins Office of Research Administration to coordinate agreements and budgets for community physicians in private practice so they can conduct studies within the walls of the community hospitals.

Brian Gragnolati

Mark Sulkowski

Daniel Ford

Gail Daumit

Gail Daumit Succeeds Daniel Ford as Vice Dean for Clinical Investigation

In June 2021, Gail L. Daumit, a professor in the departments of medicine and psychiatry and behavioral sciences, as well as the Samsung Professor of Medicine at The Johns Hopkins University, became vice dean for clinical investigation at the Johns Hopkins University School of Medicine.

Among other responsibilities, she became the institutional official in charge of protection of human subjects for Johns Hopkins Medicine and the Institutional Review Board (IRB) committees.

Daumit succeeded Daniel Ernest Ford, the David M. Levine Professor of Medicine, who held the post since 2005 and remains an internal medicine clinician and professor, as well as director of the Institute for Clinical and Translational Research (ICTR), established under his direction in September 2007.

Daumit earned her undergraduate degree at the University of Pennsylvania, her medical degree at Emory University and a master of health science in epidemiology at the Johns Hopkins Bloomberg School of Public Health.

She completed her residency in internal medicine primary care at Massachusetts General Hospital, and joined Johns Hopkins in 1996 as a Robert Wood Johnson Clinical Scholar and general internal medicine fellow.

Daumit's research focuses on developing innovative ways to improve the physical health of people with mental illness. She has obtained continuous National Institutes of Health (NIH) funding for this work since 2000, and has published more than 100 peer-reviewed articles. Her behavioral weight loss intervention trial in people with serious mental illness, ACHIEVE, won both the Society of General Internal Medicine (SGIM)'s Best Published Research Paper of the Year Award and the Society for Clinical Trials Trial of the Year Award in 2013.*

In a number of important roles, including associate director of the Welch Center for Prevention, Epidemiology and Clinical Research, and research director of the Division of General Internal Medicine in the Department of Medicine, Daumit has built a reputation as a dedicated and thoughtful clinician and researcher, a caring educator and mentor, and a dynamic and collaborative leader.†

Her predecessor Ford has developed a similarly stellar reputation during a long ca-

reer at Johns Hopkins, starting with his 1982 arrival to complete the Osler Medicine residency. Except for a brief period as a National Institutes of Health fellow, he has served on the Johns Hopkins faculty ever since. He received his master's degree in public health from the Johns Hopkins Bloomberg School of Public Health in 1986.

Ford is a professor of medicine and psychiatry in the Johns Hopkins University School of Medicine and a professor of epidemiology and health policy and management in the Bloomberg school.

His research focuses on understanding the relationships between depression and chronic medical conditions, particularly coronary artery disease, and how to improve care for patients with medical comorbidity. He was one of the first investigators to publish data documenting depression as a risk factor for myocardial infarction and stroke, and uses information technology to improve care of patients with depression and tobacco abuse.‡

---

* "Dr. Daumit Awarded Best Published Research Paper," Johns Hopkins Medicine press release, April 23, 2014, https://www.hopkinsmedicine.org/gim/news/2014_News_Items/4-23-14.html.

† "Daumit to Step Down as Vice Chair for Clinical Translational Research," *Medicine Matters*, June 25, 2021, https://medicine-matters.blogs.hopkinsmedicine.org/2021/06/daumit-to-step-down-as-vice-chair-for-clinical-translational-research/.

‡ "Daniel E. Ford, M.D.," Johns Hopkins Medicine webpage, https://www.hopkinsmedicine.org/about/leadership/biography/daniel-ford.

Allegheny Health Network

Johns Hopkins Medicine signed one of its first partnerships with an outside organization in May 2014 when it forged a research and care collaboration with Allegheny Health Network (AHN), a system of eight Pennsylvania hospitals in and around Pittsburgh, owned by Highmark Inc.

The agreement focused on oncology, giving AHN patients access to Johns Hopkins-led clinical trials, and allowing AHN physicians to consult with oncologists at the Johns Hopkins Sidney Kimmel Comprehensive Cancer Center. AHN patients with particularly complex cases could also receive care at the Kimmel Cancer Center.

Next came a master collaboration, signed Dec. 22, 2014, between Johns Hopkins Medicine and AHN. This broader agreement empowered the organizations to collaborate on research and education; seek efficiencies through supply chain initiatives and joint ventures; and use complex data analysis to improve care and outcomes.

"In this ever-changing health care environment, the goal is to provide better care for better value for patients, caregivers and payers," said Gragnolati in a press release at the time. "It makes sense to find synergies with other entities with similar missions and different strengths."

Tony Farah, chief medical officer for AHN, described the agreement as "a unique opportunity for our organization to more closely collaborate with one of the world's most accomplished and respected health care institutions so we can enhance the scope and capabilities of the services we offer to the western Pennsylvania region."[37]

The agreement was developed with the guidance of Robert Kasdin, senior vice president, chief operating officer and chief financial officer of Johns Hopkins Medicine; and John Colmers, senior vice president for health care transformation and strategic planning for Johns Hopkins Medicine.

It included joint review and care for complex cases in fetal therapy — an area in which Johns Hopkins offers rare expertise, said Ahmet Baschat, founder and director of the Johns Hopkins Center for Fetal Therapy.

In June 2017, Johns Hopkins announced an expansion of its oncology affiliation with AHN and Highmark, covering health care and insurance for people in western Pennsylvania, West Virginia and Delaware.

It supported the care of patients with rare and complex adult and pediatric cancers, as well as some organ transplant patients. It also widened the portfolio of cancer clinical trials available to AHN's patients, including some that would be conducted at AHN facilities.

In addition, AHN and Johns Hopkins experts would work together on Grand Rounds medical education presentations about patient case studies, and the Kimmel Cancer Center and AHN would collaborate on research projects related to genome sequencing and precision medicine, using the genetic code of cancer to guide treatment.

"We are honored to be working with the oncologists at Allegheny toward our shared goal of providing the best cancer care for their patients," said Rothman.

How the Johns Hopkins Center for Fetal Therapy Saved the Tingle Triplets

Katie Tingle was 16 weeks pregnant when she and her husband Shawn arrived at the Johns Hopkins Center for Fetal Therapy in the fall of 2014. Two months had passed since the couple had learned Katie was carrying triplets, a pair of identical girls and a fraternal boy.

Katie's appointment with Ahmet Baschat, the center's director, began like so many of her previous visits with other specialists, with a painstakingly long look via ultrasound at the health of the three fetuses in her womb.

This time, the focus of the ultrasound was an imbalance of fluid levels between the two identical girls. At one point, Baschat pointed out on the scan the place where that imbalance was playing out.

"There were these two little circles," Katie Tingle says. "One looked like a golf ball. The other looked like a softball. It was very extreme."

The imbalance was caused by twin-to-twin transfusion syndrome, or TTTS. Left untreated, it would likely have caused the death of both twins. At Baschat's urging, Tingle had fetoscopic surgery on Oct. 7, 2014.

Using a tiny camera for the minimally invasive procedure, called laser ablation, Baschat located and closed all of the blood vessels in the vascular equator, separating one twin's part of the placenta from the other's.

Triplets Lily, Grace and Mason were born Feb. 10, 2015, weighing between 3 pounds, 12 ounces, and 4 pounds, 11 ounces. They spent most of their first month in the neonatal intensive care unit at Johns Hopkins and were released together on March 3.*

The Center for Fetal Therapy, launched by Baschat in 2014, brings together fetal, maternal and pediatric teams to diagnose and treat a variety of complex medical conditions before birth, including in-utero surgeries when necessary for spina bifida repair, severe diaphragmatic hernias, and TTTS.

"For fetal therapy specifically, you have to be able to take care of the mother and the fetus," said Baschat in a 2015 Johns Hopkins Medicine video about the center. "And once the baby is born, the child. Johns Hopkins Hospital is actually one of the very few institutions where all of the services are in the same hospital at the highest level of acuity care. So you have the highest level of acuity for maternal care, for fetal care, and for neonatal and pediatric care. And that's rare."†

---

* Duffy, Jim, "No Womb for Error," *Hopkins Medicine* magazine, Fall 2015, https://www.hopkinsmedicine.org/news/publications/hopkins_medicine_magazine/features/fall-2015/no-womb-for-error.

† "Saving Muriel: Johns Hopkins Center for Fetal Therapy," YouTube, Johns Hopkins Medicine channel, https://www.youtube.com/watch?v=eghDW9jdGA8.

"Additionally, we are pleased to now also be Highmark's preferred referral partner, including across Highmark's BlueCard network in western Pennsylvania, as well as West Virginia and Delaware."[38]

Johns Hopkins Clinical Research Network

The arrangement meant that AHN had joined the Johns Hopkins Clinical Research Network (JHCRN), established in 2009, to bring both diversity and access to multisite research trials.

The JHCRN is part of the Johns Hopkins Institute for Clinical and Translational Research (ICTR), a member of a national consortium to improve the way biomedical research is conducted across the country. JHCRN, which began in 2007 with an alliance with the Anne Arundel Medical Center in Annapolis, is funded by the National Institutes of Health (NIH) Clinical and Translational Sciences Award (CTSA) program.

The Johns Hopkins Clinical Research Network (JHCRN), established in 2009, is a regional translational research organization that links Johns Hopkins Medicine with community health systems. This 2017 photo shows leaders of the Johns Hopkins Clinical Research Network and the Allegheny Health Network.

By 2018, JHCRN included AHN; the entire Johns Hopkins health system; Anne Arundel Medical Center; Inova Health System in northern Virginia; Peninsula Regional Medical Center in Salisbury, Maryland; and Reading Health System in Reading, Pennsylvania. In 2021, WellSpan Health, in south central Pennsylvania, joined the network.

The health systems all use Epic, which helps with data sharing, giving Johns Hopkins researchers access to an estimated nine million potential study participants in Maryland, Virginia, Washington, D.C., Pennsylvania and Florida hospitals.

Research partnerships are not new, but the JHCRN is different from most. With other types of collaborations, each institution secures and manages its own funding, and investigators share leadership. With the clinical research network, Johns Hopkins takes the lead on funding and research for all network member institutions, explained Adrian Dobs, director of Johns Hopkins' research network.

As an example, Johns Hopkins gastroenterologist Francis Giardiello and infectious disease specialist Cynthia Sears used the network to secure the 2,000 study subjects promised in their successful National Cancer Institute grant proposal, a study of the links between cancer and the biofilms in the colons of some individuals.

Using the network, they collected information from colonoscopy patients at three sites: a private practice affiliated with Reading Health System; an outpatient clinic at Green Spring Station run by Johns Hopkins faculty; and a Johns Hopkins outpatient clinic in White Marsh.

"The clinical research network gave us the confidence in applying for the grant that we could accumulate the sample size that we needed," said Giardiello, noting that an earlier proposal for 1,000 subjects was rejected. "It also smoothed things out from an administrative standpoint because we have a single consent form and a single database."[39]

The PaTH Network

Another research network under the ICTR umbrella is PaTH, which is partially funded by the Patient-Centered Outcomes Research Institute (PCORI) and is one of 33 networks that comprise PCORI's National Patient-Centered Clinical Research Network (PCORnet).

PCORI is a Washington, D.C.-based nonprofit, nongovernmental funding agency interested in comparative clinical effectiveness and patient-centered outcomes research that Congress authorized in the Patient Protection and Affordable Care Act of 2010. PCORI aims to determine which health care options are best for patients and care providers by promoting high-integrity, evidence-based information learned through research guided by patients, caregivers and the broader health care community.

PaTH includes Geisinger Health System, The Johns Hopkins University, Johns Hopkins Health System, Penn State College of Medicine, Penn State Milton S. Hershey Medical Center, Temple Health System, Lewis Katz School of Medicine at Temple University, the University of Pittsburgh, UPMC and UPMC Health Plan, The Ohio State University, The Ohio State University Wexner Medical Center, University of Michigan, and Michigan Medicine.

PaTH research uses information from patients to guide decisions and actions. Areas of research include healthy lifestyles and weight, a heart rhythm abnormality called atrial fibrillation, and lung scarring that may be genetic, known as idiopathic pulmonary fibrosis.[40]

Research program coordinator Lilly Su, right, listens to a patient in the PaTH story booth.

One project example was a PaTH story booth, set up at the Johns Hopkins Outpatient Center in 2019, which traveled among PaTH institutions, collecting stories from patients and caregivers.

The goal was twofold: to create a trove of patient and family caregiver stories for others who are going through similar situations, and to give clinicians information that helps provide better patient-centered care.

"Story Booth gives us an opportunity to engage with patients and their caregivers to gain insights and better understand their experiences with health care," said Johns Hopkins principal investigator Cheryl Dennison Himmelfarb, the Sarah E. Allison Professor of Nursing, and vice dean for research at the Johns Hopkins School of Nursing. "We aim to apply our learning to improve patient-centered approaches to health care and research," says Himmelfarb, who has joint appointments at the schools of medicine and public health.[41]

WellSpan Health

Another agreement with a Pennsylvania health system began in August 2016, when the Johns Hopkins Children's Center forged a deal with WellSpan Health, giving doctors at WellSpan York Hospital 24/7 access to Johns Hopkins pediatric surgeons.

"WellSpan's collaboration with the Johns Hopkins Children's Center will bring a cadre of specialized pediatric surgeons to the York/Adams community, ensuring that many children in our communities who need pediatric surgical care will no longer need to be transported to hospitals outside our area," said David Turkewitz, chairman and education coordinator of pediatrics and director of pediatric emergency medicine for WellSpan York Hospital, in a WellSpan press release.

"It is safer for the child, and it will allow parents to better focus on the needs

Vered Stearns   **Kenneth J. Cohen**

of their child, rather than worry about how to get to an unfamiliar hospital, where to park, where to stay, where to eat and how to navigate the hospital," he added. "This is a win for our patients and our community."[42]

The following year, in June 2017, WellSpan Health and the Johns Hopkins Kimmel Cancer Center announced a clinical collaboration that brought expanded access to clinical trials for patients of WellSpan's coordinated network of Pennsylvania cancer centers, including WellSpan Adams Cancer Center in Gettysburg, WellSpan Ephrata Cancer Center, WellSpan Sechler Family Cancer Center in Lebanon, WellSpan York Cancer Center and Cherry Tree Cancer Center in Hanover.

Under the agreement, Johns Hopkins specialists provided second opinion services to WellSpan cancer doctors. And the Johns Hopkins Kimmel Cancer Center served as a seamless referral destination for patients with rare and complex cases requiring advanced treatment technologies.

"We have worked with WellSpan's strong regional cancer program for many years, and we are enthusiastic about formalizing an agreement that will give WellSpan's clinicians and patients wider access to the unique expertise of the Johns Hopkins Kimmel Cancer Center team and clinical trials," said Kenneth J. Cohen, director of strategic planning and integration, and professor of oncology and pediatrics at the Johns Hopkins Kimmel Cancer Center.[43]

In September 2018, the Johns Hopkins Kimmel Cancer Center and WellSpan Health together announced the launch of a grant program to support research that would leverage Epic to create cancer-related health services products.

The WellSpan–Johns Hopkins Cancer Health Services Research Fund would fund two grants per year, for up to $100,000 in direct costs for each project in one year.

Award winners, known as WellSpan Scholars, would further the field of health services research by asking questions relevant to cancer and to those patients who are at risk for, who have, or who are survivors of the disease.

"This new cancer research fund is another example of the wonderfully collaborative oncology relationship between WellSpan and Johns Hopkins Kimmel Cancer Center," said Douglas Arbittier, vice president of oncology services, WellSpan Health. "Not only will this allow our WellSpan physicians to participate in research with a world-renowned academic institution, but the research will be used to enhance care for cancer patients in our communities."[44]

The first grants, awarded in April 2019, went to a weight-loss study in postmenopausal women with early-stage breast cancer, with Vered Stearns, co-director of the Breast Cancer Program at Johns Hopkins, as the principal investigator, and Jillian Smith, a surgical oncologist from WellSpan who specializes in breast cancer, as the co-principal investigator; and a study about adjuvant endocrine therapy, led by Karen Smith, breast cancer medical oncologist with Johns Hopkins.[45]

How a Johns Hopkins Doctor Made John Boehner Cry

Jessica Bienstock

Johns Hopkins doctors are often praised by grateful patients, but rarely does that praise entail the speaker of the U.S. House of Representatives choking back tears as he describes how a "miracle worker" saved the life of a colleague's baby.

At about 23 weeks of pregnancy, Rep. Jaime Herrera Beutler, a Republican representing Washington state, learned that her baby had Potter's Syndrome, a rare condition in which the kidneys do not form properly and so cannot produce the fetal urine that accounts for most amniotic fluid.

Previously, the condition was considered uniformly fatal, with most babies dying in utero or within hours of birth, and none surviving long term.

Two doctors told the congresswoman and her husband, Daniel, that the condition was untreatable. The couple then found Jessica Bienstock, a maternal-fetal medicine physician at Johns Hopkins.

Bienstock developed an unconventional regimen that called for weekly injections of saline that took the place of the missing amniotic fluid.

The baby, Abigail Rose Beutler, was born July 15, 2012, three months early and weighing just two pounds, 12 ounces, able to breathe on her own and basically healthy despite her malfunctioning kidneys.

She was immediately put on dialysis, and stayed with the regimen until February 2016, when doctors at Stanford University Medical Center transplanted a kidney donated by her father. She is now the healthy older sister to two healthy siblings.

The Congressional tribute took place a week after Abigail celebrated her first birthday.

House members rose to their feet and applauded when Herrera Beutler stood with Abigail in her arms. "If she's a happy, healthy miracle, Dr. Bienstock is the miracle worker who helped give the gift of help and life to this family," Boehner said, according to an article in *USA Today*.*

"When I met with Representative Beutler, I told her that while I couldn't promise that the amnioinfusion treatment would work, I could promise her that we would advance the science around caring for families facing a diagnosis of bilateral renal agenesis," said Bienstock in 2021. "Because of her bravery, we now have a multicenter trial going on around the country to help other babies with this diagnosis."†

* Camia, Catalina, "'Miracle Worker' and Baby Win Over Boehner and the House," *USA Today*, July 23, 2014, https://www.usatoday.com/story/news/politics/onpolitics/2014/07/23/boehner-miracle-baby-herrera-beutler/81555718/.

† Email from Jessica Bienstock to Karen Nitkin, Nov. 8, 2021.

One University and the Collaborations It Created

In 2015, speed dating came to Johns Hopkins.

Participants sat across from each other at long tables in Turner Auditorium. Everyone had three minutes to talk about their interests and three minutes to listen. Then they switched seats to repeat the process with someone new.

By the end of the evening, several matches had been made. Contact information was exchanged, along with promises to continue the relationships.

The people at the event weren't looking for romantic partners — they were forging research collaborations.

The two speed dating events, both in early 2015, attracted about 25 people each. The first brought together researchers from the School of Medicine's Department of Pathology and the Whiting School of Engineering. The second introduced pathologists to engineers and scientists at the APL.

The events were part of a new Johns Hopkins University initiative called One University, which encourages research and interactions across Johns Hopkins divisions.[46]

One University was part of a larger program called Ten by Twenty, unveiled by

Speed-dating participants from the Department of Pathology and the Applied Physics Laboratory have three minutes to describe their work to potential collaborators.

A Soaring Solution

One collaboration that came out of the speed dating session between the Department of Pathology and the Johns Hopkins University Applied Physics Laboratory (APL) led to creation of a courier system that used drones to bring blood samples to diagnostic laboratories.

Pathologist Timothy Amukele developed the idea from conversations with medical students. He realized that it would work in places like rural Kitale, Kenya, which had about 40 health clinics but just one diagnostic laboratory capable of testing blood for HIV, tuberculosis, diabetes, pregnancy and other conditions.

The clinic workers who drew blood from patients would send samples to that lab by commercial buses, which travel as far as 30 miles over bumpy roads that can break down the cells of the blood, making analysis impossible.

"The district director tells me a lot of times the samples are not useful or never arrive," said Amukele, who spent about one-third of each year in Africa, studying the number and quality of diagnostic labs.

At the speed dating event, he learned that engineers at APL could help him achieve his goal of creating a courier system with dozens of unmanned aerial vehicles (UAVs), popularly known as drones, to transport blood samples.

A few days later, Amukele drove to APL's headquarters in Laurel, Maryland, where he met with five UAV engineers. Amukele knew little about UAVs, and the APL staffers knew even less about blood tests in low-resource settings. But once they were in the same room, ideas flew.

The engineers would focus on creating a UAV system to meet Amukele's requirements for simple operation and collision-free, accurate flight. Amukele would travel to the region to learn more about its landscape and diagnostic needs.*

In 2015, Amukele won

$100,000 in the first round of Discovery Awards, along with APL's Robert Chalmers, the project lead; Joseph Moore, who was in charge of flight dynamics and modeling; and Robert Bamberger, who conducted field tests.†

On Sept. 26, 2017, the investigators reported in the American Journal of Clinical Pathology that they had set a new delivery distance record for medical drones, successfully transporting human blood samples across 161 miles of Arizona desert. Throughout the three-hour flight, they reported, the on-board payload system maintained temperature control, ensuring the samples were viable for laboratory analysis after landing.‡

In 2018, Amukele won a $75,000 Catalyst Award to continue his work.

Pathologist Timothy Amukele, left, teamed with Robert Chalmers and other Applied Physics Laboratory engineers in 2015 to create a drone courier system for transporting blood to diagnostic laboratories in Africa.

---

*   Nitkin, Karen, "Taking Collaboration to New Heights," *Dome*, June 1, 2015, https://www.hopkinsmedicine.org/news/articles/takingcollaboration-to-new-heights.

†   "Johns Hopkins Discovery Awards: 2015 Awardees," Johns Hopkins University webpage, Johns Hopkins University Research, https://research.jhu.edu/major-initiatives/discovery-awards/2015-awardees/.

‡   "Study Sets New Distance Record for Medical Drone Transport," Johns Hopkins Medicine press release, Sept. 12, 2017, [consolidate this with rest of footnote on next page] https://www.hopkinsmedicine.org/news/media/releases/study_sets_new_distance_record_for_medical_drone_transport.

Daniels in May 2013, which outlined four goals and 10 priorities for the remainder of the decade.

The goals were:

- One University — forging collaborations across disciplines
- Individual excellence — supporting faculty, students and staff
- Commitment to our communities — enriching ties to Baltimore, the nation and the world
- Institution building — building an even stronger university

One University's priorities included investments and support for collaborations across disciplines and schools.[47] Supporting those priorities, two new awards programs were launched in 2015 by the president, provost and deans of the university.

- The Discovery Fund provides awards of up to $150,000 for studies led by investigators from at least two of the university's schools or affiliates, who are poised to arrive at important discoveries or creative works.[48]
- Catalyst Awards of up to $75,000 support the promising research and creative endeavors of early-career faculty, with the goal of launching them on a path to a sustainable, rewarding academic career.[49]

Also promoting collaboration and innovation, the School of Medicine Discovery Fund Synergy Awards give grants of up to $100,000 for collaborative research involving at least one investigator in the School of Medicine.

Rising to the Challenge: 'Astounding' Numbers

The Johns Hopkins University and Johns Hopkins Health System together launched their Rising to the Challenge campaign in May 2013 with the most ambitious goal of any Johns Hopkins fundraising campaign to date: to raise $4.5 billion. The launch followed a "quiet phase" that began in January 2010 and had already received pledges of $1.94 billion.[50]

When the campaign ended on Oct. 11, 2018, with a celebration at the O'Connor Recreation Center on the university's Homewood campus, it counted gifts totaling a staggering $6.015 billion in support of students, faculty, research and discovery, and clinical care.

"The numbers in this campaign are astounding," Daniels wrote in a message sent to the Johns Hopkins community that morning. "1,600 volunteers; more than 878,000 individual gifts from 279,000 generous donors; $1.68 billion in new endowment funding, including 267 endowed professorships and 157 endowed undergraduate scholarships; $610 million in graduate and undergraduate student financial aid; and more than $2.8 billion in support of medical research and programs. In the end, we raised a total of $6.015 billion."[51]

Contributions ranged from small but meaningful gifts, such as a few hundred dollars from an 11-year-old boy to support a piano scholarship, to $350 million from billionaire alumnus and former New York City Mayor Michael Bloomberg.

Bloomberg's January 2013 gift, the largest ever to the university, brought his lifetime support of Johns Hopkins past $1 billion, a landmark of philanthropy for any university that he was believed to be the first to reach.[52]

About $100 million would be used to provide need-based financial aid for undergraduate students. The remaining $250 million would support creation of Bloomberg Distinguished Professorships, a cadre of 50 world-class interdisciplinary

Ronald J. Daniels addresses guests at a 2018 "Rising to the Challenge" event on the Johns Hopkins University campus.

professorships aimed at addressing major world problems. "The Bloomberg Professors will bridge the university's divisions, conduct and stimulate innovative research that crosses traditional disciplinary boundaries, and train a new generation of native 'interdisciplinarians,'" notes the university website.[53]

The Bloomberg Distinguished Professorships are the centerpiece of the university's focus on strengthening its "capacity for faculty-led interdisciplinary collaboration" as outlined in Daniels' Ten by Twenty vision plan.[54]

In December 2021, The Johns Hopkins University announced it would recruit an additional 50 Bloomberg Distinguished Professors to the program, primarily to work in faculty-developed interdisciplinary clusters, conducting transformational research in fields such as artificial intelligence, the health effects of climate change, racial equity, neuroscience, and pandemic preparedness and response.

"Our most pressing global challenges span many fields and geographies—and to take them on, we need more talented researchers who can do the same," said Bloomberg, whose additional investment made the next phase of the program possible.[55]

New Leader for the Department of Medicine

On August 15, 2014, Myron (Mike) Weisfeldt stepped down from the post he had held since 2001. Mark Anderson became the next William Osler Professor of Medicine, director of the Department of Medicine in the Johns Hopkins University School of Medicine and physician-in-chief of The Johns Hopkins Hospital.

By coincidence, both men were cardiologists and both were from the Midwest — Anderson from Rochester, Minnesota, and Weisfeldt from Milwaukee.

Anderson had been recruited for the prestigious role from the University of Iowa Carver College of Medicine, where he had been chair of the Department of Medicine since 2009 and director of the Cardiovascular Research Center since 2012.

"I was happy where I was, so it took a little reckoning," he told *Aequanimitas* shortly after the move. "But I knew I'd be filled with regret if I didn't take this opportunity. It's a chance to participate in something truly outstanding and to contribute to academic medicine in a way that I couldn't at any other place."[56]

Anderson was a 1981 honors graduate in biology from Macalester College in St. Paul, Minnesota, and had earned his Ph.D. in physiology and his M.D. from the University of Minnesota. He completed his internal medicine residency and fellowships in cardiology and clinical cardiac electrophysiology at Stanford before joining the faculty at Vanderbilt in 1996.

At Vanderbilt, he rose through the academic ranks, growing a National Institutes of Health-funded research laboratory and becoming the Betty and Jack Bailey Professor of Medicine. He directed the cardiology and clinical cardiac electrophysiology fellowship training programs and served as the director of the Clinical Arrhythmia Service.

In October 2005, he moved to the University of Iowa Carver College of Medicine as the director of the cardiology division; in 2009 he was named

Mark Anderson

Weisfeldt Honored for Diversity and Equity Efforts

In 2021, the American Heart Association (AHA) chose Myron (Mike) Weisfeldt for its annual Watkins-Saunders Award, which recognizes doctors who fight to overcome health and community disparities in Maryland.

The AHA award was named for two legendary cardiovascular specialists in Maryland — cardiac surgeon Levi Watkins Jr. and cardiologist Elijah Saunders — who worked to improve health equity while at Johns Hopkins and the University of Maryland, Baltimore, respectively.

Among Weisfeldt's health equity initiatives, he led the development of the Sickle Cell Center for Adults at The Johns Hopkins Hospital and the Medicine–Pediatrics Urban Health Residency Program*. He also improved diversity among the faculty, postdoctoral fellows and residents in the Department of Medicine.

When Weisfeldt became chairman of the department in 2001, underrepresented minorities made up no more than 3% of staff in any specialty. Weisfeldt started a diversity council, enlarged

Johns Hopkins' interview pool for prospective residents, and personally met with potential minority hires.

By 2004, 13% of new assistant professors in the department were underrepresented minorities.

"I have always believed that by discriminating against talented people from minority communities, we are depriving medicine and science of great people and great teachers and great products," he said.

A Diversity Visiting Professorship, established in 2003, brings prominent physicians, scientists and national leaders to the Department of Medicine for Grand Rounds, opportunities to discuss issues related to minorities in academic medicine, and meetings with leadership, house staff and fellows. The professorship is now called the Myron L. Weisfeldt Distinguished Visiting Professorship in Diversity.

"He's one of the earliest proponents of increasing women and minorities in medicine, cardiology in particular," said Roger Blumenthal, director of the Johns Hopkins Ciccarone Center for the Prevention of Cardiovascular Disease, who completed his residency and fellowship under Weisfeldt and nominated him for the Watkins-Saunders award.

"He made it more interesting, more desirable for people to get involved in cardiology and showed that academic medicine was a great career path for people who weren't well-represented in the past."†

Weisfeldt previously was honored in 2008, when the Association of Professors of Medicine gave him its Diversity Award for his work improving diversity within the Department of Medicine.‡

Myron Weisfeldt

---

*   "JHTV's Weisfeldt Honored by American Heart Association for Diversity Efforts," Johns Hopkins Technology Ventures webpage, News, https://ventures.jhu.edu/news/mike-weisfeldt-aha-diversity-johns-hopkins/.

†   "JHTV's Weisfeldt Honored by American Heart Association for Diversity Efforts."

‡   Weisfeldt portrait collection page.

---

chairman and department executive officer of internal medicine there.

Anderson's research focuses on the role of the protein CaMKII in heart failure and cardiac arrhythmias, a cause of sudden cardiac death. He has published more than 160 peer-reviewed journal articles, book chapters and book reviews, and has been an invited speaker throughout the United States and in more than a dozen nations.[57]

Lights, Camera, Hopkins

A still from a scene filmed at the Johns Hopkins Kimmel Cancer Center for Cancer: The Emperor of All Maladies.

Johns Hopkins hospitals, clinicians, researchers and patients were featured in a PBS documentary series about cancer and the centuries-long effort to understand and treat it.

Cancer: The Emperor of All Maladies, which premiered March 30, 2015, was based on Siddhartha Mukherjee's powerful Pulitzer Prize-winning book of the same name and executive-produced by famed documentarian Ken Burns.

Film crews were embedded at the Johns Hopkins Kimmel Cancer Center for 18 months to capture the hopeful and sometimes heartbreaking narratives that appeared in the three-part, six-hour series. The crews also gathered footage at the pediatric oncology wing of The Charlotte R. Bloomberg Children's Center, the Lung Cancer Program at Johns Hopkins Bayview Medical Center, and the Pancreas Multidisciplinary Cancer Clinic at The Johns Hopkins Hospital, as well as at the Charleston (West Virginia) Area Medical Center.*

Mukherjee is a cancer specialist at Columbia University, but director Barak Goodman said the production team chose Johns Hopkins for one of its locations because it wanted a hospital on the forefront of cancer treatment and research.

Johns Hopkins also had experience working with film crews as the location for two other documentaries: *Hopkins 24/7*, a documentary of life at the hospital that aired on ABC in 2000; and its sequel, *Hopkins*, a six-part follow-up that ran in 2008 and won a Peabody Award the following year.†

Johns Hopkins employees were treated to a preview of *The Emperor* series and a panel discussion on March 23 at the Chevy Chase Auditorium at the Sheikh Zayed Tower of The Johns Hopkins Hospital.

Speakers included Mukherjee and Goodman, as well as Bert Vogelstein and Kenneth Kinzler, co-directors of the Ludwig Center at the Kimmel Cancer Center. William Nelson, director of the Kimmel Cancer Center, was moderator.‡

---

* Rienzi, Greg, "Johns Hopkins Has Role in New Ken Burns Cancer Documentary, *Johns Hopkins University Gazette*, March/April 2015, https://hub.jhu.edu/gazette/2015/march-april/currents-cancer-emperor-of-maladies/.

† Rienzi, Greg, and Minkove, Judy F., "Cancer: Emperor of All Maladies," *Dome*, March 2015, https://www.hopkinsmedicine.org/news/publications/dome/dome_march_2015/cancer_emperor_of_all_maladies.

‡ "PBS Documentary-Cancer: The Emperor of All Maladies," Johns Hopkins Medicine webpage, News and Events, https://www.hopkinsmedicine.org/kimmel_cancer_center/news_events/conversations_on_cancer/pbs_documentary.html.

(In October 2022, Anderson started a new job at the University of Chicago as Executive Vice President for Medical Affairs, Dean of the Division of the Biological Sciences and Dean of the Pritzker School of Medicine.)[58]

Weisfeldt had been at Johns Hopkins for much of his education and career. He arrived on the Baltimore campus as an undergraduate in 1960, earning his bachelor's degree in 1962 and his medical degree in 1965.

After stints at Columbia-Presbyterian Medical Center, the National Institutes of Health and Massachusetts General Hospital, he returned to Johns Hopkins in 1972 as assistant professor of medicine and director of the Peter Belfer Laboratory for Myocardial Research.

He followed Richard Ross as the director of cardiology, holding the job from 1975 to 1991, when he moved to Columbia University's College of Physicians and Surgeons as the Samuel Bard Professor of Medicine and chair of the department of medicine.

He came back to Johns Hopkins in 2001 to become William Osler Professor and director of the department of medicine, a post he held until 2014. In 2021, he praised Anderson for his "remarkable leadership appointments" and said his successor "strengthened the opportunities for growth of young and exciting faculty with creative ideas and accomplishment."[59]

Weisfeldt was president of the American Heart Association (AHA) from 1989 to 1990. His cardiology contributions include his work on the first use of an implantable defibrillator and the initial studies of clot-busting drugs in the treatment of heart attack. For these achievements, he received the AHA's Award of Merit in 1989 and its Gold Heart Award in 1996.[60]

When he stepped down from the Department of Medicine, he became medical director of Johns Hopkins Technology Ventures, where he has served as its medical consultant since 2018.

Symposium Puts More Prosthetics Within Reach

This photo of 5-year-old Griffin Matuszek shows something extraordinary.

As dad Steve Matuszek looks on, Griffin picks up a favorite superhero toy using a colorful prosthetic hand. The amazing part is that the hand cost about $50 to make on a 3-D printer, compared to about $50,000 for a traditional prosthesis.

Griffin, born with a tiny left palm and undeveloped fingers, was among about 500 attendees at a Sept. 28, 2014, Johns Hopkins Medicine symposium called Prosthetists Meet Printers.

Johns Hopkins trauma surgeon Albert Chi enlisted the volunteer organization e-NABLE; the Kennedy Krieger Institute; and other leaders in medicine and industry, who joined forces to put more durable, less constrictive and much less expensive prosthetic hands within the grasp of thousands of children.*

---

*   "Give These Folks a Hand," *Dome*, November 2014, https://www.hopkinsmedicine.org/news/ publications/dome/dome_november_2014/give_ these_folks_a_hand.

Steve Kravet

Colleen Christmas

Scott Wright

Samuel (Chris) Durso

Clinical Excellence

In an academic medicine institution like Johns Hopkins, the traditional road to promotion is one that travels through a stack of published studies and a calendar filled with lectures and presentations.

Clinical excellence, which stands alongside research and education in the tripartite Johns Hopkins Medicine mission, didn't, until recently, have the same power to propel a career to its next level.

The leaders at Johns Hopkins Medicine set out to change that, first by honoring excellent clinicians through the Miller Coulson Academy of Clinical Excellence (MCACE), created in 2009; and then, in 2015, with the Johns Hopkins Medicine Clinical Awards for Physicians and Care Teams, awarded by the Office of Johns Hopkins Physicians.

The MCACE began to take shape in 2006, when Sarah Miller Coulson and Frank Coulson provided an unrestricted gift of $1 million to the Center for Innovative Medicine (CIM) at Johns Hopkins Bayview Medical Center.

The center's director is rheumatologist David Hellmann, a widely respected clinician and educator who had created the CIM in 2005 to make medicine a public trust.

At the time of the Miller Coulson gift, Hellmann had been ruminating on a question that Anne Miller, Sarah's mother, had posed years earlier, namely, "David, why aren't there more Dr. Tumultys?" The late Philip Tumulty had been Anne Miller's doctor and had represented the pinnacle of clinical excellence at Johns Hopkins for decades.

Hellmann decided that the most appropriate use of the Miller Coulson gift would be to try to answer Anne Miller's question.

He recognized that for this work to hold any water at Johns Hopkins, it would have to be meticulous and thorough, he said. Therefore, he used the Miller Coulson gift to fund a portion of the effort of four faculty members — Scott Wright, Samuel (Chris) Durso, Steve Kravet and Colleen Christmas — in determining how clinical excellence should be defined, measured and rewarded. Together they reviewed the literature, visited top medical centers, and interviewed many deans and chairs of clinical departments.

They concluded that clinical excellence was defined by professionalism, communication skills, diagnostic acumen, depth of knowledge, ability to negotiate the health care system, a passion for patient care, and service as a role model to medical trainees.

As a reflection of their scholarly approach, they published several manuscripts — including one in *The Mayo Clinic Proceedings* — that characterized a conceptual model for clinical excellence. Ultimately, this team designed and launched the MCACE — the first academy of its kind in the country. The academy's mission is to "recognize and promote clinical excellence for the benefit of the patients and communities that we serve."[61]

The MCACE inaugurated its first members with little fanfare in 2009. At first,

David Hellmann

In August 2020, David Hellmann stepped down from his roles as chairman of the Department of Medicine at Johns Hopkins Bayview Medical Center and vice dean for the Johns Hopkins Bayview campus, while continuing to lead the Center for Innovative Medicine and remaining co-chair of the Clinical Excellence Promotion Committee.*

Hellmann, the Aliki Perroti Professor of Medicine, is known for his warmth and inspiring leadership as a teacher, physician and builder of programs. His accolades include the George J. Stewart and C. Lockard Conley teaching awards from the Maryland Chapter of the American College of Physicians, and the Jane F. Desforges Distinguished Teaching Award from the American College of Physicians. He has been selected by medical students seven times to participate in their graduation, to deliver the Hippocratic Oath or as marshal or hooder.

Hellmann came to Johns Hopkins as a medical student and stayed on to complete his internal medicine residency on the Osler Medical Service in 1980. He returned to Johns Hopkins as a member of the faculty in 1986, after training in rheumatology

and spending several years on the faculty at the University of California, San Francisco.

Hellmann is fond of the phrases "medicine is a public trust," and "know your patient as a person," both emblematic of his commitment to improving the quality of life for patients and the community.

He has served in numerous important roles, including as clinical director for the Division of Rheumatology; director for the Osler residency program in the Department of Medicine; and executive vice chairman of the Department of Medicine. In 2000, he became chairman of the Department of Medicine at Johns Hopkins Bayview. Five years later, he was appointed vice dean for the campus.

Hellmann, whose clinical care and research focuses on vasculitis, is frequently sought out for his diagnostic acumen and clinical judgment. He has published more than 125 manuscripts and nearly 100 book chapters, and has served on numerous leadership and editorial boards, including the Board of Directors of the American Board of Internal Medicine Foundation and, for seven years, as editor of the journal Medicine.

In 2004, he established the

David Hellmann

Center for Innovative Medicine (CIM), which has nurtured clinicians and fostered ideas that have improved JHM and medicine as a whole. Among the most notable of these efforts was the founding of the Miller Coulson Academy of Clinical Excellence.

The Miller Coulson Academy, which debuted in 2008, formed the philosophical and operational underpinnings of the Clinical Excellence Task Force and, more recently, the Clinical Excellence promotion pathway.†

---

* "David Hellmann Stepping Down as Chairman of the Department of Medicine and Vice Dean at Johns Hopkins Bayview Medical Center," *Alumni News*, Aug. 19, 2020, https://www.hopkinsmedicine.org/news/publications/alumni-news/david-hellmann-stepping-down-as-chairman.

† "David Hellmann Stepping Down as Chairman of the Department of Medicine and Vice Dean at Johns Hopkins Bayview Medical Center."

---

membership was limited to clinicians at Bayview; within a few years, the program had expanded to include The Johns Hopkins Hospital. Ultimately, Wright became the director of the academy; Durso became chair of the Department of Medicine at Bayview; Kravet became president of Johns Hopkins Community Physicians; and Christmas became director of the Primary Care Leadership Track.

All four of the Miller Coulson Scholars now have endowed professorships. Fittingly, Wright is the Anne G. and G. Thomas Miller Professor of Medicine.

Expanding the reach of the academy was an important step toward evaluating and rewarding clinical excellence throughout Johns Hopkins Medicine, Wright explained in 2014: "Currently, if faculty members spend two-thirds of their time in patient care, that doesn't even get looked at or considered when they are up for promotion."

Academy members devote 20 or more hours a year to advance the academy's mission; they write papers and give talks about clinical excellence. They also serve as role models for medical students who can sign up for week-long rotations within the academy "to see and learn the habits of a clinically excellent physician," said Hellmann.[62]

As a few examples, the academy launched the G. Thomas Miller Coaching Program in 2013 to connect master clinicians with new faculty for clinical mentoring, and created the Frank L. Coulson Award for Clinical Excellence, which honors outstanding doctors in training in all residency programs.

In 2018, the academy launched CLOSLER.org, a website that makes it possible for people all over the world to learn about excellence in patient care from great role models.

At the 10-year anniversary of the awards, in 2018, 81 clinicians had been inducted. "These men and women — in specialties ranging from oncology to gerontology to ophthalmology — truly are standout clinicians, providing compassionate care to every patient and family, hour after hour, day after day. As individuals, they are truly a force for good; by coming together to collaborate within the Miller Coulson Academy of Clinical Excellence, their impact is nothing short of breathtaking," noted Hellmann and Wright in a publication marking the anniversary.[63]

Johns Hopkins Medicine Clinical Awards

Introduced in 2015, the Johns Hopkins Medicine Clinical Awards for Physicians and Care Teams is open to providers who practice in Johns Hopkins Medicine facilities, Johns Hopkins-employed or not, and recognizes multiple aspects of excellent patient care through six awards:

- Armstrong Award for Excellence in Quality and Safety
- Excellence in Service and Professionalism
- Best Consulting Physician
- Innovations in Clinical Care
- Clinical Collaboration and Teamwork
- William A. Baumgartner Physician of the Year

All Johns Hopkins Medicine hospitals and Johns Hopkins Community Physicians select their winners after nomination and local committee review.[64]

The awards ceremonies, held annually in the Chevy Chase Auditorium (though remotely during the COVID-19 pandemic), are organized by the Office of Johns Hopkins Physicians.

OJHP, helmed by Bill Baumgartner until his June 2018 retirement and now run by Jonathan Efron, was created in 2011 as part of the JHM 3.0 reorganization.

It supports all JHM physicians — faculty in the school of medicine as well as the thousands of other Johns Hopkins doctors who provide excellent care in community hospitals, outpatient clinics and affiliate organizations — ensuring all have access to the same tools and training, resources and information, including the honor of being recognized for clinical excellence.

Bill Baumgartner

Jonathan Efron

Clinical Excellence Promotions Track

The awards gave excellent clinicians the recognition they deserved, but it wasn't enough. In 2019, Johns Hopkins Medicine announced it was building on the Miller Coulson Academy with the launch of a new clinical excellence promotions track that rewards faculty members who demonstrate clinical excellence with patients. "Hopkins has always had a tripartite mission — research, education and clinical care; however, our traditional promotional track has not equally valued all missions," said Cynthia Rand, senior associate dean for faculty. "The Clinical Excellence promotion pathway, which started in 2019, is just as rigorous as the more traditional scholarship-focused career path. The single track has favored people who have been historically inclined toward research and peer-reviewed publication. We want people to advance for all our missions."[65]

The track is as rigorous as the current scholarship-focused career path, but could be a better fit for faculty members who have a primary focus on clinical work.

Hellmann co-leads the Clinical Excellence Promotions Committee, along with psychiatrist Meg Chisolm and ophthalmologist Sharon Solomon, both Miller Coulson Academy members. Chisolm credits leaders of the Miller Coulson Academy for establishing the criteria to measure clinical excellence, including a "360-degree" review. "They set the bar very high," she said.[66]

In December 2020, Rosalyn Stewart, director of the Johns Hopkins After-Care Clinic and director of the Johns Hopkins Hospital Substance Use Consultation Service, became the first faculty member to be promoted to full professor on the new track.

"It's been a very moving experience," said Chisolm. "People on our committee were brought to tears to see that great clinicians at Johns Hopkins — many of whom had trained hundreds of outstanding clinicians who went on to get promoted at other institutions — are finally getting recognized."[67]

Cynthia Rand   Meg Chisholm

Sharon Solomon   Rosalyn Stewart

Renovations and Expansions: Bringing Historic Buildings into the Future

A ribbon-cutting ceremony on Oct. 15, 2014, celebrated renovations to the Nelson/Harvey building in The Johns Hopkins Hospital.

The investment was part of a sweeping redevelopment that in the next few years would bring major additions and updates to Johns Hopkins Bayview Medical Center and Sibley Memorial Hospital.

The refurbished Nelson/Harvey Building, located on the northeast side of the hospital's campus, was outfitted with 136 private rooms with sleeping accommodations for family members, as well as family respite areas on each floor, updated software and equipment for caregivers, and a communication system that eliminated the disruptive paging system of the past.[68]

"We want patients to have first-class care, and that starts the moment they step

Russell Andrew Nelson   **Abner McGehee Harvey**

on our campus," Peterson said ahead of the event. "A lot of thought and care has gone into this project, and it shows."[69]

The two-year project also gave the building finishes similar to those in The Charlotte R. Bloomberg Children's Center and the Sheikh Zayed Tower, which opened in May 2012, after one of the largest hospital construction projects in U.S. history.[70]

The Nelson and Harvey buildings had opened in 1977, with the Nelson building devoted to patient care and Harvey to teaching.[71] They were named for Russell Nelson, a 1937 graduate of the school of medicine who served from 1952 to 1972 as hospital president, and A. McGehee Harvey, a 1934 graduate of Johns Hopkins' medical school, who was the longest-serving director of the Department of Medicine, heading it from 1946 to 1973.

In a December 2014 *Dome* column, Peterson said he knew both men, "albeit casually," and added:

"I'm confident that they would be immensely pleased by the rebirth of their namesake structures, which now gives us a building that will last another 30 or 40 years."[72]

Bayview's Big Upgrades

In rapid succession, Johns Hopkins Bayview Medical Center celebrated three major openings in early 2015. The Center for Translational Molecular Engineering opened in January, followed by a new Sidney Kimmel Cancer Center the following month and, in March 2015, a new North Pavilion with an adult emergency department and dedicated pediatric emergency department.

The opening of a Sidney Kimmel Cancer Center brought radiation oncology to Bayview for the first time. (The Kimmel Cancer Center at The Johns Hopkins Hospital was dedicated May 4, 2002, following a $150 million gift in November 2001 from businessman Sidney Kimmel.)[73]

The main focus of the new location was thoracic oncology — lung cancer, esophageal cancer and mesothelioma. "We consolidated all of our thoracic cancer services onto the Hopkins Bayview campus to provide better patient-centered care,"

The Johns Hopkins Bayview Medical Center opened its North Pavilion for patients at 3 a.m. on March 25, 2015.

Salwa Khan, associate director, Pediatric Hospital Medicine Program

Kimmel Cancer Center at Johns Hopkins Bayview

explained Julie Brahmer, director of the Thoracic Oncology Program. "Having all of these disciplines in one location is a great benefit to patients."

Beyond its focus on lung cancer, the center was equipped to offer expert surgical, medical and radiation oncology care in all types of cancer, in addition to palliative care.

The $26 million facility housed 21 infusion bays, 20 clinical exam rooms, a CT simulator and a linear accelerator for radiation treatments. Having radiation oncology on campus means patients could go to one place for all of their cancer treatments.[74]

In March 2015, a new North Pavilion opened at Bayview, featuring an adult emergency department designed to expedite care. For example, each new patient could be seen quickly by an initial provider at a first-look station before proceeding to the large family lounge.

The $40.1 million building also houses a pediatric center that combines a pediatric emergency department — the first at Bayview — and a pediatric inpatient unit. The center is staffed 24/7 by a multidisciplinary team trained in the medical, emotional and developmental needs of children. It also provides specialized equipment in a range of sizes, as well as technology and monitoring systems designed specifically for children.[75]

"Johns Hopkins Bayview continues to grow and expand to better serve our

Richard Bennett

Victor Broccolino Retires from Howard County General Hospital

Victor Broccolino

Steven Snelgrove

Shafeeq Ahmed

After 24 years as the gregarious and ambitious president and CEO of Howard County General Hospital (HCGH), Victor Broccolino retired on Jan. 15, 2014.

During his time as president, Broccolino was both the face and the heart of the institution. He more than doubled the hospital's size to 233 beds, added or expanded myriad services and programs, and led its 1998 merger with Johns Hopkins Medicine — the first time Johns Hopkins had joined with a private, nonprofit hospital.[*]

HCGH, founded in 1973 as a 59-bed, short-stay hospital, is the only hospital in the fast-growing county. Joining Johns Hopkins allowed it to expand and become more integrated in the community. Among other things, the hospital tripled the size of its emergency department in 2002, and added a 167,000-square-foot Medical Pavilion, which opened in 2009.[†]

Broccolino, born and raised in Baltimore, majored in business administration at the University of Baltimore and earned a master's degree in business administration from Loyola College. He worked in the banking industry for several years before becoming controller at Franklin Square Hospital, controller at Bon Secours Health Corp. and then CEO at Bon Secours from 1986 to 1989 before moving to HCGH.[‡]

"Through his outgoing personality and commitment to the community, coupled with a tremendous respect for the potential associated with the promise of medicine that Johns Hopkins offers, he has built a strong community hospital that is well-positioned to succeed in the quickly changing world of health reform," Peterson said when Broccolino announced his retirement.[§]

The next HCGH president was Steven Snelgrove, who stepped into the role on Jan. 16, 2014. He had most recently served as president of Wake Forest Baptist Health's Lexington Medical Center in Lexington, N.C. Snelgrove had joined the Wake Forest University Baptist Health System in 1989 as director of facility planning and construction.[¶]

He remained until November 2020, when M. Shafeeq Ahmed, who had joined HCGH as chief medical officer/vice president of medical affairs in July 2016, became interim president and then president of the hospital in June 2021.

Before joining Johns Hopkins, Ahmed served in numerous positions of health care leadership, including as chief medical officer and chief operating officer at two hospitals within the Baystate Health system in Massachusetts.

Before joining Baystate, Ahmed was a lieutenant commander in the U.S. Naval Medical Corps and served as an attending obstetrician/gynecologist at the Naval Hospital – Cherry Point, North Carolina, where he was also the president of the medical staff and chief of obstetrics/gynecology.

He received his medical and bachelor's degrees from Boston University, and completed a residency in obstetrics and gynecology at Baystate Medical Center.

He later completed a health care leader fellowship at the Massachusetts Healthcare Leadership College. Additionally, Ahmed holds a master of business administration degree from the Isenberg School of Business at the University of Massachusetts – Amherst. Ahmed is a board-certified fellow of the American Congress of Obstetricians and Gynecologists. Between 2014 and 2020, Becker's Hospital Review named Ahmed one of the "100 Hospital and Health System Chief Medical Officers to Know."[**]

*   Salagnik, M. William, and Hedgpeth, Dana, "Hopkins Buys Howard Hospital $142 Million Deal Continues Medical Consolidation Trend; No Staff, Service Cuts Seen," *Baltimore Sun*, March 19, 1998, https://www.baltimoresun.com/news/bs-xpm-1998-03-19-1998078009-story.html.

†   "Historical Timeline," Howard County General Hospital webpage, https://www.hopkinsmedicine.org/howard_county_general_hospital/about/timeline.html.

‡   Pichaske, Pete, "Broccolino, the Face of Howard County General, Retires after 24 years as CEO," *Baltimore Sun*, Jan. 9, 2014, https://www.baltimoresun.com/maryland/howard/columbia/ph-ho-cf-broccolino-0109-20140109-story.html.

§   "Broccolino, the Face of Howard County General, Retires after 24 years as CEO."

¶   Nitkin, Karen, "A New Member of the Johns Hopkins Family," *Dome*, Jan/Feb 2014, https://www.hopkinsmedicine.org/news/publications/dome/dome_januaryfebruary_2014/a_new_member_of_the_johns_hopkins_family.

**   "Shafeeq Ahmed Appointed President of Howard County General Hospital," Johns Hopkins Medicine news release, Howard County General Hospital, June 29, 2021, https://www.hopkinsmedicine.org/news/newsroom/news-releases/shafeeq-ahmed-appointed-president-of-howard-county-general-hospital.

community," Richard Bennett, Bayview's president, told the *Baltimore Sun*. "Our new adult emergency department was designed to streamline throughput and provide more efficient care. We're particularly excited to open a combined pediatric emergency department and pediatric inpatient unit to provide coordinated care for our youngest patients."[76] (Bennett retired as Bayview's president on Dec. 31, 2022.)

Sibley Steps into the Future

Sibley Memorial Hospital also benefitted from major expansions and modernizations around this time.

The Washington, D.C., campus added a medical building called Building A in 2011, the year after it became part of Johns Hopkins Medicine, with physician offices, outpatient surgery and imaging centers, and a parking garage. A radiation oncology center opened on campus in 2012.

In 2015, the hospital opened a new emergency department with 28 new treatment rooms; CT and MRI capabilities within and adjacent to the new ED; and decontamination showers to handle exposure to radiation, chemical or biological contaminants.[77]

The ED was created with efficient design, streamlined processes and new technology to minimize delays and maximize patient comfort and care, with the goal of reducing wait time for arriving patients. For example, a six-room FastTrack area was created for treatment of minor problems.[78]

In 2016, Sibley dedicated a new state-of-the-art hospital, known as Building B, with 200 private patient rooms, and the Johns Hopkins Kimmel Cancer Center, with 34 private infusion rooms and an inpatient oncology unit.

The 475,000-square-foot hospital expansion also contained the only pediatric radiation oncology program in Washington, D.C., operated in collaboration with Children's National Hospital; two floors of women's and infants' services; and an orthopaedic unit with a rehabilitation clinic.[79]

The expanded maternity unit boasted 18 labor, delivery and recovery rooms, equipped for high-risk and multiple birth deliveries; three C-section rooms; 50 spacious private postpartum rooms with 55-inch flat screen HDTVs and sleep sofas; and the Silberman Family Special Care Nursery with 18 private rooms, including six rooms for twins, state-of-the-art mobile and central fetal monitoring and round-the-clock neonatologists and neonatal nurses.[80]

The new orthopaedic unit has 50 all-private rooms with 55-inch flat screen HDTVs and sleep sofas; and the Morris and Gwendolyn Cafritz Foundation Clinic, a large state-of-the-art rehabilitation clinic.[81]

"As we opened the doors to the new building, we stepped into a new era of patient care, comfort and innovation and marked a major milestone in the history of our hospital," said Richard Davis, Sibley president and CEO at the time.[82]

A 2016 expansion of Sibley Memorial Hospital included 200 private patient rooms and an open and bright lobby, home to the Interplay statue by sculptor John Safer.

Richard Davis

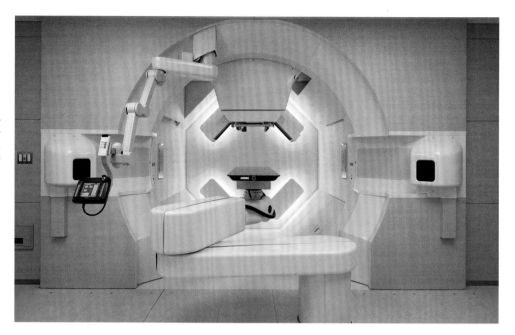

The Johns Hopkins Proton Therapy Center, which opened at Sibley Memorial Hospital on Oct. 1, 2019, provides targeted treatments that spare nearby healthy cells.

Johns Hopkins Proton Therapy Center

On Oct. 1, 2019, Johns Hopkins Medicine opened a proton center at Sibley Memorial Hospital, ushering in a new era in patient care.

The Johns Hopkins Proton Therapy Center, with experts from the Johns Hopkins Kimmel Cancer Center, gives adult and pediatric patients access to highly targeted cancer treatment that spares nearby healthy tissues and organs, reducing potential side effects including the risk of recurrence. It welcomes patients from the region, from across the country and even around the world.

Unlike traditional radiation, which uses X-rays to destroy cancer cells, proton therapy uses subatomic particles with about 2,000 times more mass, explained Matthew Ladra, assistant professor of radiation oncology and molecular radiation sciences and director of pediatric radiation oncology, Johns Hopkins Kimmel Cancer Center at Sibley Memorial Hospital.

Conventional X-ray therapy allows 30–40% of the dose to pass through the targeted tumor. But protons stop at the edge of the tumor, preventing unnecessary damage to adjacent healthy cells and tissue. As a result, there is less damage to delicate structures such as the brain, heart and spinal column. The risk of secondary cancers from treatment also decreases.

Matthew Ladra

Akila Viswanathan

"We know exactly how to control the energy and speed of protons, so they can deposit energy that kills the cancer and really no dose goes anywhere else," said Theodore DeWeese, at the time vice dean for clinical affairs and president of the Clinical Practice Association at Johns Hopkins Medicine, and since July 2022 interim Dean/CEO of Johns Hopkins Medicine.

The new 80,000-square-foot facility houses state-of-the-art proton therapy equipment, as well as next-generation imaging technologies such as dual energy CT-guided treatment that reduces the range of error, and the latest innovation in biomatrix MRI, designed to target tumors in places like the lung and liver that move due to respiration.

Lloyd Minor

Jonathan Lewin

Julie Freischlag

Jay Brooks Jackson

Gordon Tomaselli

W.P. Andrew Lee

Colleen Koch

Robert S. D. Higgins

A Next Generation of Leaders

Johns Hopkins Medicine has a storied history of creating new generations of academic medicine leaders. Here are some of the people who left Johns Hopkins during the time frame of this book to become leaders of other institutions.

Lloyd B. Minor, former director of the Department of Otolaryngology-Head and Neck Surgery and then provost and senior vice president for academic affairs for The Johns Hopkins University, has been the Carl and Elizabeth Naumann Dean of the Stanford University School of Medicine since December 2012.

Jonathan Lewin, who was Martin W. Donner Professor and director of the Russell H. Morgan Department of Radiology and Radiological Sciences at the Johns Hopkins University School of Medicine and radiologist-in-chief at The Johns Hopkins Hospital, moved to Emory in 2016, becoming executive vice president for health affairs at Emory University, executive director for Emory's Woodruff Health Sciences Center, and president, chief executive

officer, and chair of the board of directors of Emory Healthcare.

Julie Freischlag, who was chief of vascular surgery and the first female surgeon-in-chief of The Johns Hopkins Hospital, went on to become the first female chief of vascular surgery at the University of California, Los Angeles, and in 2017 became CEO of Atrium Health Wake Forest Baptist, dean of Wake Forest University School of Medicine, chief academic officer of Atrium Health and professor of vascular and endovascular surgery.

Jay Brooks Jackson, an internationally recognized researcher in HIV diagnostics, prevention and treatment, who was Baxley Professor and director of pathology, became vice president for medical affairs at the University of Minnesota in 2014, and has been University of Iowa vice president for medical affairs and the Tyrone D. Artz Dean of the Carver College of Medicine since November 2017.

Gordon Tomaselli, a Johns Hopkins cardiologist for 32 years, including 11 as director of the Division of Cardiology at the Johns

Hopkins University School of Medicine, became the Marilyn and Stanley M. Katz Dean at Albert Einstein College of Medicine on July 1, 2018.

W.P. Andrew Lee, a world leader in hand transplantation at the Johns Hopkins University School of Medicine and chair of the Department of Plastic and Reconstructive Surgery, became executive vice president for academic affairs, provost, and dean of the University of Texas Southwestern Medical School in February 2019.

Colleen Koch, who was director of anesthesiology and critical care medicine for the school of medicine, has been dean of the University of Florida College of Medicine since January 2021.

Robert S.D. Higgins, who was surgeon-in-chief at The Johns Hopkins Hospital, director of the Department of Surgery and senior associate dean for diversity and inclusion, became president of Brigham and Women's Hospital and executive vice president at Mass General Brigham, roles he assumed in December 2021.

The center's experts are using this advanced technology, supported by the latest research, to deliver individualized care.

"These advanced imaging facilities give our expert, disease-focused physicians exceptional ability to use very precise anatomy to plan proton therapy," said Akila Viswanathan, director of the Department of Radiation Oncology and Molecular Radiation Sciences.

A large mechanical arm called a gantry can move the beam 360 degrees around the patient, treating the tumor from several angles as it destroys malignant cells layer by layer.

A typical course of proton therapy is five days a week for several weeks. The treatment itself takes just a few minutes, plus time for preparation and positioning. It's usually painless, and most patients immediately return to daily activities.[83]

Specialized Pediatric Care

The Johns Hopkins Proton Therapy Center is one of few in the world — and the only one in the Washington, D.C., region — with a dedicated pediatric team that specializes in caring for young oncology patients. It delivers care in collaboration with Children's National Hospital and the Johns Hopkins pediatric oncology department.

Proton therapy can be especially valuable for children, said Ladra, because it minimizes the excess radiation that can impair growth, cognition, fertility, vision or hearing.

"If a child has a tumor located on one side of the brain, protons can effectively eliminate any radiation from traveling to the other side of the brain," Ladra explained. "Spinal tumors just behind the heart, tumors next to the eye, or tumors in the muscles next to reproductive organs are all situations where proton therapy shines and can reduce the chance of having any negative impact on those organs."[84]

Healthy at Hopkins

In September 2014, Johns Hopkins Medicine launched a Healthy Beverage Initiative, a first step in an ambitious and far-reaching Healthy at Hopkins program that would eventually include more nutritious cafeteria choices, walking competitions and fitness classes.

Under the new Healthy Beverage policy, the cafeterias on Johns Hopkins Medicine campuses added sugar-free choices to vending machines, beverage stations and retail outlets. Sugar-sweetened drinks were sold in sizes of 12 ounces or less.

Rich Safeer, who joined Johns Hopkins as medical director of the Employee Health Program in 2012 and became chief medical director of employee health and well-being in 2018, said the change followed years of work raising awareness that beverages with high sugar content were not healthy.

"As part of our mission to improve the health of the community and the world, we must lead by example," said Pamela Paulk, who was senior vice president for human resources for Johns Hopkins Medicine and the

Angelo Mojica, senior director of food and culinary services, in Bayview's Courtyard Cafe, which offers incentives for healthy food choices.

Johns Hopkins Health System. "Our goal is to provide the education and tools that will help our campus communities make healthier choices."[85]

At the same time, campus cafeterias began overhauling their menu to provide more healthful options. The Courtyard Café at the Johns Hopkins Bayview Medical Center was configured into food stations that included Lemongrass (Asian), Super Bowl (build a meal in a bowl), The Greens Project (salads and soups), Prime Grill (veggie and meat burgers) and Rocco's Brick Oven Pizza.

The cafe also offers plenty of vegan options, such as high-protein lentil penne pasta and soy "veggie shreds," which are featured at several of the locations.[86]

In 2016, the Healthy at Hopkins program was expanded to include an online portal for all staff members of The Johns Hopkins Hospital, the Johns Hopkins Bayview Medical Campus, the Johns Hopkins Health System Corporation, Johns Hopkins Community Physicians, Johns Hopkins HealthCare, and Johns Hopkins Medicine International.

Through the portal, employees receive daily well-being tips and access self-paced programs that address a variety of health challenges, such as high blood pressure, stress and diabetes. The portal also offers a library of Johns Hopkins Medicine health information, organizes the Healthy at Hopkins rewards program, and provides a calendar of health-related events.

"The Healthy at Hopkins portal is a one-stop resource for information on nutrition and exercise, logging vital statistics, tracking physical activity and measuring progress," summed up an article in *Hopkins Insider*.[87]

For the portal's popular Race the Globe steps challenge, which began in 2017, employees compete to see which entity logs the most average steps over four weeks. (The Johns Hopkins Hospital won the first year, with a staff average of 215,294 steps, or 7,689 per day.)[88] In 2021, 20% of eligible employees participated, said Safeer.[89]

Safeer's initiatives were guided by the Worksite Health ScoreCard of the U.S. Centers for Disease Control and Prevention, a measurement tool with 125 questions used to assess how well organizations were doing at implementing strategies to improve the health and well-being of their employees.

In 2016, its first year using the tool, Johns Hopkins Medicine scored 180 points out of 264, falling below the industry benchmark. The following year, entities across the enterprise improved their health programs, increasing their scores by 23 points and exceeding the industry benchmark, as well as the large employer benchmark. Since then, the

Marvin Denson, director of food and culinary services, at The Johns Hopkins Hospital in 2014.

Richard Safeer, Kris Lukish and Redonda Miller gave awards to The Johns Hopkins Hospital's top three step-takers in the 2017 Race the Globe competition: David Furukawa, Leslie O'Connell and Maria Ronces Durante.

scores have only gone higher, said Safeer.

Aligning with CDC ScoreCard, every hospital across the health system has at least one lactation support room. Sibley Memorial Hospital and Suburban Hospital both have tranquility rooms where employees can relax and relieve stress. Howard County General Hospital hosts yoga classes and has a Recharge room with exercise equipment. Johns Hopkins All Children's Hospital hosts a tobacco cessation support group.

A blood pressure campaign gave thousands of Johns Hopkins employees important information about their health; and wellness carts with healthful food and drinks also provide information about health and well-being resources.[90]

"It's not about the numbers," said Safeer. "It's about the employer caring enough to help employees make healthier choices. We spend most of our waking hours together at work. You can't underestimate the influence that the workday has on our health."[91]

Reconnecting with Joy

Safeer would soon join a new Office of Well-Being, formed to help workers reconnect with the joy and purpose of their work. Lee Daugherty Biddison, associate professor in pulmonary and critical care medicine, would lead the office as chief wellness officer.

The Office of Well-Being was created in response to growing concerns nationwide about burnout among health care providers — a serious problem that was robbing health care providers of joy and purpose, while increasing the risk of error.

"We will not survive if we don't take better care of ourselves and the people who work here," said Paul Rothman, dean of the medical faculty and CEO of Johns Hopkins Medicine, when he announced the office's launch during his annual State of Johns Hopkins Medicine address on Oct. 2, 2018.[92]

A decade earlier, a 2009 study led by researchers from the Johns Hopkins University School of Medicine and the Mayo Clinic found that 40% of surgeons out of 7,905 surveyed reported burnout, and 9% said they had made a serious medical error in the previous three months.

"People have talked about fatigue and long working hours, but our results indicate that the dominant contributors to self-reported medical errors are burnout and depression," said Charles M. Balch, a professor of surgery at the Johns Hopkins University School of Medicine and one of the study's leaders.[93]

Nationally, the problem appeared to be getting

Deborah Dang, Lee Daugherty Biddison and Rich Safeer, shortly after the launch of the Office of Well-Being in November 2018.

worse, with one 2015 study finding that 54.4 % of U.S. physicians reported professional burnout in 2014, up from 45.5 % in 2011.[94]

Johns Hopkins Medicine leaders heeded the alarm. In the fall of 2016, Paul Rothman convened a Joy in Medicine task force with the goals of identifying barriers to professional satisfaction; developing strategies and recommendations to preserve and enhance joy; and preparing and presenting a white paper to Rothman.

The task force, chaired by Janice Clements, who was vice dean for faculty, and William Baumgartner, then senior vice president of the Office of Johns Hopkins Physicians, looked at five areas:

- clinical and faculty support
- Epic workflow
- culture and work-life balance
- regulatory and other required training
- clinical operations

Rothman discussed the effort in his column in the winter 2019 issue of *Hopkins Medicine* magazine:

> *Nationwide, more than four in 10 physicians have symptoms of burnout, according to a recently published survey. In some high-stress specialties, such as critical care and neurology, up to two-thirds of doctors are burned out. Burnout is never a good outcome, but it is particularly distressing in this context because most people who become clinicians do so because they are very passionate about their work in the first place.*
>
> *The results of burnout are serious: They can include depression, alcoholism and drug abuse, and simply leaving the profession. Evidence indicates that other members of the health care team, including nurses and social workers, are also overburdened and exhausted. Biomedical researchers also face increasing pressure, as funding has flatlined or decreased, and competition for grants has risen.[95]*

Biddison was co-chair of the Working Group on Culture and Work-Life Balance and lead author of the task force report, released in December 2017. The report included a recommendation to create a wellness office to help workers "reconnect with joy," in the words of Johns Hopkins Hospital President Redonda Miller. [96]

"People are spending more time doing paperwork and less time with patients," Biddison said. "Regardless of our role, we didn't go into health care to do paperwork all the time. We went into health care because we want to help people."[97]

Deborah Dang, director of nursing for The John Hopkins Hospital, became senior director of nursing for the Office of Well-Being, and Safeer continued promoting health and well-being for all employees. Together with Biddison, they worked to ease the stresses and administrative burdens

Janice Clements

Neda Gould

that keep clinicians from doing the work that's important to them.

Dang was a member of the dean's task force as co-chair of the Regulatory and Other Required Training group. As a health services researcher, she studied factors in the nursing work environment that affected patient outcomes and nurse engagement. Through a multiyear research grant, she studied interprofessional disruptive behavior and created the Center for the Practice of Collaborative Leadership to build capacity from bedside to boardroom in personal mastery, resilience and positive psychology.[98]

Appropriately, as the new Office of Well-Being rolled out, so did Dragon Medical One, a transcription software that could turn spoken notes into text that went directly into the electronic medical record, saving considerable time for clinicians.

The system, available starting Nov. 14, 2018, added text within Epic when the user spoke into a microphone on a mobile device. Clinicians used it to take notes during their patient encounters, providing the double benefit of letting patients hear the notes, while reducing the time spent adding and editing them — time that might otherwise be required during evenings and weekends.

"Dragon Medical One in Epic permits real-time, quality transcription," said internist Joseph Cofrancesco Jr. "I can close my note after seeing a patient and not have this hanging over me."[99]

"For our busy clinicians, this speech recognition software will be crucial in so many ways," said Rothman. "It will help them work more effectively, and therefore allow them to spend more time connecting with patients, which not only improves care but makes each interaction more rewarding. This really is a powerful tool for increasing joy in medicine."[100]

The office was launched only a short time before the COVID-19 pandemic began, but in that time it helped facilitate many positive changes, including helping to establish meditation and lactation support rooms at various hospitals, offering employee resources for help with financial planning and child care, and hosting mindfulness classes led by clinical psychologist Neda Gould.

The office would become even more active and necessary during the pandemic, helping employees cope with the fears, traumas and stresses of a health crisis that seemed to grind on without end.

ENDNOTES

1 Jacobson, Joan and Smith, Linell, "Choosing Treatment Wisely," *Dome*, Jan. 1, 2014, https://www.hopkinsmedicine.org/news/publications/dome/dome_januaryfebruary_2014/choosing_treatment_wisely.

2 Brody, Howard, "Perspective: Medicine's Ethical Responsibility for Health Care Reform — The Top Five List," *New England Journal of Medicine*, Jan. 28, 2010, https://www.nejm.org/doi/full/10.1056/NEJMp0911423.

3 Colla, Carrie H., et. al., "Physician Perceptions of Choosing Wisely and Drivers of Overuse," *The American Journal of Managed Care*, May 2016, https://www.ajmc.com/view/physician-perceptions-of-choosing-wisely-and-drivers-of-overuse.

4 Karen Nitkin telephone interview with Roy Ziegelstein, Oct. 28, 2021.

5 Ziegelstein interview, Oct. 28, 2021.

6 Larochelle, Marc R., et. al., "Reducing Excess Cardiac Biomarker Testing at an Academic Medical Center," *Journal of General Internal Medicine*, June 28, 2014, https://www.ncbi.nlm.nih.gov/pmc/articles/PMC4238205/.

7 "Choosing Treatment Wisely," https://www.hopkinsmedicine.org/news/publications/dome/dome_januaryfebruary_2014/choosing_treatment_wisely.

8 Nitkin, Karen, "High Value Conference Aims to Improve Quality and Affordability of Health Care," *Dome*, Nov. 27, 2019, https://www.hopkinsmedicine.org/news/articles/high-value-conference-aims-to-improve-quality-and-affordability-of-health-care.

9 DuVernay, Christina, "FAQs About the Medicare Waiver," *Dome*, June 1, 2015, https://www.hopkinsmedicine.org/news/articles/faqs-about-the-medicare-waiver.

10 Wallach, Rachel, "Medicare Waiver," *Dome*, March 1, 2014, https://www.hopkinsmedicine.org/news/publications/dome/dome_march_2014/medicare_waiver.

11 Brown, Patricia, "Helping Patients Avoid Readmission," *Managed Care Partners*, Aug. 28, 2018, https://www.hopkinsmedicine.org/news/articles/helping-patients-avoid-readmission.

12 Sowers, Kevin, "Berkowitz Named Chief Population Health Officer and VP of Population Health for JHM," *Medicine Matters*, June 1, 2020, https://medicine-matters.blogs.hopkinsmedicine.org/2020/06/berkowitz-named-chief-population-health-officer-and-vp-of-population-health-for-jhm/.

13 "Johns Hopkins Medicine Alliance for Patients Achieves Outstanding Quality Score in 2015," *BestPractice*, Nov. 1, 2016, https://www.hopkinsmedicine.org/office-of-johns-hopkins-physicians/best-practice-news/johns-hopkins-medicine-alliance-for-patients-achieves-outstanding-quality-score-in-2015.

14 "Johns Hopkins Accountable Care Organization Achieves Perfect Quality Reporting Score," Johns Hopkins Medicine press release, Dec. 22, 2015, https://www.hopkinsmedicine.org/news/media/releases/johns_hopkins_accountable_care_organization_achieves_perfect_quality_reporting_score.

15 Staff report, "Strength—and Savings—in Numbers," *Dome*, March 1, 2014, https://www.hopkinsmedicine.org/news/publications/dome/dome_march_2014/strengthand_savingsin_numbers.

16 "Care Coordination Model," Johns Hopkins Medicine webpage, Johns Hopkins Medicine Alliance for Patients, https://www.hopkinsmedicine.org/alliance-patients/model-of-care.html.

17 "Innovating and Integrating for Higher Value Care: An Update on the Johns Hopkins Medicine Alliance for Patients, *Managed Care Partners*, Aug. 28, 2018, https://www.hopkinsmedicine.org/news/articles/innovating-and-integrating-for-higher-value-care.

18 "Johns Hopkins Accountable Care Organization Achieves Perfect Quality Reporting Score," Johns Hopkins Medicine press release, Dec. 22, 2015, https://www.hopkinsmedicine.org/news/media/releases/johns_hopkins_accountable_care_organization_achieves_perfect_quality_reporting_score.

19 "Care Transformation Organization (CTO)," Johns Hopkins Medicine webpage, Johns Hopkins Medicine Alliance for Patients, https://www.hopkinsmedicine.org/alliance-patients/care-transformation-organization.html.

20 Karen Nitkin Zoom interview with Paul Rothman, April 4, 2022.

21 Karen Nitkin telephone interview with Antony Rosen, Oct. 15, 2021.

22 Nitkin, Karen, "The Power and Promise of Big Data," *Dome*, June 5, 2014, https://www.hopkinsmedicine.org/news/publications/dome/dome_june_2014/the_power_and_promise_of_big_data.

23 "Active Care: Predicting Outcomes that Inform Prostate Cancer Patients," Johns Hopkins Medicine webpage, Technology Innovation Center, https://tic.jh.edu/work/active-care.

24 Duffy, Jim "Individualized Health Through Big Data," *Hopkins Medicine* magazine, Spring 2015, https://hub.jhu.edu/magazine/2015/spring/individualized-health-through-big-data/.

25 "Oncospace Inc. Launches New Predictive Planning Solution for Radiation Oncology," Oncospace press release, Sept. 15, 2021, https://oncospace.com/pr-2021-09-15/.

26 Oncospace press release, Sept. 15, 2021.

27 "Project Wonder" demo video created by APL, 2021.

28 Antony Rosen email to Karen Nitkin, Oct. 20, 2021.

29 Karen Nitkin interview with Antony Rosen, Oct. 15, 2021.

30 "Top 100 Worldwide Universities Granted U.S. Utility Patents for 2020," National Academy of Inventors website, https://academyofinventors.org/top-100/.

31 Schwartz, David, "IP Watchdog Ranks Top U.S. Universities for Start-Up Creation," *Tech Transfer eNews Blog*, May 19, 2020, https://techtransfercentral.com/2020/05/19/ip-watchdog-ranks-top-u-s-universities-for-start-up-creation/.

32 "Report of the Committee on the Innovation Ecosystem," Johns Hopkins University, May 2014, https://ventures.jhu.edu/wp-content/uploads/2019/10/Innovation-Report.pdf.

33 "Christy Wyskiel," Johns Hopkins University webpage, Johns Hopkins Technology Ventures, https://ventures.jhu.edu/christy-wyskiel-bio/.

34 Jacobs, Danny, "Bringing Johns Hopkins Discoveries to the World," *Dome*, Nov. 15, 2019, https://www.hopkinsmedicine.org/news/articles/bringing-johns-hopkins-discoveries-to-the-world.

35 Nitkin, Karen, "Johns Hopkins Opens a Bigger, Brighter Space for Startups," *Dome*, April 25, 2017, https://www.hopkinsmedicine.org/news/articles/johns-hopkins-opens-a-bigger-brighter-space-for-startups.

36 "Bringing Johns Hopkins Discoveries to the World."

37 "Allegheny Health Network, Highmark Health and Johns Hopkins Medicine Announce Collaboration," Johns Hopkins Medicine press release, Dec. 22, 2014, https://www.hopkinsmedicine.org/news/media/releases/allegheny_health_network_highmark_health_and_johns_hopkins_medicine_announce_collaboration.

38 "Johns Hopkins Medicine Expands Collaboration with Allegheny Health Network and Highmark," Johns Hopkins Medicine press release, June 13, 2017, https://www.hopkinsmedicine.org/news/media/releases/johns_hopkins_medicine_expands_collaboration_with_allegheny_health_network_and_highmark.

39 Nitkin, Karen, "Alliance Expands Researh Potential," *Dome*, March 1, 2018, https://www.hopkinsmedicine.org/news/articles/alliance-expands-research-potential.

40 "Welcome to the PaTH Clinical Research Network," PaTH Network webpage, https://www.pathnetwork.org/.

41 Nitkin, Karen, "Story Booth Invites People to Tell Their Healthcare Experiences," *BestPractice*, March 11, 2019, https://www.hopkinsmedicine.org/office-of-johns-hopkins-physicians/best-practice-news/story-booth-invites-people-to-tell-their-healthcare-experiences.

42 "WellSpan Health and the Johns Hopkins Children's Center Announce New Pediatric Surgery Collaboration in York County," WellSpan Health press release, Aug. 30, 2016, https://www.wellspan.org/news/story/wellspan-health-and-the-johns-hopkins-childrens-center-announce-new-pediatric-surgery-collaboration-in-york-county/N4789.

43 "Johns Hopkins Kimmel Cancer Center and WellSpan Collaborate to Fight Cancer," Johns Hopkins Medicine press release, June 26, 2017, https://www.hopkinsmedicine.org/news/media/releases/johns_hopkins_kimmel_cancer_center_and_wellspan_collaborate_to_fight_cancer.

44 "WellSpan and Johns Hopkins Kimmel Cancer Center Announce New Cancer Research Fund," WellSpan Health press release, Sept. 17, 2018, https://www.wellspan.org/news/story/wellspan-and-johns-hopkins-kimmel-cancer-center-announce-new-cancer-research-fund/N4931.

45 Webster, Jenn, "Wellspan Health Realizes New Benefits as Collaboration with Johns Hopkins Sidney Kimmel Comprehensive Cancer Center Grows," *South Central Pennsylvania MDNews*, April 25, 2019, https://southcentralpennsylvania.mdnews.com/wellspan-health-realizes-new-benefits-collaboration-johns-hopkins-sidney-kimmel-comprehensive-cancer.

46 Nitkin, Karen, "Taking Collaboration to New Heights," *Dome*, June 1, 2015, https://www.hopkinsmedicine.org/news/articles/taking-collaboration-to-new-heights.

47 "One University," Johns Hopkins University webpage, 10X2020 Progress Report 2015, https://10x2020progress.jhu.edu/priorities/one-university/.

48 "Johns Hopkins Discovery Awards," Johns Hopkins University webpage, Johns Hopkins University Research, https://research.jhu.edu/major-initiatives/discovery-awards/.

49 "Johns Hopkins Catalyst Awards," Johns Hopkins University webpage, Johns Hopkins University Research, https://research.jhu.edu/major-initiatives/catalyst-awards/.

50 O'Shea, Dennis, "Rising to the Challenge," *Hopkins Medicine* magazine, Summer 2013, https://hub.jhu.edu/magazine/2013/summer/rising-campaign-launch/.

51 Hub staff report, "Johns Hopkins Concludes its 'Rising to the Challenge' Capital Campaign," *The Hub*, Oc. 12, 2018, Thttps://hub.jhu.edu/2018/10/12/rising-to-the-challenge-campaign-closing/.

52 Hub staff report, "Michael R. Bloomberg commits $350 million to Johns Hopkins for Transformational Academic Initiative," *The Hub*, Jan. 26, 2013, https://hub.jhu.edu/2013/01/26/bloomberg-gift/.

53 "Bloomberg Distinguished Professorships," Johns Hopkins University webpage, Research, https://research.jhu.edu/bloomberg-distinguished-professorships/.

54 Brooks, Kelly, "Four New Bloomberg Distinguished Professors Named at Johns Hopkins, *The Hub*, July 8, 2015, https://hub.jhu.edu/2015/07/08/bloomberg-distinguished-professors-fanzo-ha-li-yuille/.

55 Hub staff report, "Johns Hopkins University's Bloomberg Distinguished Professorships Program to Double in Size," *The Hub*, Dec. 14, 2021, https://hub.jhu.edu/2021/12/14/bloomberg-distinguished-professorships-program-expands/.

56 "Introducing Mark Anderson," *Aequanimitas*, Fall 2014, https://www.hopkinsmedicine.org/news/publications/aequanimitas/aequanimitas_fall_2014/introducing_mark_anderson.

57 "Mark Edward Anderson, M.D., Ph.D.," Johns Hopkins Medicine webpage, https://www.hopkinsmedicine.org/profiles/details/mark-anderson.

58 *UChicago News*, "Mark Anderson, Renowned Cardiac Expert and Medical Leader, Named to Lead University of Chicago Medicine," July 19, 2022, https://news.uchicago.edu/story/mark-anderson-renowned-cardiac-expert-and-medical-leader-named-lead-university-chicago.

59 Email from Mike Weisfeldt to Karen Nitkin, Nov. 8, 2021.

60 "Myron Weisfeldt," The Johns Hopkins Medical Institutions webpage, Portrait Collection, https://portraitcollection.jhmi.edu/portraits/weisfeldt-myron.

61 "The Miller Coulson Academy of Clinical Excellence at Johns Hopkins: Celebrating 10 Years of Exceptional Patient Care," *MCACE 10 Year Magazine*, 2018, https://www.hopkinscim.org/wp-content/uploads/2018/05/MCACE-10-Year-Magazine-PDF-Web-2.pdf.

62 Nitkin, Karen, "Academy Awards," *Dome*, June 2014, https://www.hopkinsmedicine.org/news/publications/dome/dome_june_2014/academy_awards.

63 *MCACE 10 Year Magazine*.

64 "Announcing the Winners of the Second Annual Johns Hopkins Medicine Clinical Awards," *BestPractice*, Dec. 1, 2016, https://www.hopkinsmedicine.org/office-of-johns-hopkins-physicians/best-practice-news/announcing-the-winners-of-the-second-annual-johns-hopkins-medicine-clinical-awards.

65 Nitkin, Karen, "Clinical Excellence Track Will Provide a New Path to Promotion," *BestPractice*, June 13, 2019, https://www.hopkinsmedicine.org/office-of-johns-hopkins-physicians/best-practice-news/clinical-excellence-track-will-provide-a-new-path-to-promotion.

66 "A New Pathway to Promotion for Excellent Clinicians," *Breakthrough*, Center for Innovative Medicine, Holiday 2019, https://www.hopkinscim.org/breakthrough/holiday-2019/a-new-pathway-to-promotion-for-excellent-clinicians/.

67 "Faculty Promotions and Additions," *Medicine Matters*, April 2, 2021.

68 "Media Advisory: The Johns Hopkins Hospital Celebrates Completion of Hospital Renovation Project," Johns Hopkins Medicine media release, https://www.hopkinsmedicine.org/news/media/releases/media_advisory_the_johns_hopkins_hospital_celebrates_completion_of_hospital_renovation_project.

69 "The Johns Hopkins Hospital Adds More Private Patient Rooms with Recent Renovation," Johns Hopkins Medicine media release, https://www.hopkinsmedicine.org/news/media/releases/the_johns_hopkins_hospital_adds_more_private_patient_rooms_with_recent_renovation.

70 "The Johns Hopkins Hospital Adds More Private Patient Rooms with Recent Renovation."

71 Grauer, Neil A., *Leading the Way*, Johns Hopkins Medicine in Association with The Johns Hopkins University Press, 2012, p. 107.

72 Peterson, Ronald R., "Makeover Milestones, *Dome*, December 2014, https://www.hopkinsmedicine.org/news/publications/dome/dome_december_2014/major_makeover_milestones.

73 *Leading the Way*, p. 203.

74 Baker, Sara, and Tong, Karen, "New Buildings Bring Next Generation of Patient Care," *Dome*, May 2015, https://www.hopkinsmedicine.org/news/articles/new-buildings-bring-next-generation-of-patient-care.

75 "New Buildings Bring Next Generation of Patient Care."

76 Cohn, Meredith, "Bayview Unveils New Facilities for Emergency Care, Cancer and Imaging," *Baltimore Sun*, March 3, 2015, https://www.baltimoresun.com/health/bs-hs-bayview-hospital-20150303-story.html.

77 "Expansion at Sibley Memorial Hospital," *Dome*, October 2016, https://www.hopkinsmedicine.org/news/articles/expansion-at-sibley-memorial-hospital.

78 "Emergency Department," *Sibley Magazine*, Sibley on Health Fall 2016, https://www.hopkinsmedicine.org/news/publications/sibley_on_health/sibley_on_health_fall_2016/emergency_department.

79 "Expansion at Sibley Memorial Hospital."

80 "Women's and Infant Services," *Sibley Magazine*, Sibley on Health Fall 2016, https://www.hopkinsmedicine.org/news/publications/sibley_on_health/sibley_on_health_fall_2016/womens_and_infants_services.

81 "Institute for Bone and Joint Health," *Sibley Magazine*, Sibley on Health Fall 2016, https://www.hopkinsmedicine.org/news/publications/sibley_on_health/sibley_on_health_fall_2016/institute_for_bone_and_joint_health.

82 Davis, Richard O. "President's Message: Welcome to Your New Sibley," *Sibley Magazine*, Sibley on Health Fall 2016, https://www.hopkinsmedicine.org/news/publications/sibley_on_health/sibley_on_health_fall_2016/presidents_message.

83 Nitkin, Karen, "Johns Hopkins Opens National Proton Center at Sibley Memorial Hospital," *Dome*, Sept. 26, 2019, https://www.hopkinsmedicine.org/news/articles/johns-hopkins-opens-national-proton-center-at-sibley-memorial-hospital.

84 "Johns Hopkins Opens National Proton Center at Sibley Memorial Hospital."

85   "Johns Hopkins Medicine Announces Healthy Beverage Initiative," Johns Hopkins Medicine press release, Oct. 10, 2014, https://www.hopkinsmedicine.org/news/media/releases/johns_hopkins_medicine_announces_healthy_beverage_initiative.

86   Minkove, Judy F., "Cafeteria Makeover: Where Healthy Means Tasty," *Dome*, Jan. 28, 2019, https://www.hopkinsmedicine.org/news/articles/cafeteria-makeover-where-healthy-means-tasty.

87   "A New Healthy at Hopkins Online Portal and Rewards Program," *Hopkins Insider*, Sept. 29, 2016, https://www.hopkinsmedicine.org/news/articles/a-new-healthy-at-hopkins-online-portal-and-rewards-program.

88   Price, Stephanie, "The Johns Hopkins Hospital Wins the Race the Globe Challenge," *Hopkins Insider*, July 13, 2017, https://www.hopkinsmedicine.org/news/articles/the-johns-hopkins-hospital-wins-the-race-the-globe-challenge.

89   Email from Rich Safeer to Karen Nitkin, Nov. 10, 2021.

90   Email from Rich Safeer to Karen Nitkin, Nov. 8, 2021.

91   "Making Work a Healthier Place," *Hopkins Insider*, Jan. 9, 2018, https://www.hopkinsmedicine.org/news/articles/making-work-a-healthier-place.

92   Nitkin, Karen, "Well-Being Office Plans to Make Hopkins Easier," *BestPractice*, Nov. 6, 2018, https://www.hopkinsmedicine.org/office-of-johns-hopkins-physicians/best-practice-news/well-being-office-plans-to-make-hopkins-easier.

93   "Burned Out Surgeons More Likely to Commit Errors," *The JHU Gazette*, Nov. 11, 2009, https://gazette.jhu.edu/2009/11/30/burned-out-surgeons-more-likely-to-commit-errors/.

94   Shanafelt, Tait D., et. al., "Changes in Burnout and Satisfaction with Work-Life Balance in Physicians and the General US Working Population Between 2011 and 2014," *Mayo Clinic Proceedings*, December 2015, https://pubmed.ncbi.nlm.nih.gov/26653297/.

95   Rothman, Paul, "Bringing Joy Back into Medicine," Post-Op Column, *Hopkins Medicine* magazine, Winter 2019, https://www.hopkinsmedicine.org/news/publications/hopkins_medicine_magazine/forum/winter-2019/bringing-joy-back-into-medicine.

96   DuVernay, Christina, "Wellness Office to Promote Joy and Work-Life Balance," *BestPractice*, Feb. 8, 2018, https://www.hopkinsmedicine.org/office-of-johns-hopkins-physicians/best-practice-news/wellness-office-to-promote-joy-and-worklife-balance.

97   "Well-Being Office Plans to Make Hopkins Easier."

98   Deborah Dang email to Karen Nitkin, November 16, 2021.

99   Nitkin, Karen, "Tips and Tricks for Getting the Most out of Dragon Medical One, *Dome*, Dec. 28, 2018, https://www.hopkinsmedicine.org/news/articles/tips-and-tricks-for-getting-the-most-out-of-dragon-medical-one.

100 Nitkin, Karen, "Dragon Medical One Brings Voice Recognition to Epic," *Hopkins Insider*, Nov. 16, 2018, https://www.hopkinsmedicine.org/news/articles/hed-dragon-medical-one-brings-voice-recognition-to-epic.

Chapter Three

Inclusion and Community (2015–2017)

O N APRIL 19, 2015, Freddie Carlos Gray Jr. died from a spinal cord injury sustained in Baltimore City in a police transport van.

The tragedy — and its violent aftermath — reverberated through Baltimore City, the nation, and Johns Hopkins Medicine, an institution that has been deeply entwined with Baltimore from its start.

The events prompted a renewed focus on diversity and inclusion across Johns Hopkins Medicine, as well as new programs, such as BLocal and Y2CONNECT, aimed at improving the lives and health of residents in Baltimore's least-advantaged neighborhoods.

In addition, these years saw the ascent of Redonda Miller as the first female president in the history of The Johns Hopkins Hospital, as well as a new school of medicine training program specifically for trainees seeking primary care careers.

The institution's unwavering dedication to clinical excellence was underscored with the creation of the Armstrong Institute Center for Diagnostic Excellence, the Bloomberg-Kimmel Institute for Cancer Immunotherapy, a pain management clinic to avert opioid dependence, a center for transgender health, and breakthroughs in organ transplants for people with HIV or hepatitis C.

Baltimore Uprisings and Aftermath

Gray, a 25-year-old Black man, sustained a fatal injury on April 12, after police stopped and searched him in his Sandtown-Winchester neighborhood, found a pocket knife, and put him in the van, shackled but with no seat belt. He died a week later.

Gray's death brought Baltimore's long-simmering racial tensions to a boil. Over several days and nights of protests that worsened after Gray's April 27 funeral, businesses and vehicles were damaged, police officers were injured, and hundreds of people were arrested. Gov. Larry Hogan deployed Maryland National Guard troops and declared a state of emergency for Baltimore that lasted through May 6.

Even before the fires were extinguished and the television crews went home, Johns Hopkins leaders got to work, sending a survey to faculty and staff seeking ideas to heal the city "we love and serve," as Dean/CEO Paul Rothman described in a November 2015 *Dome* column.

< **The Johns Hopkins Center for Transgender Health opened in 2017, providing gender-affirming care for Kez Hall, Dariel Peay and Dylan Ballerstadt, shown here left to right, and many others.**

"A consensus emerged," he wrote. "We need to apply our Johns Hopkins resources and ingenuity to address the vast gaps in our city between the haves and the have-nots. One person wrote, 'If we can figure out how to bypass a brain aneurysm, we can figure out how to connect the 'two Baltimores' and make them one.'"[1]

Healthy Baltimore 2020, a 2016 blueprint for health from the city's health department, cited studies showing that residents in Baltimore's wealthier neighborhoods lived, on average, 20 years longer than those in poor neighborhoods, largely because of factors including the physical and emotional harms caused by poverty, gun violence, substance use disorder and inadequate access to nutritious food.[2]

"We aim to improve health, but we recognize that this is not enough," wrote Leana Wen, the city's health commissioner at the time. "We must specifically call out the historical policies that have promoted structural racism and injustice."[3]

BUSINESSES PARTNERING FOR BALTIMORE

HopkinsLocal and BLocal

Johns Hopkins University and Medicine, which consistently rank among the region's largest private-sector employers, were already hubs for jobs, for neighborhood revitalization, for philanthropic support and for public health partnerships.

It was time to do more.

In September 2015, Johns Hopkins University and Medicine launched HopkinsLocal, a community investment program that "offers new approaches to build on Johns Hopkins' existing programs to invest in communities," said a *Hub* article. [4]

A *Dome* article described HopkinsLocal's three-pronged approach:

BUILD: Johns Hopkins will work to expand participation from certified minority, women and other disadvantaged businesses across its portfolio of construction projects. It will implement a local hiring policy to require contractors to make a good-faith effort to hire local residents for new jobs that result from construction projects.

HIRE: The university and health system will increase employment of city residents while supporting the growth and retention of local and underrepresented employees. The two entities will aim to ensure that 40% of new hires in targeted jobs are from selected ZIP codes that are in need of economic opportunities. They will also focus on recruitment of Baltimore residents for positions at all levels and on expanding partnerships with organizations that help identify and prepare individuals for careers with Johns Hopkins.

BUY: Johns Hopkins will increase spending with local businesses, including those owned by minorities and women, by as much as $6 million over the next three years. It will increase outreach to local and disadvantaged businesses to engage them in the competitive bidding process and support employees to buy local by providing a directory of prescreened vendors from which to purchase goods and services.[5]

In April 2016, HopkinsLocal became part of a larger effort called BLocal, involving Johns Hopkins and 24 other Baltimore-based businesses.

"HopkinsLocal is our comprehensive approach to leverage Johns Hopkins' economic power to do more to build, buy, and hire locally," said Ronald J. Daniels, president of The Johns Hopkins University. "Building on that promise, BLocal aims to help bolster a local economy, not on a project-by-project basis, but through a collective, deep-seated change approach."[6]

The idea began when a group of Johns Hopkins trustees led by Mike Hankin, president and CEO of Brown Advisory, asked how they could join in Johns Hopkins' efforts.

"As Mike reached out to other Hopkins trustees leading major companies in Baltimore, it quickly became apparent that many businesses in the city are committed to expanding opportunities and looking for ways to provide an even stronger economic engine for the people of Baltimore," Ronald R. Peterson, president of The Johns Hopkins Hospital and Health System, told the *Hub*.[7]

Together, business leaders pledged $69 million to support local and minority-owned, women-owned, and disadvantaged businesses over the following three years.[8]

BLocal far exceeded that goal. In its first three years, it grew to 28 partners, who spent more than $280 million, hired 1,729 Baltimore residents and, provided paid internships for 1,576 Baltimore youth, said a 2019 *Hub* article.[9]

Ronald J. Daniels

Hiring and Promoting

Job-training programs were essential to the Johns Hopkins mission of improving opportunities for Baltimore residents. BLocal partners launched BUILD College, a 13-week education program that helps local, women-owned, and minority-owned construction companies learn how to grow their businesses.[10]

In 2016, the health system's Department of Finance created a Supply Chain Academy in partnership with Baltimore City Community College.

For six hours a day over eight weeks, high school graduates learned the complexities of large-scale shipping, receiving and distribution operations, and other aspects of supply chain management. At the end of the course, Johns Hopkins and other employers interviewed the graduates for jobs.

Graduation from the institute was intended to serve as the first step on a career path, rather than mere training for an entry-level job.[11]

Two years later, the Johns Hopkins Medicine Department of Finance launched a similar program, this one for health system employees and focused on the fundamentals of billing, coding, collections and other financial functions vital to the well-being of the institution.

To qualify for the 12-week, twice-a-week Revenue Cycle Management Academy, class members must work for the Johns Hopkins Health System and be recommended for the course by their managers. Graduates can apply for open positions in the finance department.[12]

Michael D. Hankin

Dancing with the Hopkins Stars

In this 2016 photo, Hope Byers and Christopher Terndrup's bold, barefooted performance earned them the top award at the United Way event.

Though it was planned before the uprisings, Dancing with the Hopkins Stars, modeled after the popular *Dancing with the Stars* television show, generated enthusiasm and support for the annual Johns Hopkins United Way campaign.

The first dance-off, on May 26, 2015, featured 20 Johns Hopkins faculty and staff members who had been rehearsing for months for their big night on the stage in Turner Auditorium.

Human resources Senior Vice President Pamela Paulk wowed the crowd when she was flipped by her partner, transplant surgeon Dorry Segev. The two took home the first of two trophies at the end of the night, for audience favorite. The second went to molecular biologist Lucio Gama and nursing instructor JoAnn Ioannou, who won for raising the most money, an impressive $11,666, for The Journey Home, a United Way program aimed at ending homelessness.

Other highlights included the opening number, in which Landon King, executive vice dean, and administrator Kat Ries danced "The Hustle," a performance filled with exciting drops and lifts. Certified medical assistant Katie Marks and oncologist Mike Carducci also performed a tango, in which Marks slithered across the stage before dropping into a perfect split.

The first year, the event raised $52,000 for the United Way.* Over the years, Johns Hopkins has been a major contributor to United Way, said Peterson. The chair of the United Way of Central Maryland in 2021-2022 was Charles B. Reuland, executive vice president and chief operating officer of the Johns Hopkins Health System. Patricia M.C. Brown, then senior vice president of Johns Hopkins Medicine and president of Johns Hopkins HealthCare, was chair in 2013-2014.

* "Dancing with the Hopkins Stars Raises $52,000 for United Way," *Biomedical Odyssey*, June 10, 2015, https://biomedicalodyssey.blogs. hopkinsmedicine.org/2015/06/dancing-with-the-hopkins-stars-raises-52000-for-united-way/.

Community Conversations

Paul Rothman, dean of the medical faculty and CEO of Johns Hopkins Medicine, greets participant Naomi Bennett during a 2017 community conversation at the Henderson-Hopkins school.

Dean/CEO Paul Rothman knew that one key to building trust and alliances between Johns Hopkins and East Baltimore was to meet with community members and do more listening than talking.

In 2017, he began a series of East Baltimore Community Conversations to learn about East Baltimore's struggles and how Johns Hopkins could help. "In my five years in Baltimore in this position, I've come to love this city," he said during that first meeting, on June 15, 2017, at the Henderson-Hopkins School.

The event, hosted by Rothman and coordinated by Johns Hopkins University's Office of Government and Community Affairs, had about 75 participants, including city and school officials, as well as leaders of faith organizations and community groups.[13]

Meetings in 2021 and 2022 were held over Zoom because of the COVID-19 pandemic.[14]

Because of the COVID-19 pandemic, the 2021 and 2022 community conversations were held over Zoom. In 2021, clockwise from top left, were Michael Preston, director of East Baltimore community affairs for The Johns Hopkins University and Johns Hopkins Medicine; Sherita Golden, vice president and chief diversity officer for Johns Hopkins Medicine; Panagis Galiatsatos, pulmonary and critical care medicine physician; and Dean/CEO Paul Rothman.

Opportunities for Young People

Since the inception of YouthWorks in the mid-1990s by the Baltimore City Mayor's Office of Employment Development, Johns Hopkins had provided significant numbers of high school students summer job opportunities, particularly at The Johns Hopkins Hospital.

But from those community meetings and other conversations, it became clear that introducing *more* young people to careers in health care would provide new job opportunities while creating a pipeline for a more diverse workforce of clinicians and researchers.

Johns Hopkins responded by creating or expanding several internship and summer jobs programs for Baltimore students.

One of the most exciting was P-TECH (Pathways in Technology Early College High School), launched in 2016 in partnership with the Paul Laurence Dunbar High School, near The Johns Hopkins Hospital. Students in P-TECH graduate from high school with a no-cost associate's degree by augmenting their regular high school courses with community college classes.

Piloted in a New York City school in 2011, the P-TECH model calls for high schools, community colleges and corporate partners to work together crafting a curriculum aimed at a specific set of skilled jobs identified by the corporate partner.

When Daniels toured the original P-TECH high school in Brooklyn in 2011, he imagined how the model might work in Baltimore.

Five years later, working closely with Gov. Hogan, Johns Hopkins launched the state's first P-TECH program at Dunbar High School.

Dunbar's P-TECH students can select education tracks in health care information technology, nursing, physical therapy and respiratory services. Johns Hopkins provides one-on-one mentors to students, offering school and career advice, as well as encouragement.

"Over the past two years, I have been thrilled with the resolute commitment we've seen from students, families and volunteers — all aimed at moving these young people more quickly toward higher education and good jobs," Daniels said in 2018. "This is exactly what we envisioned when we started P-TECH."[15]

Another initiative, the Summer Scholars program, launched in 2008, aims to increase diversity in academic medicine and science by providing structured summer internship opportunities to high school students, mostly residents of Baltimore City. It was expanded in the summer of 2016, with 400 high school students working as summer interns at Johns Hopkins, up from 300 in 2015.

The program includes internships in brain sciences, aging and biomedical research, as well as the Medical Education Resources Initiative for Teens (MERIT) program, started in 2010 by Johns Hopkins medical students Tyler Mains and Mark Wilcox.[16]

MERIT accepts sophomores from high schools throughout Baltimore, who devote summers and school-year Saturdays to activities that prepare them for health careers. They shadow professionals, work in research labs, study health disparities, prepare for the SATs and learn leadership skills.

In this 2018 photo, Alexia Smith (left) of the Johns Hopkins University Office of Community Affairs, walks with Dunbar High School students (from left) Caitlun Clark, Nayonna Earl, Simone Wade and Alvin Winn, who are enrolled in Maryland's Pathways in Technology Early College High School program.

Western High School rising senior Savannah Tripp adjusts pH levels in a Johns Hopkins lab, guided by MERIT mentor Oliver Rogers, in 2015.

Y2CONNECT

Arik Marcell

A website and free smartphone app, launched in November 2016, gave Baltimore City young people access to information about local organizations and youth-focused topics, including job readiness and mental health services.

The creation of Y2CONNECT was led by Arik Marcell, associate professor in the Johns Hopkins University School of Medicine's Department of Pediatrics, and funded by the U.S. Centers for Disease Control and Prevention, the Secretary's Minority AIDS Initiative Fund and the Thomas Wilson Foundation.

"Since moving to Baltimore in 2001, I have recognized a need for a comprehensive resource guide for youth in this city, so I've been focused on creating one ever since," said Marcell. "Our main goal as youth-serving professionals is to help connect youth across all dimensions of their lives, from addressing their basic life skills to health care needs, in order to assist them in becoming successful adults."*

---

*   "Johns Hopkins Launches Website and App to Connect Baltimore City Youth to Health Resources," Johns Hopkins Medicine media release, Nov. 1, 2016, https://www.hopkinsmedicine.org/news/media/releases/johns_hopkins_launches_website_and_app_to_connect_baltimore_city_youth_to_health_resources.

To make all this possible, MERIT, which counts Johns Hopkins Bayview Medical Center and the Johns Hopkins University School of Medicine as its two largest partners, relies on about 250 volunteer health care workers, educators and scientists across the city.[17]

Hiring Outside the Box

In the months following the Baltimore uprising, Johns Hopkins could draw upon a strong history of recognizing that some of the best employees don't come to job interviews with polished resumes or expensive suits.

Paulk encouraged her human resources team to think expansively about hiring.

During her tenure, she said, Johns Hopkins conducted training programs at homeless shelters in Baltimore, and hired more than 100 people as a result.[18]

She also advocated for hiring previously incarcerated people, and Johns Hopkins has hired hundreds of ex-offenders since the late 1990s. Per university hiring policy, job applicants do not check a box indicating if they have been convicted of a crime. Instead, any criminal history is reviewed after the applicant is chosen, and considered with factors such as age at time of conviction and attempts at rehabilitation.[19]

Collie Thomas, for example, joined Johns Hopkins in 2015 after serving 10 years for the fatal stabbing of her boyfriend. According to a *Pew Trust* article,

Camille Range, Talayah Jackson, Thomas Houston, Veronica Vela and Ebony McMorris pose with the Ward Infinity pitch award in December 2021. The team planned to use the prize money to improve food access in Washington, D.C.

Ward Infinity

In 2017, Sibley Memorial Hospital launched a program called Ward Infinity, which supports community advocates, entrepreneurs and social enterprises in the District of Columbia's Wards 7 and 8.

The program grew out of conversations between Sibley leaders and residents who were interested in improving health and well-being for people in the district's underinvested communities through initiatives that would improve access to affordable housing, nutritious food and other social determinants of health.

"We wanted to set up an initiative that would create community-driven solutions," said Marissa McKeever, Sibley's director of government and community affairs and a key developer of the program. "By listening to the community and not just coming in with our own agenda, we are fostering real collaboration and encouraging people to be change agents."* (On July 15, 2022, McKeever began a new role as chief of staff and senior adviser to the school of medicine.†)

Each year, Sibley selects five teams, each led by a community member, that work closely with Sibley's Innovation Hub and receive coaching and guidance from leading industry designers and entrepreneurs, plus up to $25,000 to help develop solutions related to health and wellness.‡

One 2019 Ward Infinity participant, Tambra

Raye Stevenson, a doctoral student at American University School of Communication, said the program helped her team develop a scalable signature program, Women Advancing Nutrition, Dietetics and Agriculture (WANDA), which is building a pipeline and platform for women and girls to become healthful food advocates and leaders in their families and communities.

"The Ward Infinity program was instrumental in reaffirming that I am the change that my community needs," said Stevenson. "Also, it set our team on a trajectory to make a greater impact and build partnerships with fellow innovators to create a culture of health in our community."§

Marissa McKeever

* Morris, Michel, "Ward Infinity Program Helps Foster Community-Driven Solutions to DC Health Issues," *Community Health Stories*, Sept. 1, 2021, https://www.hopkinsmedicine.org/news/articles/ward-infinity-program-helps-foster-community-driven-solutions-to-dc-health-issues.

† DeWeese, Theodore, "Marissa McKeever Named SOM Chief of Staff and Senior Adviser to JHM Interim Dean and CEO," message to colleagues from the Johns Hopkins University School of Medicine, June 29, 2022.

‡ Wright, Sarafina, "Residents East of the River Team with Hospital, *Washington Informer*, Aug. 23, 2017, https://www.washingtoninformer.com/residents-east-of-the-river-team-with-hospital/.

§ "Ward Infinity Program Helps Foster Community-Driven Solutions to DC Health Issues."

Thomas grabbed the knife for protection and accidentally killed him as he rushed toward her.[20]

When Thomas was released in 2013, her daughter advised her to join a re-entry program she'd seen on Facebook, Turnaround Tuesday. Through that connection, she got a job at Johns Hopkins, where she worked as a peer recovery coach until 2022, when she joined Turnaround Tuesday as an outreach specialist.

The nonprofit's board is co-chaired by Peterson, now president emeritus of the Johns Hopkins Health System. Redonda Miller is also a member of the board.

"We gave Turnaround Tuesday the opportunity to test the program with us and people did very well," said Peterson in 2022. "Today, more and more organizations are considering granting opportunity to returning citizens."[21]

In 2014, Paulk was recognized at the White House as a Champion of Change for her work and advocacy in the hiring of ex-offenders. "For over 10 years, Paulk and Peterson have been leaders in the push to give qualified ex-offenders a second chance at a job and a life," said a 2014 press release about the honor.

"First and foremost, this is a good business decision," said Paulk at the time. "You've got great workers who truly are pleased to have a second chance. These are good, loyal, solid workers. And I have the numbers to prove it." She cited a 2009 study at Johns Hopkins of almost 500 ex-offender hires that showed a retention rate after 40 months that was better than a matched group of nonoffenders.[22]

Diversity and Inclusion

"Diversity is not an obligation but an asset."

So wrote Rothman and Peterson in 2015, in Johns Hopkins Medicine's first diversity and inclusion annual report.[23]

The institution has a long history of encouraging diversity and inclusion.

From the start, it accepted patients of all backgrounds — unusual at the time and especially in segregated Baltimore. Black patients had access to care at The Johns Hopkins Hospital, but were treated in segregated wards.

The first medical school class, in 1893, had three female students at a time when women were generally excluded from medical education or careers.[24]

Yet the school of medicine did not have an African-American on faculty until 1966, when Roland Smoot became part-time faculty and was granted admitting privileges by The Johns Hopkins Hospital. Smoot, who died in 2006, went on to become the first African-American assistant professor in 1974 and was instrumental in significantly expanding the number of underrepresented medical students at Johns Hopkins.[25]

James Page

What Do UIM and UIS Mean?

The Association of American Medical Colleges defines UIM as underrepresented in medicine. The term refers to "racial and ethnic populations that are underrepresented in the medical profession relative to their numbers in the general population."

In 2021, the term referred to Black, Mexican-American, Native American (American Indian, Alaska Native and Native Hawaiian), and mainland Puerto Rican populations. But since the focus of the definition is not on specific ethnicities, this allows the status of groups to change if they gain or lose representation in the medical profession.*

A similar term, UIS, refers to people who are underrepresented in science or engineering compared to the general population, including women, people with disabilities, and three racial and ethnic groups: Blacks, Hispanics, and American Indians or Alaska Natives, according to the National Science Foundation.†

---

*   Ogunyemi, Dotun, "Unique Populations," Association of American Medical Colleges (AAMC) webpage, Diversity and Inclusion, https://www.aamc.org/professional-development/ affinity-groups/gfa/unique-populations.
†   "Women, Minorities and Persons with Disabilities in Science and Engineering," webpage, National Science Foundation, https://www.nsf.gov/statistics/2017/nsf17310/digest/ introduction/.

Robert Lee Gamble, the school of medicine's first African-American graduate, earned his M.D. in 1967.

Faculty efforts to increase the diversity of the medical school class gained steam in the late 1980s and onward, with prominent faculty members including cardiac surgeon Levi Watkins, pediatric neurosurgeon Ben Carson and disparities expert and internist Lisa Cooper actively recruiting and mentoring minority medical students, residents and fellows.[26]

A more formal and unified approach began when James Page joined Johns Hopkins Medicine on Aug. 4, 2014, as its first vice president of diversity and inclusion.

Page, who had been assistant vice president and chief diversity officer at Cincinnati Children's Hospital Medical Center, was tasked with creating Johns Hopkins Medicine's Office of Diversity and Inclusion, a first for the institution.

The office would support the existing diversity and inclusion efforts that were taking place in pockets throughout Johns Hopkins Medicine, while bringing a unified strategy and approach to the mission of recruiting, retaining and supporting people who are underrepresented in medicine, science, nursing and health care administration, said Rothman.[27]

The mission would take on even more urgency after the April 2015 death of Freddie Gray and subsequent uprisings in Baltimore.

One of the new office's first acts was to create mandatory formal search committees for every leadership position across the health system, with committee members required to complete unconscious bias training.

"We are in the process of building a diverse workforce capable of relating to our patients and speaking their language, both literally and figuratively," wrote Rothman in a May 2016 *Dome* column.

"This is not just about fairness," he continued. "Diversity in medicine has measurable benefits. Studies show that students trained at diverse schools are more comfortable treating patients from a wide range of ethnic backgrounds. When the physician is the same race as the patient, patients report higher levels of trust and satisfaction. The visits even last longer — by 2.2 minutes, on average. When patients enter our hospitals, they want to see staff members and physicians who resemble them."[28]

In July 2019, Sherita Golden took the helm of the office as the new vice president and chief diversity officer at Johns Hopkins Medicine.

Within months, Golden, the Hugh P. McCormick Family Professor of Endocrinology and Metabolism, was playing a key role in helping to inform the institution's response to a COVID-19 pandemic that disproportionately harmed Black, Hispanic/Latinx and Native American communities, as well as a national social justice movement.

Alejandro Garcia

Supporting Research for New Underrepresented Faculty

In 2014, Rothman's office launched a Strategic Planning Recruitment and Retention Program to increase the number of underrepresented school of medicine faculty.

The program allotted $250,000 per year to support the research efforts of new underrepresented in medicine faculty. Support and mentoring are essential components of the program.

By 2018, the program had recruited or retained 14 faculty members, including Alejandro Garcia, hired in 2016. Garcia, who grew up north of New York City, earned his undergraduate degree from Cornell University and his M.D. from Weill Cornell Medical College, then completed a general surgery residency at the Columbia University Medical Center.

The recruitment and retention funding he received at Johns Hopkins, he said, gave him time to study how to reduce kidney damage in pediatric patients who use a heart-lung treatment called extracorporeal membrane oxygenation, or ECMO. "If you spend all your time operating, you don't get the opportunity to advance the field to the next level," he said.*

*   Nitkin, Karen, "Investments Add Diversity to Labs and Clinics," *Dome*, Jan. 8, 2018, https://www. hopkinsmedicine.org/news/articles/ investments-add-diversity-to-labs-and-clinics.

Levi Watkins Jr.

Levi Watkins Jr.

Levi Watkins Jr., who died in 2015 at age 70, was a pioneer in both cardiac surgery and civil rights. He implanted the first automatic heart defibrillator in a patient and was instrumental in recruiting underrepresented in medicine students to the Johns Hopkins University School of Medicine.

Watkins was born in Parsons, Kansas, and moved with his family to Alabama, where he attended the Dexter Avenue Baptist Church, led by Martin Luther King Jr. He played an active role in the Montgomery bus boycotts in the 1950s and became the first African-American admitted to Vanderbilt University School of Medicine, in 1966.

After earning his medical degree in 1970, he began a surgical internship at the Johns Hopkins University School of Medicine and, after a research stint at Harvard Medical School, became The Johns Hopkins Hospital's first Black chief resident in cardiac surgery.

From his earliest days at Johns Hopkins, he played a pivotal part in changing the institution's role in medical education.

In 1975, at the request of the newly appointed school of medicine dean, Richard Ross, Watkins and a fellow African-American faculty colleague, ophthalmologist Earl Kidwell, a 1973 school of medicine graduate, launched a concerted nationwide drive to recruit talented underrepresented in medicine students who were interested in studying medicine.

Within a few years, Johns Hopkins was attracting Black students from all over the nation who were convinced by Watkins that Johns Hopkins wanted them. The success of the Johns Hopkins minority recruitment campaign soon made it a model imitated by other medical schools.

In 1980, Watkins gained renown for implanting the first automatic heart defibrillator in a patient suffering from repeated, life-threatening episodes of ventricular fibrillation, or irregular heartbeats. Such a procedure is now commonplace, saving untold lives annually.

In 1982, he created the annual Martin Luther King Jr. Commemoration at Johns Hopkins, which over the years welcomed luminaries including Maya Angelou, Stevie Wonder, James Earl Jones, Bishop Desmond Tutu, Rosa Parks and Coretta Scott King as speakers at the popular event.*

In 1983, he was appointed to the medical school's admissions board, and the recruitment, retention, and graduation rates for minorities steadily climbed. Each year, Watkins hosted a reception for Black students, house staff and faculty, which grew from 10 or so attendees in the 1970s to more than 100 by the mid-1990s.

In 1991, Watkins was promoted to full professor of cardiac surgery and vice dean for postdoctoral programs and faculty development. In that role, he established the nation's first postdoctoral association at Johns Hopkins, another pioneering effort that has been emulated elsewhere. He retired in 2013.

---

*   *Hub* staff report, "Levi Watkins Jr., Pioneering Hopkins Cardiac Surgeon and Civil Rights Activist, Dies at 70," *The Hub*, April 12, 2015, https://hub.jhu.edu/2015/04/12/levi-watkins-jr-obit/.

Days of Service

On Jan. 26, 2018, more than 400 Johns Hopkins employees fanned out across Baltimore, Washington, D.C., and St. Petersburg, Florida.

Wearing bright yellow T-shirts with the image of Martin Luther King Jr. and the words "Keep the Dream Alive," they prepared meals, planted gardens, painted murals and enthusiastically took on other projects for 21 nonprofit organizations in those cities.

They were part of the inaugural Johns Hopkins Martin Luther King Jr. Day of Service, which has become an annual event organized by the Johns Hopkins Office of Work, Life and Engagement and the Johns Hopkins Medicine Office of Diversity, Inclusion and Health Equity. *

The second year, more than 500 employees volunteered, this time in blue T-shirts with the words "The Time is Always Right to Do What is Right."†

"This day of service is just another way Johns Hopkins demonstrates our commitment to the many communities that we are part of," said Inez Stewart, senior vice president of human resources for Johns Hopkins Medicine.‡

The popular event was put on hold in 2021 and 2022 because of COVID-19.§

---

*   Smith, Patrick, "Inaugural Day of Service Honors MLK Jr.'s Legacy," *The Hub*, Jan. 31, 2018, https://hub.jhu.edu/at-work/2018/01/31/inaugural-day-of-service/.
†   Sachs, Katelynn, "The Time is Always Right to Do What is Right," *Dome*, Jan. 30, 2019, https://www.hopkinsmedicine.org/news/articles/the-time-is-always-right-to-do-what-is-right.
‡   "The Time is Always Right to Do What is Right."
§   "MLK Day of Service," Johns Hopkins Medicine webpage, Our Diverse Community, https://www.hopkinsmedicine.org/diversity/our-diverse-community/mlk-day-of-service.html.

Robert Higgins: Surgeon with a Vision

One significant hire in 2015 was Robert Higgins, a renowned cardiothoracic and heart-lung transplant surgeon who joined Johns Hopkins that July as surgeon-in-chief at The Johns Hopkins Hospital, William Stewart Halsted Professor of Surgery, and director of the Department of Surgery at the Johns Hopkins University School of Medicine.

Higgins, the first Black physician to chair any department in the school of medicine, was hired April 6 — literally days before the Baltimore uprisings. From the start, he was a strong mentor and advocate for people who are underrepresented in medicine (UIM) — a role that was formalized in July 2020 when he was appointed senior associate dean for diversity and inclusion for the medical school.

"We want to be representative of the community we serve," said Higgins in 2017. "A critical part of my job is to recruit excellent candidates from all backgrounds, especially those who are women and underrepresented minorities."[29]

Higgins came from The Ohio State University, where he was professor and chairman of the Department of Surgery, as well as surgeon-in-chief and director of the Comprehensive Transplant Center at Wexner Medical Center. He succeeded Julie Freischlag, who in March 2003 became the first female and sixth person to serve as William Stewart Halsted Professor, surgeon-in-chief at The Johns Hopkins Hospital and director of the Department of Surgery.

His interest in medicine was sparked at a young age by his father, an internist

Robert Higgins

A meeting of WISH (Women in Surgery at Hopkins). Back row, left to right, Anuradha Kar; Robert Higgins; Macey Henderson; Gina Adrales; Tanjala Purnell; Jacqueline Garonzik Wang and Alisa Coker. Front row, left to right, Gloria Fleming; Nita Ahuja; Caitlin Hicks; Melissa Camp; and Margaret Arnold.

In 2017, division chiefs Gina Adrales, of minimally invasive surgery; Jennifer Lawton, of cardiac surgery; and Nita Ahuja, of surgical oncology, stand in front of an archival photo showing Elizabeth Hurdon, center, the first woman on the medical staff of The Johns Hopkins Hospital.

and the first African-American physician to practice in Charleston, S.C. He died in a car accident when Higgins was 5.

"Growing up, I always had this idea that my father had unfinished business and that if I pursued medicine, I'd be able to finish his work," said Higgins. "Fast forward 50 years, and I've been blessed with so many opportunities, and I've had the privilege to work with so many great people."[30]

When Higgins took the job, no woman had ever served as chief of one of the Department of Surgery's divisions. By 2017, women headed three out of seven divisions. Gina Adrales led minimally invasive surgery; Nita Ahuja led surgical oncology; and Jennifer Lawton was chief of cardiac surgery.[31]

Higgins also created an endowed memorial lectureship in honor of Watkins, the cardiac surgery and civil rights pioneer who died on April 11, 2015. In 2020, he launched the Levi Watkins Mentorship Program in collaboration with JHM Human Resources and the Office of Diversity, Inclusion and Health Equity, an initiative to help UIMs advance their careers through mentorship and special projects.[32]

In December 2021, Higgins left Johns Hopkins to become president of Brigham and Women's Hospital and executive vice president at Mass General Brigham.[33]

After Higgins left, Andrew Cameron became surgeon-in-chief and professor of surgery for The Johns Hopkins Hospital. Cameron, the son of renowned Johns Hopkins surgeon John Cameron, was born at The Johns Hopkins Hospital and raised in Baltimore. He earned his undergraduate degree at Harvard College and his medical degree at Johns Hopkins, joining the Johns Hopkins faculty in 2006 as head of the liver transplant program.[34]

Gender Equity

On Oct. 14, 2015, Johns Hopkins Medicine celebrated a milestone: More than 200 women had attained the highest academic rank of full professor at Johns Hopkins.

Rothman and Janice Clements, vice dean of the faculty, spoke about the school of medicine's strides toward gender equity in an event at the Chevy Chase Bank Conference Center at The Johns Hopkins Hospital.[35]

The first woman to become a full professor in the school of medicine was Florence Sabin, who earned her M.D. at Johns Hopkins in 1900 and became a full professor in 1917. The second was pediatric cardiologist Helen Taussig in 1959. By 1979, just seven women in the hospital's 90-year history had been full professors.

In 2003, the school promoted its 100th woman to that rank: Judith Karp, now professor emerita of oncology and the former director of the Johns Hopkins Kimmel Cancer Center's leukemia program.

Helen Brooke Taussig

Janice Clements

Barbara Fivush

Julie Freischlag

It took more than a century to reach the milestone of 100 female professors and only 12 additional years to reach the 200 mark, a testament to the institution's commitment to gender equity, said pediatrics professor Barbara Fivush in 2015. "We view this as a huge shared accomplishment: So many of our earlier women professors paved the way for other women to achieve this rank and honor," she said.[36]

Fivush was one of many leaders who fought for gender equity at Johns Hopkins. In 2005, when an internal report showed that women faculty in the school of medicine lagged behind men in salaries, promotions and leadership roles, she fought — successfully — to change the status quo.[37]

In 2008, she teamed with Clements to create the Office of Women in Science and Medicine (OWISM), which supports female faculty through education, advocacy, mentoring and networking.[38] In 2011, Fivush became the institution's first associate dean for women in science and medicine; she became senior associate dean in 2017.[39]

By 2021, just six years after the 200-female-professor milestone, the number of women who were full professors in the school of medicine passed 315.[40]

"I am very proud of the work of the OWISM," Fivush said in 2021. "With the support from the dean's office, the impact and progress in gender equity at the school of medicine has been remarkable."[41]

Redonda Miller Breaks Glass Ceiling at The Johns Hopkins Hospital

On July 1, 2016, Miller made history when she became the 11th president of The Johns Hopkins Hospital and the first female president since it was founded in 1889.

Johns Hopkins employees learned of the appointment in a May 12, 2016, message from Peterson, who had held the title of president of The Johns Hopkins Hospital for 19 years. Peterson had announced that January that he would step down from that particular role at the hospital while remaining president of the health system and executive vice president of Johns Hopkins Medicine.

"Redonda's extraordinary combination of exceptional medical prowess, years of progressive administrative experience, and the well-earned respect of senior clinical and administrative leadership will serve us all well," Peterson wrote. "Her deep

Redonda Miller, left, takes the keys to The Johns Hopkins Hospital from Ronald Peterson.

Lisa Cooper – Vice President for Health Care Equity

Lisa Cooper became the institution's first vice president for health care equity on Jan. 1, 2016, providing the vision for the health system's subsequent plans to address health care disparities in high-risk populations.

"As our nation struggles to improve population health while managing health care spending, moving from health disparities to health equity has emerged as a top priority here at Johns Hopkins Medicine and across the country," wrote Paul Rothman and Ron Peterson in a letter to colleagues announcing the appointment.*

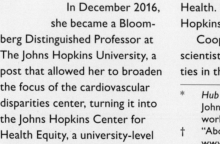

Cooper, the James F. Fries Professor of Medicine in the Division of General Internal Medicine, was director at the time of the Johns Hopkins Center to Eliminate Cardiovascular Health Disparities, which she and her colleagues in the Welch Center for Prevention, Epidemiology and Clinical Research created in 2010 with a grant from the National Institutes of Health.

Lisa Cooper

In December 2016, she became a Bloomberg Distinguished Professor at The Johns Hopkins University, a post that allowed her to broaden the focus of the cardiovascular disparities center, turning it into the Johns Hopkins Center for Health Equity, a university-level initiative which she founded and leads.

The center promotes equity in health and health care for socially at-risk populations through advancing scientific knowledge, training scholars, building academic-community partnerships, promoting sustainable changes in practice and policy, and raising public awareness of health inequities. (The Bloomberg Distinguished Professor program, described in Chapter Two, has

created a cadre of world-class interdisciplinary professorships aimed at addressing major world problems.†)

In April 2020, she added yet more responsibilities when she became director of the Johns Hopkins Urban Health Institute, leading a team focused on advancing health equity in Baltimore City.‡

Health disparities have been a lifelong interest for Cooper. She was born and grew up in Africa, attended an international school in Liberia, and saw how poverty and privilege shape each person's health and life trajectory.

After earning her bachelor's degree from Emory University and her medical degree from the University of North Carolina at Chapel Hill, Cooper completed her internship and residency in internal medicine at the University of Maryland School of Medicine and the Baltimore VA Medical Center.

Inspired to dig deeper into health equity issues, Cooper came to the Johns Hopkins University School of Medicine for her postdoctoral fellowship, at the same time pursuing a master's degree at the Johns Hopkins Bloomberg School of Public Health. She joined the Johns Hopkins faculty in 1994.§

Cooper was one of the first scientists to document disparities in the quality of relationships

between physicians and patients from socially at-risk groups. She designed interventions targeting physicians' communication skills, patients' self-management skills, and health care organizations' ability to address needs of populations experiencing health disparities.

She is the author of more than 250 publications and has been the principal investigator of more than 20 federal and private foundation grants. She has also been a devoted mentor to more than 75 individuals seeking careers in medicine, nursing and public health.¶

She has received many national and international honors for her work, including a MacArthur Fellowship (also known as a "genius grant") in 2007 and election in 2008 to the National Academy of Medicine. In 2014, she received the Herbert W. Nickens Award from the Association of American Medical Colleges for outstanding contributions to promoting justice in medical education and health care equity in the United States.**

In September 2021, President Joseph R. Biden appointed her to the President's Council of Advisors on Science and Technology (PCAST), which advises the president on developments related to science, innovation and technology, including health and medicine. ||

*   *Hub* staff report, "Lisa Cooper Named VP for Health Care Equity at Johns Hopkins Medicine," *The Hub*, Dec. 8, 2015, https://hub.jhu.edu/at-work/2015/12/08/lisa-cooper-health-care-equity/.

†   "About Us," Johns Hopkins Center for Health Equity, webpage, https://www.jhsph.edu/research/centers-and-institutes/johns-hopkins-center-for-health-equity/about-us/.

‡   "About Us," Johns Hopkins Urban Health Institute, webpage, https://urbanhealth.jhu.edu/about/.

§   *Hub* staff report, "Health equity researcher Lisa Cooper named 25th Bloomberg Distinguished Professor," *The Hub*, Dec. 7, 2016, https://hub.jhu.edu/2016/12/07/lisa-cooper-bloomberg-distinguished-professor/.

¶   "Lisa Cooper, M.D.," Johns Hopkins Urban Health Institute webpage, About our Team, https://urbanhealth.jhu.edu/about/our-team/.

** "Lisa Cooper Named VP for Health Care Equity at Johns Hopkins Medicine."

|| "Johns Hopkins Health Disparities Researcher Lisa Cooper Appointed by Biden to President's Council of Advisors on Science and Technology," published Sept. 22, 2021, Johns Hopkins Bloomberg School of Public Health, https://publichealth.jhu.edu/2021/johns-hopkins-health-disparities-researcher-lisa-cooper-appointed-by-biden-to-presidents-council-of-advisors-on-science-and-technology.

Jacqueline Schultz – President of Suburban Hospital

On Aug. 15, 2016, Jacqueline "Jacky" Schultz became president of Suburban Hospital, after serving as interim president since October 2015.

Schultz joined Suburban in 2005 as senior vice president of patient care and chief nursing officer, and also served as senior vice president for patient care, senior vice president for clinical operations, and executive vice president and chief operating officer.

She earned her bachelor's degree from Cornell University and her master's degree in nursing from Seton Hall University. Before joining Johns Hopkins, she had worked in patient care leadership roles at Saint Barnabas Health Care System and Monmouth Medical Center, both in New Jersey*.

After Schultz retired Sept. 4, 2020, the interim president, LeighAnn Sidone, described her con-

tributions as "immeasurable."

Sidone, also vice president of nursing and chief nursing officer, listed some of Schultz's accomplishments in the hospital's annual nursing report:

"She was instrumental in helping the hospital to achieve Magnet recognition from the American Nurses Credentialing Center and oversaw the recent campus transformation and opening of the 300,000 square-foot state-of-the-art North Building. Under Jacky's leadership, Suburban received national recognition across its service lines, including from *U.S. News and World Report*, the Emergency Nurses Association, the American College of Cardiology, the American Heart Association, the American Association of Critical-Care Nurses, The Leapfrog Group, *Bethesda Magazine* and more."†

Jacky Schultz in 2014 and in her early days as a nurse

LeighAnn Sidone

---

*   "Jacqueline Schultz Named President of Suburban Hospital," *New Directions*, Sept. 28, 2016, https://www.hopkinsmedicine.org/news/publications/johns_hopkins_medicine_suburban/new_directions_fall_2016/jacqueline_schultz_named_president_of_suburban_hospital.

†   "Jacky Schultz Retires from Suburban Hospital," *Suburban Hospital Nursing Annual Report 2020*, Jan. 1, 2021, https://www.hopkinsmedicine.org/news/articles/jacky-schultz-retires-from-suburban-hospital.

---

understanding and appreciation of The Johns Hopkins Hospital's culture and her working knowledge of the Maryland financial rate-setting system make her extremely well suited to lead at this time in our history."[42]

Peterson and Rothman had co-chaired the search committee for the new president. They didn't have to look far. Miller, a native of Ohio, had been at Johns Hopkins since arriving for medical school in 1988.

After completing her internship and residency in internal medicine at The Johns Hopkins Hospital, she served as an assistant chief of service in 1996. The following year, she joined the Johns Hopkins University School of Medicine faculty as an assistant professor of medicine. In 2004, she earned her M.B.A. from The Johns Hopkins University (at what is now the Carey Business School), and in 2006, she was promoted to associate professor.

After spending the early part of her career in academic medicine, during which she served as associate program director of the Osler Medical Residency Program, and later as assistant dean for student affairs for the school of medicine, Miller shifted her focus to hospital administration. She initially served as vice chair of clinical operations for the Department of Medicine, and then as vice president of medical affairs for The Johns Hopkins Hospital and senior vice president of medical affairs for the health system.

During her tenure as president, The Johns Hopkins Hospital has maintained

Deborah Baker – Senior Vice President for Nursing

Deborah Baker

On July 1, 2016, Deborah Baker became senior vice president for nursing at Johns Hopkins Health System. She took the newly created post while serving concurrently as the vice president of nursing and patient care services for The Johns Hopkins Hospital, an interim role she had held since September 2015.

As senior vice president for nursing, Baker partners with chief nursing officers at the other Johns Hopkins hospitals to develop and deliver strategic planning and vision related to the care of patients and the shift toward a full-service, integrated health system.

Baker earned her bachelor's, master's and doctoral degrees at the Johns Hopkins University School of Nursing, and held a variety of teaching and leadership roles since joining the staff of The Johns Hopkins Hospital as a clinical nurse in 1992.

Baker has devoted her nursing career to patient- and family-centered care, with posts that have included: director of nursing for Surgery, for Wilmer Ophthalmology and the Comprehensive Acute Care Rehabilitation Unit at The Johns Hopkins Hospital (2008–2015); assistant director of nursing for advanced practice and nurse practitioner in the Division of Surgical Oncology at The Johns Hopkins Hospital (2005–2008); and manager of the Department of Surgery NP/PA teams (2003–2005).*

One of her most important and challenging tasks has been recruiting and retaining nurses amid shortages that have grown worse during the COVID-19 pandemic.

In an interview in September 2020, she said one important strategy for ensuring adequate nursing staff to care for patients with COVID-19 has been to train interested nurses in critical care, where shortages were most acute.

"This cross-training is now in place across the Johns Hopkins Health System," she said. "And it's something that will remain."†

---

*   "Deborah J. Baker, D.N.P., C.R.N.P," About Johns Hopkins Medicine, webpage, Leadership, https://www.hopkinsmedicine.org/about/leadership/biography/deborah-baker.

†   "Deb Baker: Reflections of a CNO During COVID-19," *Johns Hopkins Nursing*, Sept. 2, 2020, https://magazine.nursing.jhu.edu/2020/09/deb-baker-reflections-of-a-cno-during-covid-19/.

---

its ranking among the top hospitals in the nation on the *U.S. News & World Report* Honor Roll and earned its fourth consecutive Magnet designation for nursing excellence from the American Nurses Credentialing Center.[43]

Miller's time as president has been marked by enhanced emphasis on the hospital's efforts to address the social determinants of health in East Baltimore. The initiatives she championed include an expansion of the hospital's justice-involved hiring program and the creation of a partnership between 10 city hospitals, Baltimore City, and a local nonprofit to provide supportive housing for individuals experiencing homelessness.

In the years following her appointment, Miller has continued her clinical practice, saying that seeing patients keeps her grounded and helps her decision-making.[44]

Miller is active in several community and professional organizations and has held leadership positions for various American Hospital Association and United Way initiatives. In 2020, she was inducted into both the National Academy of Medicine and the Maryland Chamber of Commerce Business Hall of Fame.

The Life and Legacy of Henrietta Lacks

Former Baltimore resident Oprah Winfrey returned to the city in September 2016 to film scenes of *The Immortal Life of Henrietta Lacks* in and around The Johns Hopkins Hospital.

The HBO/Harpo movie, based on Rebecca Skloot's bestselling 2010 book of the same name, debuted in April 2017. It told the true story of Lacks, a Black woman who died of cervical cancer at The Johns Hopkins Hospital on Oct. 4, 1951, never knowing that her remarkable cell line would contribute to some of the world's most important and lifesaving research. Winfrey played Henrietta's daughter, Deborah Lacks.

Henrietta Lacks, who lived in the segregated Baltimore community of Turner Station, was a 31-year-old mother of five when she died of cervical cancer at The Johns Hopkins Hospital, despite receiving the best treatment available at the time.

During her treatment, some of her discarded pathology tissue was sent to George Gey's lab. For years, Gey, a prominent cancer and virus researcher, had been collecting discarded pathology tissue from all patients who came to The Johns Hopkins Hospital with cervical cancer. Cells isolated from each tissue sample eventually died — except the cells from Lacks, which not only lived but doubled every 20 to 24 hours.

Henrietta Lacks

In the ensuing years, scientists used that cell line — named HeLa from the first two letters of her first and last names — to develop polio and COVID-19 vaccines, and to understand cancer, polio, the human genome and more.[45]

However, the book and movie raised questions about whether Lacks had given permission for her cells to be used for research, even though patient consent as it is understood today did not exist back then, and Johns Hopkins never owned the rights to the HeLa cell line and never sold or profited from the discovery or distribution of the HeLa cell line.[46]

George Otto Gey

As it often does, Johns Hopkins Medicine chose to embrace the complexities of a difficult situation. The path forward, leaders decided, involved listening, learning, honoring and giving.

Henrietta Lacks Symposiums, Scholarships and a New Building

On Oct. 2, 2010, eight months after *Immortal Life* was published, Johns Hopkins Medicine hosted its first Henrietta Lacks Memorial Lecture in Turner Auditorium. The lecture, which is free and open to the public, became an annual event, hosted by the Johns Hopkins Institute for Clinical and Translational Research (ICTR). In its first year, it attracted about 600 attendees, including about 20 members of the Lacks family.

Dan Ford, at the time vice dean for clinical investigation, oversaw the program

Students from four Baltimore high schools attended the Henrietta Lacks High School Symposium in 2016.

James J. Potter

in cooperation with members of the Lacks family. It included a presentation on the scientific relevance of HeLa cells by Johns Hopkins faculty member James J. Potter; a presentation on the history of HeLa cells by Roland Pattillo, professor of obstetrics and gynecology at Morehouse School of Medicine; and a talk by *Immortal Life* author Rebecca Skloot.

It also included announcements about two new annual scholarships:

- The Henrietta Lacks East Baltimore Health Sciences Scholarship, sponsored by the Johns Hopkins Health System, gives up to $10,000 per year to support promising graduates of Paul Laurence Dunbar High School; and
- The Henrietta Lacks Award for Community–University Collaboration, sponsored by the Johns Hopkins Urban Health Institute, provides a $15,000 annual award to recognize outstanding, pre-existing community–university collaborations in Baltimore.[47]

The first annual Henrietta Lacks High School Day took place in March 2014. During those visits, students from surrounding schools engage in biomedical and ethics discussions with faculty members and sometimes hear from descendants of Henrietta Lacks. Many labs invite students to see working stem cell labs, where they often view HeLa cells and learn about their contributions.[48]

These programs provided education and opportunity. In 2018, Rothman and Daniels announced a more concrete way to honor Lacks — they would build a new research building and name it for her.

The Henrietta Lacks Building

The announcement at the Oct. 6, 2018 Henrietta Lacks Memorial Lecture brought cheers and a standing ovation. The leaders explained that the building would be constructed at the intersection of Ashland and Rutland avenues, next to Deering Hall, which houses the Berman Institute of Bioethics. It would promote research ethics and support programs that enhance community partnerships and participation. In October 2019, Vines Architecture was chosen to lead the planning stage.[49]

"In addition to changing the course of modern medicine with the impact of HeLa cells, we have learned many important lessons from Henrietta Lacks' experience," said Rothman. "We have a responsibility to our patients, in the care we deliver, as well as when we partner with patients to pursue important scientific questions," Rothman said. "As we honor Mrs. Lacks and her family, we will take a thoughtful approach to the design of the building so that it conveys our commitment to developing trust with our patients, their families, and our community."[50]

Best in Snow

Winter Storm Jonas dumped nearly three feet of snow on the Baltimore region Jan. 22–23, 2016.

More than 3,500 staff members bunked down at The Johns Hopkins Hospital, Johns Hopkins Bayview Medical Center, Sibley Memorial Hospital, Suburban Hospital and Howard County General Hospital. Some arrived on Friday before the snow started and didn't leave until Tuesday.

Between shifts, staffers slept on cots and played in the snow. As the drifts piled up and the region came to a standstill, the hospitals bustled. From Friday afternoon to Sunday night, 60 babies were delivered, 73 surgeries were performed and memories were made.*

---

* Nitkin, Karen, "Snow Days," *Dome*, Feb. 16, 2016, https://www.hopkinsmedicine.org/news/articles/snow-days.

Innovations in Medical Education

In 2015 and 2016, the Johns Hopkins University School of Medicine launched several training programs designed to meet the interests and career paths of individual medical students, residents, fellows and graduate students. The support made it easier for trainees to pursue careers such as primary care physician or physician-scientist.

The Pathways Program

Residents and aspiring physician-scientists at The Johns Hopkins Hospital gained new opportunities to pursue career-focused training with the July 2015 launch of the Pathways Program within the Osler Medical Residency.

The innovative approach transformed the second and third years of internal medicine training into four specialized programs with individualized coursework, hands-on experiences, sophisticated exposure to content and experts, and individualized mentorship.

The four pathways are:

- The Physician-Scientist Pathway
- Global Health Pathway
- Patient Safety and Quality Pathway
- Medical Education Pathway[51]

Each was designed for residents likely to pursue careers in their chosen pathway. Selection is competitive, reflecting the resources involved and the fact that residents would receive a more focused experience instead of traditional training.[52]

Robert Brodsky, left, was head of the physician-scientist pathway in the Osler Medical Residency Training Program in 2016 and a research mentor to Jason Vaught, right, who became assistant professor of Gynecology and Obstetrics.

Physician-Scientist Training Program

Launched in 2016, this program provides specialized support to physicians who plan to devote much of their time to biomedical research.

This support for physician-scientists at the school of medicine includes helping trainees identify research mentors, providing funding for research projects and travel to conferences, helping residents and fellows apply for career development awards and research grants, and helping physicians focus their clinical training to allow for lab time during their residencies or fellowships.[53]

Primary Care Track

The Johns Hopkins University School of Medicine has long been known for placing a strong emphasis on training in medical specialties. Leaders believed it was time for primary care to get the same attention.

In January 2016, nine first-year medical students enrolled in the school's new Primary Care Leadership Track.[54]

"It is a bit of a culture shift for an academic place to think about primary care," said track director Colleen Christmas, a primary care geriatrician and the inaugural Rosemarie Hope Reid, M.D. Professor in the school of medicine."[55]

The three-year program provides specialized training and mentorship in internal medicine, geriatrics, pediatrics and family medicine, preparing future physicians to provide comprehensive care of the whole patient over the course of a lifetime.[56]

The track helps address chronic physician shortages in a vitally important field, and complements other Johns Hopkins programs offering primary care education, including residency training and a clerkship for first-year medical students.

"This program is taking it to the next level," said Steven Kravet, Johns Hopkins Community Physicians president. "We think it's important for the pipeline of primary care physicians to have the opportunity to train from medical school through residency at Hopkins and hopefully stay on in these careers."[57]

Johns Hopkins All Children's Hospital

On April 5, 2016, five years after Johns Hopkins acquired All Children's Hospital in St. Petersburg, Florida, the pediatric hospital officially changed its name to Johns Hopkins All Children's Hospital.

The hospital announced the new name and unveiled its colorful new sign at a sunny outdoor celebration that also marked All Children's 90th anniversary of providing pediatric health care and service to families.[58]

The hospital had been founded in 1926 as the American Legion Hospital for Crippled Children; after the development of the polio vaccine, it developed a new focus in pediatric care, rebranding as All Children's Hospital in 1967. In 2010, it opened a state-of-the-art 259-bed facility, with more than half of the beds devoted to intensive care-level services.

The relationship between All Children's and Johns Hopkins began around 2009, when Ron Werthman, then chief financial officer for the health system, got a phone call from an All Children's executive who wanted to explore affiliations between the two medical centers.

The discussion quickly turned to the idea of merging the two institutions, as Peterson explained in a February 2013 *Dome* column.

On April 1, 2011, in a non-cash transaction, All Children's Hospital became the only hospital in the United States outside of the Baltimore-Washington metro area to join Johns Hopkins.[59]

"What was in it for us?" Peterson asked rhetorically in the *Dome* column. "For one, George Dover, head of the Johns Hopkins Children's Center and director of pediatrics, recognized an opportunity to develop new residency and fellowship training programs and new research programs there. He also recognized the prospect for new employment opportunities for recent graduates of our own training programs. And perhaps most important of all, he recognized the opportunity to make a major difference in the care of children in that part of our country. So that fit neatly into our philosophy."[60]

As part of Johns Hopkins, All Children's in 2014 began a residency program, as well as clinical and research fellowship programs that provided even more highly specialized training.

In 2017, the first cohort of 10 doctors graduated from the residency program.

Of the six women and four men who accepted diplomas at a beachside hotel ceremony that June, two moved to Maryland for fellowships at The Johns Hopkins Hospital and three stayed at Johns Hopkins All Children's Hospital. All took with them the knowledge they had gained as pioneers.

Most hospitals rely on residents to handle the logistics of day-to-day patient care. But Johns Hopkins All Children's was already functioning smoothly as a freestanding children's hospital. That meant Raquel Hernandez, director of the Office of Medical Education at Johns Hopkins All Children's, could create a program with more opportunities for residents to tailor and optimize their learning.

"I tried to hold back tears," said Hernandez after the first graduation ceremony. "None of us knew what we were getting into. The courage and vision of the residents kept me believing we're doing the right thing."[61]

The integration also brought research capabilities to All Children's, which in 2013 became the site of the first accredited pediatric biorepository in Florida, providing an organized

Raquel Hernandez

Unveiling the Johns Hopkins All Children's Hospital sign on April 5, 2016.

Pediatric residents at All Children's Hospital in 2015: Standing left to right; Paul Gilbert, Elena Rueda-de-Leon and Alexander Kim. Seated front is Racha Khalaf.

way to store specimens in a highly protected setting, ensuring their value in research.

The same year, two precision medicine-based studies were established:

- The Prospective Inception Cohort Study (iPICS) is designed to identify key predictors of outcomes in children with a variety of acute and chronic health conditions.
- The Prospective Research on Early Determinants of Illness and Children's Health Trajectories (PREDICT) seeks to learn why some children, healthy at birth and in early childhood, go on to develop health problems.[62]

In 2019, the hospital and its nurses received Magnet designation from the American Nurses Credentialing Center (ANCC) in recognition of nursing excellence and the highest level of professionalism in nursing practice.[63]

And in 2021, *U.S. News & World Report* ranked Johns Hopkins All Children's Hospital in a tie for best pediatric hospital in Florida in its 2021-2022 rankings.[64]

Children's Center leaders through the years, left to right: George Dover, Tina Cheng, Margaret Moon and David Hackam.

Children's Center Leadership

In July 2016, George J. Dover retired after 20 years as director of the Children's Center and pediatrician-in-chief. He had presided over a period of tremendous growth, innovation and discovery, with many of his trainees going on to become national leaders in pediatric medicine.[65]

From 2014 until 2016, he co-directed the Children's Center with David Hackam, Garrett Professor of Pediatric Surgery, chief of pediatric surgery and pediatric surgeon-in-chief at The Johns Hopkins Hospital.[66]

After Dover left, Hackam shared the directorship with Tina Cheng, who became the Given Foundation Professor of Pediatrics, director of the Department of Pediatrics for the Johns Hopkins University School of Medicine and pediatrician-in-chief of The Johns Hopkins Hospital.

Hackam had grown up in Stafford, England, and obtained his bachelor's and medical degrees from the University of Western Ontario, and a Ph.D. in cell biology at the University of Toronto. He came to Johns Hopkins from the University of Pittsburgh School of Medicine, where he was the Watson Family Professor of Surgery, the associate dean for medical student research, and co-director of the Fetal Diagnosis and Treatment Center of the Children's Hospital of Pittsburgh.[67]

In 2020, Cheng joined the Cincinnati Children's Hospital Medical Center as B.K. Rachford Memorial Chair in Pediatrics, chief medical officer, and Research Foundation director.[68]

Hackam now co-directs the Children's Center with Margaret Moon, director of the Department of Pediatrics at the Johns Hopkins University School of Medicine and pediatrician-in-chief at the Johns Hopkins Children's Center.

Moon, who earned her undergraduate degree from Michigan State University, her M.D. from the Johns Hopkins University School of Medicine and her M.P.H. from the Johns Hopkins Bloomberg School of Public Health, specializes in ethics in clinical practice, ethics education, ethics in research, and urgent care pediatrics. Her clinical time takes place in the Johns Hopkins pediatric emergency department, where she teaches residents in the urgent care setting.

She is also a core faculty member of the Johns Hopkins Berman Institute of Bioethics, where she is supported as the Freeman Scholar in Clinical Ethics, and serves as an ethics member of the Johns Hopkins IRB.[69]

Construction and Grand Openings Across the Enterprise

Construction cranes and grand openings were abundant between 2016 and 2018, years that saw tremendous expansion across Johns Hopkins Medicine.

Among the new facilities:

- Sibley Memorial Hospital opened the Advanced Clinical Care Unit to provide care for complex medical and surgical patients who require specialized nursing assessments and care.
- The Johns Hopkins Hospital opened the John G. Bartlett Specialty Practice, a facility that co-locates much-needed services for patients with a range of infectious diseases, including infections following transplantation or from medical devices, HIV and viral hepatitis.
- Johns Hopkins Bayview Medical Center opened a facility at 5500 E. Lombard St., which became the new home of the Community Psychiatry Program, previously at the campus' Mason F. Lord Building.
- The Johns Hopkins Children's Center opened its first pediatric cardiac intensive care unit, as well as the first outpatient pediatric hemodialysis and peritoneal dialysis center in Maryland.
- The Johns Hopkins Department of Psychiatry and Behavioral Sciences opened a new psychiatry office in the medical pavilion next to Howard County General Hospital.[70]

Community Physicians at Remington Row

Remington Row opened on Sept. 28, 2016, in Remington, the diverse Baltimore neighborhood that brushes the southern edge of The Johns Hopkins University's Homewood campus.

Although the building's high windows and ornate brickwork evoke a

The Remington Row building is home to Johns Hopkins Community Physicians (JHCP) practices.

repurposed factory, Remington Row was built from the ground up by local developer Seawall Development, in collaboration with Johns Hopkins.[71]

The second floor of the building is devoted to three Johns Hopkins Community Physicians (JHCP) practices that moved from dated offices in the Wyman Park Building, adjacent to the Homewood campus of The Johns Hopkins University. The move not only allowed the practices to update their facilities, but also to experiment with a new model of patient-centered care, said Kravet.[72]

Each examination room is named after a city landmark or a bird portrayed in a photo on the exterior door. There are no individual providers' offices. All clinicians work together in a centralized area, a decision that promotes teamwork and aims to trim the time that patients wait.[73]

"JHCP has the special privilege of bringing Johns Hopkins Medicine into our communities," Kravet said. "Primary care is local, so we pay particular attention to opportunities to integrate into our neighborhoods in small ways and big ways."

Kravet said similar priorities characterize the more than 50 JHCP practices in Maryland, Washington, D.C., and northern Virginia, where more than 500 health care providers serve approximately 250,000 patients each year.

JHCP took shape in 2003 by integrating the physician practices at Johns Hopkins Bayview Medical Center with the Johns Hopkins Medical Services Corp., an entity that used to provide most of Maryland's primary care.

Since then, it has expanded its scope to more fully embody JHM's tripartite mission, implementing a longitudinal clerkship in the school of medicine; developing research platforms to support major research grants for the school of medicine and school of public health, and helping to launch primary care models in India and Saudi Arabia.[74]

Capacity Command Center

A bustling 5,500-square-foot command center at The Johns Hopkins Hospital opened in 2016, representing a giant leap forward for efficiency, integration and innovation.

The Judy Reitz Capacity Command Center is packed with giant monitors displaying real-time data about everything from bed availability and operating room efficiency to patient status and staffing, giving staff the information they need to make quick, informed decisions about patient placement.

It's named for Reitz, who retired in 2016 as the first woman executive vice president and chief operating officer for The Johns Hopkins Hospital.

James Scheulen, who oversees the center, saw the need for this kind of innovation from both a personal and a professional perspective. When a relative was ill and needed to be moved from a hospital in Frederick to The Johns Hopkins Hospital, Scheulen saw that it took multiple phone calls between the two hospitals to work out the details of the transfer.

James Scheulen at the opening of the Capacity Command Center, January 2016.

Community Meals

When her primary care practice moved from Wyman Park to Remington Row in September 2016, pediatrician Tina Kumra wanted to know how she could contribute to the well-being of the surrounding neighborhood.

She asked Alice Bassett-Jellema, pastor of the local Church of the Guardian Angel, who suggested a focus on nutrition.

The result was a weekly community dinner at the church, prepared by Kumra and other Johns Hopkins Community Physicians volunteers.

Kumra started the program with a $10,000 grant from the Brancati Center of the Johns Hopkins Division of General Internal Medicine, which works to improve health through community partnerships.*

About 20 residents cook and eat together at the church each week, often preparing food from Gather Baltimore, a volunteer-based program that collects surplus vegetables and fruits to sell at low cost to those in need.

"Some of the people who are most engaged are teenagers, and it's really phenomenal," Kumra said. "They're not only enjoying problem-solving by figuring out what to do with the different ingredients we have each week, but they're also leading the cooking classes. The elementary school kids follow along. If we can change habits early in life, it will have an incredible impact."†

---

* "The Brancati Center for the Advancement of Community Care," Brancati Center webpage, https://www.brancaticenter.org/.
† Smith, Linell, "Practicing Culinary Medicine," *Dome*, Oct. 3, 2017, https://www.hopkinsmedicine.org/news/articles/practicing-culinary-medicine.

Pediatrician Tina Kumra, center; medical office assistant Devona Fletcher, in blue scrubs; and other Johns Hopkins colleagues help prepare the weekly community dinner at The Church of the Guardian Angel.

As chief administrative officer for emergency medicine, he knew The Johns Hopkins Hospital could do more to proactively manage patient capacity.

The command center, designed and built in collaboration with GE Healthcare Partners, replaced the previous scattershot process, which involved phone calls, emails and even faxes between the admitting department, the physician referral line team, the ambulance system and bed management.

As a result, wait times for Johns Hopkins care have been reduced, and patient capacity has increased.

The system is one of the first of its kind; in its first few months, representatives from 50 health systems across the U.S. and from four countries visited to see it in operation.[75]

When the COVID-19 pandemic began in March 2020, the command center proved up to the task of managing the flow of patients with the new and highly infectious illness.

"We didn't need a separate plan for patients with COVID-19 because this is what we do every day," said Scheulen in March 2021. "These were patients with emergency needs just like others, but they had COVID-19. They're patients and they're being transported, and we know how to do that."[76]

Capacity Command Center

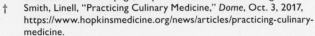

The Center for Diagnostic Excellence

Anne and C. Michael Armstrong

Sometimes, a headache is just a headache. But sometimes it's caused by health concerns like allergies, stress or worse — a stroke.

A headache is just one of the many symptoms that can lead doctors to misdiagnose a medical issue. Research shows that diagnostic errors affect roughly one in 20 adults in the U.S., or 12 million Americans a year. As many as one-third of these errors may result in serious permanent injuries, including disability or death.[77]

To address this issue, Johns Hopkins Medicine on Oct. 31, 2016, announced the opening of the Armstrong Institute Center for Diagnostic Excellence to enhance diagnostic accuracy, cut waste on unnecessary diagnostic testing and move the needle on eliminating preventable harms from diagnostic errors worldwide.

It is led by David Newman-Toker, Johns Hopkins professor of neurology, ophthalmology and otolaryngology–head and neck surgery, and an internationally recognized leader in diagnostic research and diagnostic safety.

"Misdiagnosis is incredibly frequent because medicine is incredibly hard. There's uncertainty, complexity and incomplete information all the time," said Newman-Toker. "But we can do better than we're doing right now, and our new center will lead change to make that a reality."[78]

The center, regularly featured in national media, provides resources for patients, providers, scientists and administrators working together to tackle the challenges of misdiagnosis.[79]

The Center for Diagnostic Excellence was made possible by a $5 million gift from C. Michael Armstrong, who had also provided $10 million to fund the Johns Hopkins Armstrong Institute for Patient Safety and Quality more than five years earlier.

Armstrong, chairman of Johns Hopkins Medicine's board of trustees from 2005 to 2013, had nearly died from medical errors at hospitals outside of the Johns Hopkins system.[80]

First, doctors missed signs that he had leukemia. Then he developed a serious infection post-chemotherapy. Years later, he was belatedly diagnosed with advanced cancer and given a 50-50 chance of living five years. If he survived, he vowed, he would "do something big" for patient safety. Armstrong finished treatment in 2009.

"We have been making excellent progress on patient safety and quality, but we can do better," Armstrong said when announcing the gift that launched the Armstrong Institute. "We must take our patient safety research and results to the next level to be the best."[81]

The institute opened in June 2011 with Peter Pronovost as its first director. The former Johns Hopkins anesthesiologist and critical care specialist also became senior vice president for patient safety and quality.

Pronovost, winner of a 2008 MacArthur Fellowship, was best known for developing a simple checklist of five steps for doctors to follow when placing a central-line catheter, an intervention that dramatically reduced the risk of hospital-acquired infections.

David Newman-Toker

Peter Pronovost

Time magazine in 2008 named Pronovost one of the 100 most influential people in the world.

"Michigan hospitals began implementing Pronovost's checklists in ICUs in 2003," said the *Time* article. "Within three months, hospital-acquired infections at typical ICUs in the state dropped from 2.7 per 1,000 patients to zero. More than 1,500 lives were saved in the first 18 months."[82]

Clinical Communities

As leader of the Armstrong Institute, Pronovost quickly developed four "clinical communities" that brought together clinicians from across the health system to identify and combat infections and other patient safety problems. Those first communities were grouped around intensive care units, hospital medicine (the practice of caring for acutely ill patients in hospital settings), the improvement of medication safety, and the post-anesthesia care units (PACUs).[83]

"Comprised of inter-professional members (to include patients and families), these physician-led groups foster collaboration and support peer learning," said the clinical communities website. "Key stakeholders establish quality and safety initiatives, work collaboratively within their organizations, develop project goals, targets and timelines and are accountable for measurable results. Tapping into the wisdom of those closest to the point of care encourages innovative solutions that are wiser, effective and more likely to be implemented."[84]

By 2021, Johns Hopkins Medicine had launched more than 20 such communities, including ones for diabetes, opioid management and strokes.

The institute also has a Patient Safety and Quality Leadership Academy, a nine-month multidisciplinary training program for future quality and safety leaders.

In December 2018, Allen Kachalia joined Johns Hopkins Medicine as the senior vice president of patient safety and quality and the director of the Armstrong Institute for Patient Safety and Quality.

Kachalia is a general internist whose research focuses on how law affects medical care, particularly how liability system reform and the disclosure of medical error relate to the quality and safety of health care. He came to Johns Hopkins from Brigham and Women's Hospital in Boston, where he had been chief quality officer and vice president for quality and safety since 2014.

According to the Johns Hopkins Medicine press release, "Dr. Kachalia has combined clinical expertise with innovation and investigation that has been focused on improving health care quality and safety, and has become an international leader and scholar in the field."[85]

Bloomberg~Kimmel Institute for Cancer Immunotherapy

The Bloomberg-Kimmel Institute for Cancer Immunotherapy was dedicated on March 29, 2016, with a standing room-only ceremony led by then-Vice President Joe Biden and Michael R. Bloomberg, the philanthropist, entrepreneur, Johns Hopkins alumnus and three-term mayor of New York City.

The institute supports a new and promising type of treatment that works by empowering each patient's highly individual immune system to detect and attack cancer cells. It was funded with $50 million each from longtime Johns Hopkins donors Bloomberg and Jones Apparel Group founder Sidney Kimmel, and $25 million from more than a dozen additional supporters.

Joe Biden with Michael Bloomberg at the 2016 dedication of the Bloomberg~Kimmel Institute for Cancer Immunotherapy.

The Bloomberg-Kimmel Institute for Cancer Immunotherapy, with Drew Pardoll as director, united Johns Hopkins Kimmel Cancer Center experts with immunology, genetics, microbiology and biomedical engineering experts throughout Johns Hopkins. The concentrated effort involves more than 100 scientists and clinicians.

"We believe the focused and collaborative research made possible through the institute will advance immunotherapies to the point where the immune system will ultimately be able to beat 100 percent of cancers," said Pardoll, a leading immunology researcher at Johns Hopkins. "The potential to control or cure even the most advanced, treatment-resistant cancers has been elusive until now."

The event took place in the sunny second-floor atrium of the Anne and Mike Armstrong Medical Education Building on the Johns Hopkins medical campus.

Medical students in their white coats stood on stairs flanking the room and often drew the praise and attention of speakers. After the event, Biden, who would become the nation's 46th president on Jan. 20, 2021, shook the students' hands and posed with them for selfies.

"This institute is going to perfect new therapies and bring hope to millions of people," Biden said. "I'm convinced, not only will we save millions of lives, we will re-instill in the American public the notion that anything is possible."

Bloomberg also spoke at the dedication. "Ending cancer is a dream we have all held," he said.

Kimmel spoke in a pre-recorded video. "I have always sought to invest in the best people at the best place — connecting promise to progress," he said.[86]

Andrew Cameron

Christine Durand

Niraj Desai

Dorry Segev

Nation's First HIV-Positive Liver Transplant

Andrew Cameron's cell phone rang as he drove to work on March 18, 2016. It was 6:30 a.m., the sky beginning to brighten.

Cameron, a liver transplant surgeon at The Johns Hopkins Hospital, got calls like that about 100 times a year. A New England woman had died; her grieving family wanted her organs to live on. But this time was different: The woman had human immunodeficiency virus — HIV.

For the next two days, infectious diseases specialist Christine Durand gathered information about the donor to ensure that her liver and kidneys were suitable for transplant.

On March 20, Cameron performed the nation's first transplant of an HIV-positive liver. The same day, Johns Hopkins kidney transplant surgeon Niraj Desai performed the nation's first HIV-positive kidney transplant.

Only HIV-positive patients can receive HIV-positive organs.

Although the surgeries were no more complicated than usual, "we knew there was more to it," Cameron says.

People with HIV now live long enough for their livers and kidneys to wear out — the damage accelerated by cancer, cirrhosis, hepatitis C and therapies for HIV. About 30% of people with HIV suffer kidney damage, said nephrologist Derek Fine. A transplanted kidney adds years of life and ends uncomfortable and time-consuming dialysis.

The lifesaving surgeries were possible because Dorry Segev, another Johns Hopkins transplant surgeon, pushed to reverse a 1988 federal law forbidding the use of organs from donors diagnosed with HIV. At that time, AIDS was new, poorly understood and almost always fatal.

The HIV Organ Policy Equity (HOPE) Act, which became law in November 2013, allows HIV-positive organ transplants. Earlier in 2016, under Segev's leadership, The Johns Hopkins Hospital became the first hospital approved for such surgeries.[87]

This illustration, developed for Johns Hopkins Medicine by Francesco Bongiorni, shows a red ribbon symbolizing support for people with HIV, as well as workers who continue the progress.

Nina Martinez, left, donated her HIV-positive kidney in a first-of-its-kind surgery performed by Dorry Segev, right.

Another Historic First for HIV-Positive Transplants

Three years later, on March 25, 2019, surgeons Segev and Desai made history again when they performed the first transplant of an HIV-positive kidney from a live donor to another person living with HIV.

The donor, Nina Martinez, had lived with HIV since receiving a blood transfusion when she was 6 weeks old. She was 35 at the time of the surgery.

As with the earlier HIV-positive transplants from deceased donors, the surgery itself was not medically exceptional. Its power came from its ability to help sweep away stigmas and barriers, paving the way for lifesaving kidney transplants from people with HIV to others with HIV.

"There's the perception that people [with HIV] like me, we bring death," said Martinez. "But in this moment, I get to bring life." Her brave act was just the latest development in longstanding campaigns by both Martinez and Segev to improve the lives of people with HIV.

"This is a celebration of the evolution of HIV care," said Segev. "Today it's not even a disease. It's a condition. And for people with HIV, one of the biggest stigmas has been that they have not been able to donate a kidney."[88]

Martinez had been following advances in HIV-positive transplants since the HOPE Act was signed. In March 2016, she learned that Johns Hopkins had performed the nation's first HIV-to-HIV kidney and liver transplants from a deceased donor.

Because of health concerns for potential donors, it would take three more years and additional research before the first live HIV-positive kidney donation. "We don't want an HIV-positive donor to donate a kidney and then lose the function of the remaining one," explained Segev.[89]

Hepatitis C-Positive Transplants

In 2014, nearly 20,000 Americans died of illnesses related to hepatitis C — more than from HIV, pneumonia and tuberculosis combined. Baltimore had a higher rate of infection than most U.S. cities, with an estimated 60,000 cases.

That year, the outlook for people infected with the hepatitis C virus improved dramatically when the U.S. Food and Drug Administration approved new drugs that inhibit the virus' ability to reproduce.

The treatments, which have gotten even better over time, allowed Johns Hopkins to cure thousands of people of hepatitis C.[90]

They also opened the door to an entirely new category of transplants pioneered at Johns Hopkins, from donors with hepatitis C to recipients without the virus.

Beginning in 2016, Johns Hopkins successfully transplanted 10 hepatitis C-infected kidneys into patients without hepatitis C and used the new antivirals as prophylaxis, to prevent the patients from becoming infected with the virus.

"These 10 kidneys we used are 10 kidneys that would not have been transplanted outside of this study," said Desai, senior author of the study, appropriately called EXPANDER (Exploring Renal Transplants Using Hepatitis C Infected Donors for HCV-Negative Recipients), published in *Annals of Internal Medicine*. "Figuring out how to use these kidneys is a way to do more transplants and save more lives."

No recipients ever developed any clinical signs of chronic hepatitis C infection. In addition, the kidneys themselves functioned well. At the time of the study's publication, all patients were at least a year out from their transplant and doing well, said Desai.

"This was an overwhelmingly positive study," said Durand. "We're always trying to expand what we consider acceptable for an organ donor."

Following EXPANDER, Desai, Durand and colleagues published a second study in 2021, also in the *Annals of Internal Medicine*. This one, called REHANNA (Renal Transplants in Hepatitis C Negative Recipients with RNA Positive Donors), showed that the strategy of using hepatitis C-infected kidneys for patients without hepatitis C could also be successful with shorter course prophylaxis, said Durand.[91]

Erwin and Stephanie Cooper Greenberg

Johns Hopkins Greenberg Bladder Cancer Institute

Tens of thousands of people are diagnosed each year with bladder cancer, yet it tends to receive less attention and funding compared to other cancers.

Erwin and Stephanie Cooper Greenberg helped to change that when they donated $15 million in May 2014 to establish the Johns Hopkins Greenberg Bladder Cancer Institute, the world's first academic entity dedicated solely to improving bladder cancer knowledge and treatment.

The Greenbergs, who serve on several Johns Hopkins boards, are directors of the Erwin and Stephanie Greenberg Foundation; their gift was part of a $45 million co-investment with Johns Hopkins University.[92]

David J. McConkey

David J. McConkey, a professor in the Department of Urology at Johns Hopkins Medicine, is the inaugural director of the center, which brings together researchers from across the institution and beyond to advance understanding of bladder cancer and how to treat it.

The center is a Johns Hopkins Precision Medicine Center of Excellence, benefitting from technology platforms and support that allow researchers to use larger data sets to find patterns and tailor treatments to individual patients.[93]

"Until recently, bladder cancer medicine remained essentially unchanged for 50 years," noted a fundraising document for the center. "Now we have entered a new era, heralded by the emergence of a dazzling array of new treatments: gene therapies, immunotherapies, targeted therapies, combination therapies, tumor sequencing, novel chemotherapies that utilize bioengineered materials such as nanoparticles, and likely other strategies as yet unimagined. These late-breaking treatments offer greater effectiveness and less toxicity, better long-term outcomes and improved quality of life for patients."[94]

In July 2019, the National Institutes of Health awarded researchers at the Greenberg center a $3.2 million grant to study new treatment options for early-stage bladder cancer, along with experts from the Johns Hopkins Kimmel Cancer Center, the Bloomberg-Kimmel Institute for Cancer Immunotherapy, the Brady Urological Institute and the Center for Computational Genomics at Johns Hopkins.

Early-stage bladder cancer, formally known as non-muscle invasive bladder cancer, is identified in close to 50,000 patients each year, according to the National Cancer Institute, and typically is treated with a weakened form of bacillus Calmette-Guérin (BCG) bacteria instilled directly into the bladder.

People undergoing this treatment have a high risk of cancer recurrence and typically experience side effects such as increased frequency of and discomfort with urination.

The grant supports efforts by Johns Hopkins physician-scientists to participate in a novel multidisciplinary clinical trial for this cancer, called ADAPT-BLADDER, for its ability to accommodate changes during the trial that optimize and accelerate tests of promising treatments.[95]

Johns Hopkins Center for Transgender Health

Paula Neira

Paula M. Neira remembers the 2017 phone call that became a milestone in her career: She was able to tell the transgender man seeking "top surgery" that the Johns Hopkins Center for Transgender Health could provide the gender-affirming procedure at The Johns Hopkins Hospital.

"It was the first time we offered gender-affirming surgery since the 1970s," said Neira, at the time clinical program director of the Johns Hopkins Center for Transgender Health. Neira, a nurse, lawyer and former naval officer, later became program director of LGBTQ+ Equity and Education in the Johns Hopkins Medicine Office of Diversity, Inclusion and Health Equity.

"With this new center, we aim to lead by action: to improve the ability for

Rainbow Connection

Baltimore's annual LGBTQ+ Pride parade has had an enthusiastic Johns Hopkins presence since 2014. Most years, well over a hundred representatives from Johns Hopkins Medicine, The Johns Hopkins University and the Applied Physics Laboratory march in the parade, with many wearing bright colors to form a "scrubs rainbow.*"

The parade is part of an annual Pride weekend in Baltimore that began with a small rally in 1975 and now includes a high-heel footrace and live music.†

---

* Hub staff report, Johns Hopkins Well-Represented at Baltimore's Annual Pride Celebration," *The Hub*, July 27, 2015, https://hub.jhu.edu/2015/07/27/hopkins-baltimore-gay-pride/.

† Cohen, Lauren, "The Evolution of Baltimore's Pride Festival," *Baltimore* magazine, https://www.baltimoremagazine.com/section/community/the-evolution-of-baltimores-pride-festival/.

transgender people to get health care, to provide medically needed care, and to offer care in a supportive and affirming way," said Neira in 2017.[96]

The center, which opened in 2017, represented a new chapter in Johns Hopkins Medicine's LGBTQ+ care. In the late 1960s, Johns Hopkins Medicine was the leading academic medical center providing transgender health care in the United States. Then, amid controversy about its services, the original Gender Identity Clinic closed in 1979.

In 2015, Johns Hopkins leaders pledged to resume offering comprehensive services to the transgender community. The center involved 25 to 30 professionals across many departments, including plastic surgery, urology, gynecology, endocrinology, dermatology, primary care, pediatrics, mental health, nursing and social work. Over time, the size of the team grew as the center evolved its reach across the enterprise.

In March 2022, facial plastic and reconstructive surgeon Fan Liang joined Johns Hopkins as the new director of the Center for Transgender Health. Liang had earned her undergraduate degree at Princeton University and her medical degree at Harvard Medical School before completing an integrated plastic surgery residency at the University of Southern California and a fellowship in pediatric and craniofacial plastic surgery at the University of Michigan.

By that time, The Johns Hopkins Center for Transgender Health had a team of more than 50 clinicians and staff, and had performed more than 600 gender-affirming surgeries and interacted with more than 2,800 patients.[97]

That includes people like Dariel Peay, a senior patient access supervisor in the Johns Hopkins Department of Dermatology.

Fan Liang

Paula Neira Christens 'U.S.N.S. Harvey Milk'

In a historic ceremony, Paula M. Neira, then clinical program director of the Johns Hopkins Center for Transgender Health, smashed a bottle of sparkling wine across the bow of the future *U.S.N.S. Harvey Milk* (T-AO-206) on Nov. 6, 2021.

The ship was the first to be named for an openly gay person: Harvey Milk, an iconic LGBTQ+ civil rights leader and one of the first openly LGBTQ+ people to be elected to public office in the U.S. He was a member of the San Francisco Board

of Supervisors when he was assassinated on Nov. 27, 1978. Both Neira and Milk had served as officers in the U.S. Navy.

In 2015, Neira made naval history as the first transgender Navy veteran to have her discharge documentation updated to reflect her correct name by order of the Navy.

In 2016, the Secretary of the Navy named her the co-sponsor of the *Harvey Milk*, the second ship in the *John Lewis* class of fleet replenishment oilers, used to refuel and resupply other ships.*

---

* Bacon, Johns, "Navy Christens Ship USNS Harvey Milk, After Gay Rights Activist," *USA TODAY*, Nov. 7, 2021, https://www.usatoday.com/story/news/nation/2021/11/07/navy-christens-ship-usns-harvey-milk-after-gay-rights-activist/6330296001/.

Peay's prom marked the first time she went to a high school event dressed as a woman. She wore a white pantsuit and matching high-heel pumps, and she felt terrific. "That was my way of being me," she said. "You could hear the gasps."

In October 2017, Peay, assigned as male at birth, further aligned her outward appearance with her gender identity when she had breast augmentation surgery at the center.

Peay became a co-leader of on-campus transgender awareness workshops and a facilitator for training through Safe Zone, a free education program for the Johns Hopkins LGBTQ+ community and allies.

"What surprises me is that people want to learn about how to become an ally for my community," she said. "I always assumed that people have their stereotypes and their beliefs and they stick to them. It turns out they want to know and they want to educate other people. I tell people, the moment you let others define who you are, you've given up your power. Never give up your power. I defined myself as a trans person, even when I didn't know what trans was."[98]

Opioid Prescriptions and Substance Use Disorder

By 2017, opioid abuse and overdoses were serious and growing health problems across the United States.

The Department of Health and Human Services (HHS) declared a public health emergency in April of that year, sharing the devastating statistic that 91 people a day were dying of opioid-related overdoses.[99]

Hospital prescribing practices needed to change.

A Johns Hopkins study, published that August in *JAMA Surgery*, highlighted the need for more personalized pain management to avoid overprescribing opioids and reduce risks linked to improperly stored opioids in the home.

The review of six published studies found that a substantial majority of patients used only some or none of the pills, and more than 90% failed to dispose of the leftovers in recommended ways.

"If we can better tailor the amount of opioids prescribed to the needs of patients, we can ensure patients receive appropriate pain control after surgery, yet reduce the number of extra oxycodone and other opioid tablets in many homes

Suzanne Nesbit and Mark Bicket in a clinical community of experts discussing opioid stewardship.

Dan Ashby   Peter Hill   Steven Levin

that are just waiting to be lost, sold, taken by error, or accidentally discovered by a child," said Mark Bicket, the paper's first author, who was assistant professor of anesthesiology and critical care at Johns Hopkins.[100]

To reduce prescription opioid misuse, the Armstrong Institute for Patient Safety and Quality in 2017 formed a clinical community of experts from throughout the Johns Hopkins Health System to take on the issue of opioid stewardship.

The group included doctors, nurses, nurse practitioners, pharmacists, counselors, administrators and patients representing all six Johns Hopkins hospitals and Johns Hopkins Community Physicians. It was chaired by Suzanne Nesbit, clinical pharmacy specialist for pain management, and Steven Levin, director of the regional pain management program.

Executive champions were Peter Hill, senior vice president of medical affairs for the Johns Hopkins Health System, and Dan Ashby, vice president and chief pharmacy officer.[101]

Marie Hanna

Beth Hogans

Perioperative Pain Clinic

In June 2017, the Department of Anesthesiology and Critical Care Medicine launched something new: a Perioperative Pain Clinic.

The clinic, led by anesthesiologist and critical care specialist Marie Hanna, deploys a full arsenal of pain management tools — education, psychiatry, physical therapy, and narcotic and non-narcotic medications— to help patients understand and manage surgical pain, sometimes starting several weeks before a procedure and continuing for months after they leave the operating room.[102]

Hanna has two rules for patients in the clinic. They must not get any opioids, legal or illegal, from anyone outside her clinic. And they can take opioids only when necessary — not to eliminate pain, but to provide enough relief to allow physical activity.[103]

"The whole purpose is to fight the opioid epidemic," said Hanna at the time. "I would say 80 to 90 percent of patients on opioids don't realize there are other options to control their pain, like non-narcotic medications and physical therapy."[104]

Hanna in 2021 said the Perioperative Pain Clinic had grown, and its methods have been adopted by other institutions. "Thanks to the support of the hospital leadership, we became leaders in this field," she said.

"I want to send a message of hope," she added. "The opioid crisis is not insurmountable. My team and I are inspired by every patient who overcomes their reliance on opioids after years of dependency."[105]

In addition, first-year students in the Johns Hopkins University School of Medicine are required to take a pain care medicine course. During four days of instruction, students hear from Johns Hopkins patients; physicians, including pain specialists, neurologists, rehabilitation specialists, oncologists, and surgeons; pharmacists; physical therapists; psychologists; nurse practitioners; and pain scientists.

"Our whole program is geared to teaching students the hazards of opioids and the fact that pain should be addressed comprehensively," said neurologist Beth Hogans, who developed the class in 2010 and is director of the National Institutes of Health Pain Consortium Johns Hopkins University Center of Excellence in Pain Education (CoEPE). The students learn about comprehensive pain management, including non-pharmacological approaches in the spring; the following fall they learn about the management of substance use disorders.

The curriculum was created to align with national and international recommendations for pain education in medical training, and contributed to recommendations of the American Academy of Pain Medicine, said Hogans, whose most recent pain management book is *Pain Medicine at a Glance*, intended for medical students.[106]

At the Broadway Center for Addiction, Kenneth Stoller, left, sees patients and mentors newly minted addictions psychiatrists Ann Ruble and Vinay Parekh.

Clinical Care for Substance Use Disorder

Johns Hopkins offers many treatment options for people with substance use. These include:

- The Intensive Treatment Unit (ITU): a six-bed inpatient service at The Johns Hopkins Hospital designed for rapid (two-to-four-day) medically supervised withdrawal from drugs and alcohol, along with medical stabilization.[107]
- The Motivated Behaviors Unit (MBU): an eight-bed unit at The Johns Hopkins Hospital that specializes in the treatment of patients with both substance use disorders and other psychiatric disorders.[108]
- The Johns Hopkins Broadway Center for Addiction: a clinic offering comprehensive outpatient treatment services for people with acute or chronic substance use disorders.[109]
- The Addiction Treatment Services (ATS): a comprehensive treatment center at the Johns Hopkins Bayview Medical Center for individuals suffering from substance use disorder, including co-occurring psychiatric illnesses.[110]
- The Center for Addiction and Pregnancy (CAP): located at Hopkins Bayview, with an outpatient program and an available overnight housing unit, which provides a comprehensive, coordinated, and multidisciplinary approach for mothers who are drug-dependent and their drug-affected babies.[111]
- The Addiction Medicine Unit (AMU): an 18-bed inpatient medical unit at Hopkins Bayview that provides safe medical detoxification for those addicted to alcohol or benzodiazepines.[112]
- The Mental Illness and Substance Abuse (MISA) treatment team at Hopkins Bayview, which provides outpatient services for those who have a co-occurring mental illness and a current or recent substance use problem. Individual therapy, group therapy, medication management and urine testing are a core part of the treatment structure.[113]
- The Addiction Treatment Center (ATC) at Suburban Hospital, which has been helping teens fight substance use since the 1980s, said its director, Beth Kane Davidson. The program lasts four to six months, starting with counseling sessions two to three times per week and becoming less frequent over time. Clients are expected to transition to a support group such as Narcotics Anonymous.[114]

ENDNOTES

1 Rothman, Paul B., "The Promise of Opportunity," *Dome*, November 2015, https://www.hopkinsmedicine.org/news/articles/the-promise-of-opportunity.

2 City of Baltimore and Baltimore City Health Department, "*Healthy Baltimore 2020*: A Blueprint for Health," March 2017, https://health.baltimorecity.gov/sites/default/files/HB2020%20-%20April%202017.pdf.

3 "*Healthy Baltimore 2020*: A Blueprint for Health."

4 Hub staff report, Johns Hopkins launches HopkinsLocal, an Initiative aimed at Strengthening Baltimore," *The Hub*, Sept. 30, 2015, https://hub.jhu.edu/2015/09/30/hopkins-local-launch/.

5 Callaway, Marian, "HopkinsLocal Seeks to Expand Economic Opportunities in Baltimore," *Dome*, November 2015, https://www.hopkinsmedicine.org/news/articles/hopkinslocal-seeks-to-expand-economic-opportunities-in-baltimore.

6 Reeves, Tracey, "Johns Hopkins among 25 Businesses Teaming up to Expand Economic Opportunities in Baltimore," *The Hub*, April 4, 2016, https://hub.jhu.edu/2016/04/04/blocal-baltimore-launch/.

7 "Johns Hopkins among 25 Businesses Teaming up to Expand Economic Opportunities in Baltimore."

8 "Johns Hopkins among 25 Businesses Teaming up to Expand Economic Opportunities in Baltimore."

9 Rosen, Jill, "BLocal Effort Drives Millions of Dollars into Baltimore's Economy," *The Hub*, May 15, 2019, https://hub.jhu.edu/2019/05/15/blocal-commits-50-million-more/.

10 "BLocal Effort Drives Millions of Dollars into Baltimore's Economy."

11 Smith, Patrick, "Homegrown Talent," *Dome*, April 28, 2016, https://www.hopkinsmedicine.org/news/articles/homegrown-talent.

12 Smith, Patrick, "Revenue Cycle Management Academy Trains Hopkins Staffers for the 'Other Side' of Health Care," *Dome*, Aug. 6, 2018, https://www.hopkinsmedicine.org/news/articles/revenue-cycle-management-academy-trains-hopkins-staffers-for-the-other-side-of-health-care.

13 Nitkin, Karen, "Dean Rothman Hosts Community Conversation in East Baltimore," *Dome*, June 20, 2017, https://www.hopkinsmedicine.org/news/articles/dean-rothman-hosts-community-conversation-in-east-baltimore.

14 Morris, Michel, "Community Conversation Emphasized Education and Trust," *Community Health Stories*, Jan. 14, 2021, https://www.hopkinsmedicine.org/news/articles/community-conversation-emphasizes-education-and-trust.

15 Smith, Patrick, "Pathways to Opportunity," *The Hub*, March 1, 2018, https://hub.jhu.edu/2018/03/01/ptech-students-get-college-experience/.

16 "Summer Scholars Program at Johns Hopkins Bayview," Johns Hopkins Medicine webpage, Johns Hopkins Bayview Medical Center, https://www.hopkinsmedicine.org/johns_hopkins_bayview/education_training/summer_scholars_program/.

17 Nitkin, Karen, "Creating Community Health Leaders," *Dome*, Aug. 26, 2015, https://www.hopkinsmedicine.org/news/articles/creating-community-health-leaders.

18 Pamela Paulk phone conversation with Karen Nitkin, October 21, 2021.

19 Paulk, Pamela D., "The Johns Hopkins Hospital Success in Hiring Ex-Offenders," Johns Hopkins Medicine report, September 2016, https://www.diversityincbestpractices.com/medialib/uploads/2016/09/Paulk-Presentation-Hiring-Ex-Offenders-09142016.pdf.

20 Quinton, Sophie, "Matching Ex-Offenders with Hard to Fill Health Care Jobs," *Stateline*, April 19, 2017, https://www.pewtrusts.org/en/research-and-analysis/blogs/stateline/2017/04/19/matching-ex-offenders-with-hard-to-fill-health-care-jobs.

21 Karen Nitkin telephone interview with Ronald Peterson, Jan. 10, 2022.

22 "Johns Hopkins Medicine Senior Vice President for Human Resources Honored as Champion of Change," Johns Hopkins Medicine media release, July 2, 2014, https://www.hopkinsmedicine.org/news/media/releases/johns_hopkins_medicine_senior_vice_president.

23 Diversity and Inclusion 2015 Annual Report, Johns Hopkins Medicine, https://www.hopkinsmedicine.org/diversity/_documents/diversity-annual-report2015.pdf.

24 "Women Gaining Access to Medical Education," Johns Hopkins Medicine webpage, History, https://www.hopkinsmedicine.org/about/history/women-med-ed.html.

25 "Hopkins GIM: What's New," Johns Hopkins Medicine webpage, General Internal Medicine: What's New, 2006, https://www.hopkinsmedicine.org/gim/news/2006_News_Items/1_31_06.html.

26 Edelson, Mat, "Homing in on Diversity," *Hopkins Medicine* magazine, Spring/Summer 2018 https://www.hopkinsmedicine.org/news/publications/hopkins_medicine_magazine/features/spring-summer-2018/homing-in-on-diversity.

27 Karen Nitkin Zoom interview with Paul Rothman, Nov. 23, 2021.

28 Rothman, Paul B., "Diversity in Medicine has Measurable Benefits," *Dome*, May 27, 2016, https://www.hopkinsmedicine.org/news/articles/diversity-in-medicine-has-measurable-benefits.

29 Nitkin, Karen, "The Rise of Female Surgeons at Johns Hopkins," *Dome*, Aug. 28, 2017, https://www.hopkinsmedicine.org/news/articles/the-rise-of-female-surgeons-at-johns-hopkins.

30 "Renowned Heart-Lung Transplant Expert Robert Higgins to Become Johns Hopkins' New Surgeon-in-Chief," Johns Hopkins Medicine media release, April 6, 2015, https://www.hopkinsmedicine.org/news/media/releases/renowned_heart_lung_transplant_expert_robert_higgins_to_become_johns_hopkins_new_surgeon_in_chief_.

31 "The Rise of Female Surgeons at Johns Hopkins."

32 Shapiro, Marc, "Robert Higgins Helps Set the Agenda in Diversity and Inclusion," *Dome*, Feb. 3, 2021, https://www.hopkinsmedicine.org/news/articles/robert-higgins-helps-set-the-agenda-in-diversity-and-inclusion.

33 Lagasse, Jeff, "Mass General Brigham names Robert Higgins president of Brigham and Women's Hospitals," *Healthcare Finance News*, Aug. 23, 2021, https://www.healthcarefinancenews.com/news/mass-general-brigham-names-robert-higgins-president-brigham-and-womens-hospitals.

34 "Andrew MacGregor Cameron, M.D., Ph.D.," Johns Hopkins Medicine webpage, https://www.hopkinsmedicine.org/profiles/details/andrew-cameron.

35 DuVernay, Christina, "Gender Equity Progress," *Dome*, Oct. 1, 2015, https://www.hopkinsmedicine.org/news/articles/gender-equity-progress.

36 "Gender Equity Progress."

37 "A Strong Advocate for Gender Equity," *Dome*, Sept. 16, 2011, https://www.hopkinsmedicine.org/news/publications/dome/september_2011/a_strong_advocate_for_gender_equity.

38 "Johns Hopkins Medicine at 25," Johns Hopkins Medicine webpage, https://www.hopkinsmedicine.org/25-anniversary/milestones.html.

39 "A Strong Advocate for Gender Equity."

40 "Johns Hopkins Medicine at 25."

41 Email from Barbara Fivush to Karen Nitkin, Dec. 12, 2021.

42 "Johns Hopkins Hospital Names First Female President in 127-Year History," Johns Hopkins Medicine media release, May 12, 2016, https://www.hopkinsmedicine.org/news/media/releases/johns_hopkins_hospital_names_first_female_president_in_127_year_history.

43 "Redonda Miller, M.D., M.B.A.," Johns Hopkins Medicine webpage, Leadership, https://www.hopkinsmedicine.org/about/leadership/biography/redonda-miller-md-mba.

44 Nitkin, Karen, "Q-and-A with Redonda G. Miller," *Dome*, June 28, 2016, https://www.hopkinsmedicine.org/news/articles/q-and-a-with-redonda-g-miller.

45 "Honoring Henrietta," Johns Hopkins Medicine webpage, The Legacy of Henrietta Lacks, https://www.hopkinsmedicine.org/henriettalacks/.

46 "The Immortal Life of Henrietta Lacks," Johns Hopkins Medicine webpage, The Legacy of Henrietta Lacks, https://www.hopkinsmedicine.org/henriettalacks/immortal-life-of-henrietta-lacks.html.

47 "Honoring Henrietta Lacks."

48 "Henrietta Lacks High School Day," Johns Hopkins Institute for Clinical and Translational Research webpage, https://ictr.johnshopkins.edu/community-engagement/programs/henrietta-lacks-high-school-day/.

49 "One Step Closer: Johns Hopkins Selects Architect for Early-Stage Planning of Multidisciplinary Building In Honor Of Henrietta Lacks," Johns Hopkins Medicine new release, Oct. 5, 2019, https://www.hopkinsmedicine.org/newsroom/news-releases/one-step-closer-johns-hopkins-selects-architect-for-early-stage-planning-of-multidisciplinary-building-in-honor-of-henrietta-lacks.

50 Hub staff report, "Johns Hopkins to Name Research Building in Honor of Henrietta Lacks," The Hub, Oct. 6, 2018, https://hub.jhu.edu/2018/10/06/henrietta-lacks-building-naming/.

51 "The Pathways Program," Johns Hopkins Department of Medicine webpage, Osler Medical Residency, https://www.hopkinsmedicine.org/medicine/education/hstrainingprogram/overview/pathways/index.html.

52 "The Pathways Program."

53 Minkove, Judy F., and Nitkin, Karen, "Guiding the Next Generation of Physician-Scientists at Johns Hopkins," Dome, Nov. 30, 2016, https://www.hopkinsmedicine.org/news/articles/guiding-the-next-generation-of-physician-scientists-at-johns-hopkins.

54 "Primary Care Leadership Track Rolls Out," BestPractice, Nov. 4, 2016, https://www.hopkinsmedicine.org/office-of-johns-hopkins-physicians/best-practice-news/primary-care-leadership-track-rolls-out.

55 "Primary Care Leadership Track Rolls Out."

56 "M.D. Curriculum," Johns Hopkins Medicine webpage, School of Medicine, https://www.hopkinsmedicine.org/som/education-programs/md-program/curriculum-and-degrees/md-curriculum.html.

57 "Primary Care Leadership Track Rolls Out."

58 "All Children's Hospital Celebrates 90 Years and Changes Name to Johns Hopkins All Children's Hospital," Johns Hopkins Medicine news release, April 5, 2016, https://www.hopkinsmedicine.org/news/media/releases/all_childrens_hospital_celebrates_90_years_and_changes_name_to_johns_hopkins_all_childrens_hospital.

59 Peterson, Ronald R., "A Tale of Two Hospitals," Dome column, February 2013, https://www.hopkinsmedicine.org/news/publications/dome/dome_february_2013/a_tale_of_two_hospitals.

60 "A Tale of Two Hospitals."

61 Nitkin, Karen, "First Residents Graduate from Johns Hopkins All Children's Hospital," Dome, July/August 2017, https://www.hopkinsmedicine.org/news/images_publications/johns_hopkins_dome_julyaugust2017.pdf.

62 "Our History," Johns Hopkins All Children's Hospital webpage, About Us, https://www.hopkinsallchildrens.org/About-Us/Our-History.

63 "Our History."

64 "Johns Hopkins All Children's Hospital Ranked #1 Children's Hospital in Florida," Johns Hopkins All Children's Hospital general news release, June 15, 2021, https://www.hopkinsallchildrens.org/ACH-News/General-News/Johns-Hopkins-All-Children-s-Hospital-Ties-for-Top.

65 Blum, Karen, "The Dover Decades, Hopkins Children's magazine, Summer 2016, https://www.hopkinsmedicine.org/johns-hopkins-childrens-center/_documents/_publications/hopkins_childrens_magazine_summer2016.pdf.

66 "The Dover Decades."

67 "At the Helm: David Hackam, Pediatric Surgeon-in-Chief, Co-Director Johns Hopkins Children's Center," At the Helm, Nov. 28 2017, https://www.hopkinsmedicine.org/news/articles/at-the-helm-david-hackam.

68 "Tina Lee Cheng," Johns Hopkins Medical Institutions webpage, Portrait Collection, https://portraitcollection.jhmi.edu/portraits/cheng-tina-lee.

69 "Margaret R. Moon, M.D., M.P.H.," Johns Hopkins Medicine webpage, profiles, https://www.hopkinsmedicine.org/profiles/details/margaret-moon.

70 "Year in Review 2017," Dome, Jan. 8, 2018, https://www.hopkinsmedicine.org/news/articles/year-in-review-2017.

71 Scharper, Julie, "Development Projects Bring New Energy to Baltimore's Remington Neighborhood," The Hub, Sept. 22, 2016, https://hub.jhu.edu/2016/09/22/remington-row-development-project/.

72 "Development Projects Bring New Energy to Baltimore's Remington Neighborhood."

73 "Johns Hopkins Medicine at 25."

74 "Johns Hopkins Medicine at 25."

75 "The Johns Hopkins Hospital Launches Capacity Command Center to Enhance Hospital Operations," Johns Hopkins Medicine news release, Oct. 26, 2016, https://www.hopkinsmedicine.org/news/media/releases/the_johns_hopkins_hospital_launches_capacity_command_center_to_enhance_hospital_operations.

76 "Covid-19 Story Tip: Johns Hopkins Capacity Command Center Boosts Hospital's Ability to Care for Patients During COVID-19 Pandemic," Johns Hopkins Medicine news release, March 10, 2021, https://www.hopkinsmedicine.org/news/newsroom/news-releases/covid-19-story-tip-johns-hopkins-capacity-command-center-boosts-hospitals-ability-to-care-for-patients-during-covid-19-pandemic.

77 "Johns Hopkins Opens New Center to Reduce Diagnostic Errors," Johns Hopkins Medicine news release, Oct. 31, 2016, https://www.hopkinsmedicine.org/news/media/releases/johns_hopkins_opens_new_center_to_reduce_diagnostic_errors.

78 "Johns Hopkins Opens New Center to Reduce Diagnostic Errors."

79 "Center for Diagnostic Excellence," Johns Hopkins Medicine webpage, Armstrong Institute for Patient Safety and Quality, https://www.hopkinsmedicine.org/armstrong_institute/centers/center_for_diagnostic_excellence/.

80 Nitkin, Karen; Broadhead, Lisa; Smith, Linell; Smith, Patrick, "No Room for Error," Dome, Jan. 8, 2016, https://www.hopkinsmedicine.org/news/articles/no-room-for-error.

81 "Johns Hopkins Establishes Armstrong Institute for Patient Safety and Quality," Johns Hopkins Medicine media release, May 26, 2011, https://www.hopkinsmedicine.org/news/media/releases/johns_hopkins_establishes_armstrong_institute_for_patient_safety_and_quality.

82 Kingsbury, Kathleen, "Peter Pronovost," Time magazine, The 2008 Time 100, May 12, 2008, http://content.time.com/time/specials/2007/article/0,28804,1733748_1733754_1735344,00.html.

83 Shapiro, Stephanie, "Striving for Consensus on Quality," Dome, Feb. 9, 2012, https://www.hopkinsmedicine.org/news/publications/dome/february_2012_/striving_for_consensus_on_quality_.

84 Clinical Communities, Johns Hopkins Medicine webpage, Armstrong Institute for Patient Safety and Quality, https://www.hopkinsmedicine.org/armstrong_institute/programs/get_involved/clinical_communities.html.

85 "Kachalia Promoted to Professor of Medicine," Johns Hopkins Medicine General Internal Medicine announcement, May 14, 2020, https://www.hopkinsmedicine.org/gim/news/2020_News_Items/05_14_20.html.

86 Nitkin, Karen, "Vice President Biden Speaks at Launch of Bloomberg-Kimmel Institute for Cancer Immunotherapy," Dome, March 29, 2016, https://www.hopkinsmedicine.org/news/articles/vice-president-biden-speaks-at-launch-of-bloombergkimmel-institute-for-cancer-immunotherapy.

87 Nitkin, Karen, "The Road to Hope," Dome, July/August 2016, https://www.hopkinsmedicine.org/news/articles/the-road-to-hope.

88 Nitkin, Karen, "Bringing Life, Hopkins Medicine magazine, Spring/Summer 2019, https://www.hopkinsmedicine.org/news/publications/hopkins_medicine_magazine/features/spring-summer-2019/bringing-life.

89 Nitkin, Karen, "The Organ Donor, the Surgeon, and a U.S. First for People Living with HIV," Dome, March 26, 2019, https://www.hopkinsmedicine.org/news/articles/the-donor-the-surgeon-and-the-doors-they-are-opening.

90 Smith, Patrick, "Making a Dent in Baltimore's Hepatitis C Epidemic," *Dome*, September 2016, https://www.hopkinsmedicine.org/news/articles/making-a-dent-in-baltimores-hepatitis-c-epidemic.

91 Christine Durand email to Karen Nitkin, Dec. 8, 2021.

92 "Johns Hopkins Greenberg Bladder Cancer Institute," case statement from Johns Hopkins Medicine, https://www.hopkinsmedicine.org/greenberg-bladder-cancer-institute/_docs/jhgbci%20case%20statement.pdf.

93 "Johns Hopkins Precision Medicine," Johns Hopkins Medicine webpage, https://www.hopkinsmedicine.org/precisionmedicine/.

94 "Johns Hopkins Greenberg Bladder Cancer Institute."

95 "Johns Hopkins Research Focused on Early Stage Bladder Cancer Wins $3.2 Million Federal Grant," Johns Hopkins Medicine news release, July 1, 2019, https://www.hopkinsmedicine.org/news/newsroom/news-releases/johns-hopkins-research-focused-on-early-stage-bladder-cancer-wins-32-million-federal-grant.

96 Diversity and Inclusion 2017 Annual Report, Johns Hopkins Medicine, https://www.hopkinsmedicine.org/diversity/_assets/documents/annual-reports/2017/dar_awa_web.pdf.

97 "Fan Liang, M.D.," Johns Hopkins Medicine webpage, Center for Transgender Health, https://www.hopkinsmedicine.org/center-transgender-health/fan-liang.html.

98 Nitkin, Karen, "The Joy of Being Yourself," *Dome*, Nov. 28, 2018, https://www.hopkinsmedicine.org/news/articles/the-joy-of-being-yourself.

99 "HHS Acting Secretary Declares Public Health Emergency to Address National Opioid Crisis," U.S. Department of Health and Human Services news release, Oct. 26, 2017, https://public3.pagefreezer.com/browse/HHS.gov/31-12-2020T08:51/https://www.hhs.gov/about/news/2017/10/26/hhs-acting-secretary-declares-public-health-emergency-address-national-opioid-crisis.html.

100 "Study Adds to Evidence That Most Prescribed Opioid Pills Go Unused," Johns Hopkins Medicine news release, Aug. 2, 2018, https://www.hopkinsmedicine.org/news/media/releases/study_adds_to_evidence_that_most_prescribed_opioid_pills_go_unused_.

101 Nitkin, Karen, "Health System Leaders Work Together for Better Opioid Stewardship," *Dome*, March 1, 2018, https://www.hopkinsmedicine.org/news/articles/health-system-leaders-work-together-for-better-opioid-stewardship.

102 Nitkin, Karen, "Treating Pain, Averting Addiction," *Dome*, March 1, 2018, https://www.hopkinsmedicine.org/news/articles/treating-pain-averting-addiction.

103 Nitkin, Karen, "Reverting to Physical Activity," *Dome*, Feb. 9, 2018, https://www.hopkinsmedicine.org/news/articles/returning-to-physical-activity.

104 "Treating Pain, Averting Addiction."

105 Email from Marie Hanna to Karen Nitkin, Dec. 16, 2021.

106 Email from Beth Hogans to Karen Nitkin, Dec. 20, 2021.

107 "Intensive Treatment Unit (ITU)," Johns Hopkins Medicine webpage, Psychiatry and Behavioral Sciences, https://www.hopkinsmedicine.org/psychiatry/specialty_areas/substance_abuse/intensive_treatment_unit.html.

108 "Motivated Behaviors Unit," Johns Hopkins Medicine webpage, Psychiatry and Behavioral Sciences, https://www.hopkinsmedicine.org/psychiatry/specialty_areas/substance_abuse/motivated_behaviors_unit.html.

109 "Broadway Center for Addiction," Johns Hopkins Medicine webpage, Psychiatry and Behavioral Sciences, https://www.hopkinsmedicine.org/psychiatry/specialty_areas/substance_abuse/broadway_center_for_addiction.html.

110 "Addiction Treatment Services (ATS)," Johns Hopkins Medicine webpage, Psychiatry and Behavioral Sciences, https://www.hopkinsmedicine.org/psychiatry/patient_information/bayview/medical_services/substance_abuse/addiction_treatment_service.html.

111 "Center for Addiction and Pregnancy (CAP), Johns Hopkins Medicine webpage, Psychiatry and Behavioral Sciences, https://www.hopkinsmedicine.org/psychiatry/patient_information/bayview/medical_services/substance_abuse/center_addiction_pregnancy.html.

112 "Addiction Medicine Unit," Johns Hopkins Medicine webpage, Psychiatry and Behavioral Sciences, https://www.hopkinsmedicine.org/johns_hopkins_bayview/medical_services/addiction_medicine/addiction_medicine_unit.html.

113 "Mental Illness and Substance Abuse (MISA), Johns Hopkins Medicine webpage, Psychiatry and Behavioral Sciences, https://www.hopkinsmedicine.org/psychiatry/patient_information/bayview/medical_services/substance_abuse/MISA.html.

114 "A Medical Model to Treat Opioid Addiction," 2016 Diversity Annual Report, Johns Hopkins Medicine, https://www.hopkinsmedicine.org/diversity/annual-reports/2016-annual-report/a-medical-model-to-treat-opioid-addiction.

Kevin Sowers

Chapter Four

The Kevin Sowers Years (2018–)

N FEB. 1, 2018, KEVIN W. SOWERS became the second person in history to serve as president of the Johns Hopkins Health System and executive vice president of Johns Hopkins Medicine.

Sowers, a clinical nurse by training, had been with the Duke University Health System for 32 years.

"We really wanted to continue to expand the integration of the health system, and he'd done a lot of that work at Duke," recalled Paul B. Rothman, dean of the medical faculty and CEO of Johns Hopkins Medicine. "In the end, that was what made Kevin stand out. He really valued all parts of the mission and he had experience with integration."[1]

Bringing those skills and vision to Johns Hopkins, Sowers, in collaboration with school of medicine and health system leadership, created a clinical road map that would expand outpatient offerings and strengthen referrals to specialized care.

He also decisively steered the health system through a crisis at the Johns Hopkins All Children's Heart Institute and, of course, the COVID-19 pandemic (discussed in detail in Chapter Five), while joining Rothman in celebrating milestones, including a Nobel Prize win for genetic medicine professor Gregg Semenza.

Kevin Sowers Joins Johns Hopkins

After 44 years at Johns Hopkins, Ronald R. Peterson retired at the end of 2017, becoming president emeritus of the Johns Hopkins Health System and special adviser to the dean/CEO of Johns Hopkins Medicine for one year.

For months, a search committee led by Rothman and David Hodgson, chair of the Johns Hopkins Medicine board of trustees, had conducted an extensive national quest for the next person to fill one of the top posts in academic medicine.

The search led them to Sowers, who joined Johns Hopkins Medicine from the Duke University Health System, the last eight years as president and CEO of Duke University Hospital.[2]

Sowers had earned his bachelor of science degree from Capital University School of Nursing and a master of science from Duke University School of Nursing. He began his career with Duke University Hospital in 1985 as a staff nurse in oncology.

Ronald R. Peterson's Legacy of Leadership

Ronald R. Peterson, a native of Edison, New Jersey, came to The Johns Hopkins University as a premed student in 1966. When he graduated four years later, family circumstances compelled him to get a job rather than pursue a medical education.

After working as a city school science teacher and as a part-time employee of a local kosher caterer, Peterson in 1972 entered a George Washington University graduate program in health care administration. When the master's program gave him the opportunity for an administrative residency at The Johns Hopkins Hospital in 1973, he jumped at the chance.*

He remained at Johns Hopkins the rest of his career, his ascent fueled by his remarkable mastery of the minutest details, coupled with his big-picture vision for the institution and how it could best serve patients and the community.†

After completing the administrative residency, Peterson was hired as administrator in the Johns Hopkins Department of Psychiatry. He then held several staff roles before becoming administrator of the then-financially challenged Johns Hopkins Children's Center in 1978.

Over the years, Peterson became known as a "fiscal surgeon" for his ability to manage financially challenged organizations, including the former Baltimore City Hospitals, which he dramatically revived, updating the physical campus and transforming a $7 million loss per year into a healthy bottom line. The hospital, now known as Johns Hopkins Bayview Medical Center, was acquired by Johns Hopkins in 1984 with Peterson named president.

In 1995, Peterson became executive vice president and chief operating officer of the health system. A year later, he was officially named president of The Johns Hopkins Hospital and Health System, a role he held until July 2016.

At the hospital, Peterson oversaw the opening of a new comprehensive cancer center and undertook one of the most expansive private building campaigns in Maryland history, culminating in two new state-of-the-art hospital towers for adult and pediatric patients.‡

At the health system, he was responsible for the acquisition of Howard County General Hospital, Sibley Memorial Hospital, Suburban Hospital and All Children's Hospital. On his watch, Johns Hopkins Community Physicians, which he chaired, grew logarithmically. Significant emphasis was also placed on the development of major ambulatory health and surgery centers, most notably at Green Spring Station and White Marsh.§

"While Ron has a talent for streamlining operations and consummating bold strategic projects, even more striking is the way he cares for people," wrote Rothman in a letter announcing Peterson's retirement.

"He's the type of leader who is not motivated by individual ambition but rather a desire to be the very best steward — of our tradition of excellence, our collegial community, and of course, our mission to improve lives.

"In a 2009 interview, Ron described his role this way: 'What we do is help to create the environment that enables the great, bright clinicians and scientists and staff to carry on their important work each and every day.' Year after year, with his intelligence and his trademark equanimity, he not only has built that supportive foundation but he has insulated the people of Johns Hopkins Medicine from challenging outside forces so that each of us could focus our talents on what matters most.

"For that, we all owe him a debt of gratitude."¶

* Grauer, Neil A., "A Hopkins Person Beyond Compare," *Dome*, Nov. 8, 2017, https://www.hopkinsmedicine.org/news/articles/a-hopkins-person-beyond-compare.

† "A Hopkins Person Beyond Compare."

‡ https://www.hopkinsmedicine.org/news/publications/hopkins_medicine_magazine/circling_the_dome/fall-2017/fiscal-surgeon-ronald-r-peterson-to-retire.

§ Karen Nitkin phone interview with Ron Peterson, Feb. 24, 2022.

¶ Rothman, Paul B., "Ronald R. Peterson to Retire as President of the Johns Hopkins Health System and EVP of Johns Hopkins Medicine," Dear Colleagues letter to Johns Hopkins Medicine community, June 30, 2017, https://www.insidehopkinsmedicine.org/news/leadershipcorner/2017/ronald-r-peterson-to-retire-as-president-of-the-johns-hopkins-health-system-and-evp-of-johns-hopkins-medicine.cfm.

He held several key nursing positions, including nurse educator, director of medical oncology and nurse internships, and director of a cancer care consortium and unit manager for hematology, oncology and GYN oncology. He later served as chief operating officer for Duke University Hospital and interim CEO for Durham Regional Hospital.

As Duke University Hospital's president, Sowers oversaw operations of the 1,000-bed teaching hospital, including implementation of the strategic plan, in collaboration with the physician practice group and school of medicine.

He also led several major building projects for the hospital, including an expansion of the emergency department, a modernization of operating rooms, a new cancer center and an ICU bed tower. His faculty posts included adjunct professor at the University of North Carolina Gillings School of Global Public Health and Duke University Fuqua School of Business, and clinical associate at Duke University's Graduate School of Nursing.

He served as a consultant to the Sun Yat-sen University Cancer Center in Taiwan, establishing a department of oncology nursing and educating nurses on oncology care.

Sowers was not planning to leave Duke when, one day in late 2017, he got a phone call from Rothman, asking if the two could meet for a meal in Washington, D.C.

"It was a fun lunch," Sowers recalled in 2021. "He really wanted to get to know me and spent most of the time asking questions about where I grew up and why I became a nurse. At the end of our meal he said, 'Let me tell you about the job.'"

Sowers didn't think he wanted the job, but he agreed to go to New York to meet with Hodgson, managing director of General Atlantic Partners, an international equity firm. "This person is an incredible board chair," said Sowers. "I just thought this is someone really connected to the mission and passion of Hopkins."

As he continued the recruitment process and got to know more people at Johns Hopkins Medicine, Sowers grew even more impressed with their devotion to the organization's tripartite mission of research, education and clinical care.

"I was looking for someone on the same page on issues and who also really knew how to run hospitals and health systems," said Rothman in 2022.

When the post was offered, Sowers accepted it. His partner, Anthony Evans, soon joined him in Baltimore.

"I've never seen a group so committed to the mission of this place," Sowers said in a 2021 interview. "These are people who desperately care about making a positive impact for our patients and our community. It was very authentic and it's a wonderful culture. You could feel it. At four years in, I still think it's a wonderful, magical place."[3]

When Sowers joined Johns Hopkins Medicine, he took the helm of an $8 billion academic medical center and health system with around 40,000 employees. His background was well-suited to the task ahead of him, defining and executing strategies to advance all three parts of the Johns Hopkins Medicine mission — delivering outstanding care, training the next generation of leaders, and advancing research and discovery.[4]

"Kevin is a visionary leader with a deep understanding of the role and responsibilities of academic and community health centers, as well as the challenges we face," Rothman said when he announced the hire.

"With his experience leading a prestigious academic health center, his ability to forge strong collaborative relationships, and his demonstrated knack for thriving in complex environments, Kevin will be instrumental in advancing our mission and reinforcing our commitment to the communities we serve."[5]

A New Clinical Road Map Charts a Strategic Course

In 2018, as Sowers settled into his job, he saw an opportunity to create a bold clinical road map — one that would position the health system for a robust future in a rapidly changing care environment characterized by an increased ability to care for patients in an outpatient setting.

Johns Hopkins Community Physicians opened a practice in McLean, Virginia, in February 2022.

The idea, he said in a 2021 interview, was to create an integrated care delivery system that would "sustain referral patterns, complement our current assets and build on the promise of the brand," while giving patients more opportunities to access outpatient care in convenient locations.[6]

Areas under consideration for development included Maryland's Prince Georges and Montgomery counties; Washington, D.C.; northern Virginia; and southern Pennsylvania.

Creating a strong ambulatory presence in those places would help secure a steady stream of tertiary referrals to The Johns Hopkins Hospital, Sowers said, while also freeing up hospital beds for cases requiring specialized care.[7]

"For close to 130 years, what has made Johns Hopkins Medicine has been our ability to take the most complex cases and apply our clinical expertise to caring for those patients," he said. "It has allowed our learners access to learn about the disease and its treatment while also allowing us to research those disease processes to try to improve outcomes for the next generation."[8]

Working with the boards of Sibley Memorial and Suburban hospitals, Sowers spurred creation of a National Capital Region (NCR) Executive Governance Committee with board members from Sibley and Suburban, as well as Johns Hopkins Medicine leaders, to oversee JHM's clinical strategy in that region — a move that advanced the governance and administrative structures to acknowledge the development of a region and installed the NCR's first president.

Carolyn Carpenter, who started the newly created job on Aug. 24, 2020, would lead all hospital-based and community-based ambulatory sites across the NCR region.

Carpenter brought 25 years of health care administration leadership experience to her new role, most recently as president of Sentara Norfolk General Hospital and corporate vice president for Sentara Healthcare in Virginia. Before that, she was with Duke University Health System for 20 years, in roles including chief operating officer for Duke University Hospital, associate dean and administrator for the Duke Cancer Institute, and associate vice president of Duke Health.

Carolyn Carpenter

Leadership Changes at Sibley and Suburban

Working closely with Carolyn Carpenter were new leaders at both Sibley Memorial Hospital and Suburban Hospital.

Hasan A. Zia became president and chief operating officer for Sibley in December 2020, after serving as interim president since October 2018. Before that, he had been vice president of medical affairs and chief medical officer, director of critical care and emergency surgery, and patient safety officer and medical director for quality and patient safety for the hospital. He also led implementation of the Epic electronic medical record at Sibley and Howard County General Hospital.

Zia earned his medical degree from the University of Virginia and did his surgical residency at George Washington University. He completed a fellowship in trauma surgery and surgical critical care at UVA. He is a fellow of the American College of Surgeons, and in 2017 he earned an M.B.A. in health care management from the Johns Hopkins University Carey Business School.*

Jessica Melton

Hasan A. Zia

Jessica Melton became president and chief operating officer of Suburban Hospital on April 15, 2021. The first African American to become a Johns Hopkins president, Melton joined Johns Hopkins from Sentara Norfolk General Hospital in Virginia, where she was the chief operating officer. Before that, she held leadership roles at Duke University Hospital in Durham, N.C., including vice president of medical, surgical and critical care services; vice president of emergency services and patient flow; administrative director of emergency preparedness and trauma; and director of the Duke Preparedness and Response Center.

Melton received both a bachelor of science in biology and a master of healthcare administration with concentrations in finance, community preparedness

and disaster management from the University of North Carolina at Chapel Hill. She completed an administrative fellowship at Duke University Hospital, and received Six Sigma training and certification as an executive coach through North Carolina State University. She is a member of the National Association of Health Services Executives, the American College of Healthcare Executives and the Delta Omega Honorary Society in Public Health.

Melton was nationally recognized as a 2018 Up & Comer by *Modern Healthcare* magazine and was named a Woman to Watch in Healthcare Leadership in 2016 by Deloitte. She was in the second class of Carol Emmott Fellows, which graduated in 2018.†

---

*   "Hasan Zia Named President and Chief Operating Officer of Sibley Memorial Hospital," Johns Hopkins Medicine news release, Dec. 7, 2020, https://www.hopkinsmedicine.org/news/newsroom/news-releases/hasan-zia-named-president-and-chief-operating-officer-of-sibley-memorial-hospital.

†   "Suburban Hospital Names Jessica Melton President and Chief Operating Officer," Johns Hopkins Medicine news release, March 10, 2021, https://www.hopkinsmedicine.org/news/newsroom/news-releases/suburban-hospital-names-jessica-melton-president-and-chief-operating-officer.

She was chosen for the inaugural 2017 class of the competitive, national Carol Emmott Fellowship, which is dedicated to development and promotion of female health care executives.[9]

"The clinical road map has been critically important in helping us set the path for clinical growth," she said during a July 2021 video conversation with Sowers. "We've organized it around building on our strengths that exist already in the national capital region, making sure we meet the core service needs of community hospitals, as well as understanding what our patients and community need from specific services."[10]

In February 2022, a new 8,600 square-foot Johns Hopkins Community Physicians practice opened in McLean, Virginia. The facility, with 12 exam rooms, houses a mix of adult primary care, subspecialty and ancillary services, such as lab services, EKGs and COVID-19 testing.[11]

Belward Farm

A larger project, on a Montgomery County site called Belward Farm, was finally moving forward: The 126,200-square-foot Johns Hopkins Medical Office & Surgery Center was expected to open at the site in 2024.[12]

The Johns Hopkins University had acquired the 138-acre site in 1989, and won Montgomery County approval in 1996 to develop a research campus at the location.[13] But the project was mired for years in legal disputes and changed plans.

Elizabeth Banks and her siblings, Rolland Banks Jr. and Beulah Newell, whose family had farmed the land since 1873, sold the property to Johns Hopkins for $5 million, far below market value, to use, as the contract stated, "for agricultural, academic, research and development, delivery of health and medical care and services, or related purposes only."

In 2011, the university sought permission to change its plans to a much larger proposal with 23 buildings totaling 4.7 million square feet. The new plan was approved, but it drew a legal challenge from Banks family members and community residents.[14]

In 2012, a Montgomery County court sided with Johns Hopkins,[15] a decision affirmed a year later by the Maryland Court of Special Appeals.[16]

The current plan is much smaller, calling for a first phase of a single 71-foot-high building, parking facilities, public space and other amenities including a cafe, on 11 acres of the property.[17]

The ambulatory facility was expected to house primary care, cardiology, digestive services, urology, medical oncology, radiation oncology, anesthesia, gynecology (fertility services), general surgery (breast), otolaryngology, orthopaedics, rheumatology, physical medicine and rehabilitation, neurosurgery, ophthalmology and imaging.[18]

The CMSC/North Tower Addition

In 2019, the health system unveiled the largest construction project on Johns Hopkins' medical campus since the Sheikh Zayed Tower and the Charlotte Bloomberg Children's Center opened in 2012.

The multiyear project, slated for completion in June 2026, would renovate the former Johns Hopkins Hospital Children's Medical and Surgical Center (CMSC) and construct a 12-story tower on the site of the former Brady Building, a seven-story structure that dated to 1915.[19]

The project, known as the CMSC/North Tower Addition, would have 440,770 square feet of new and renovated space, including 200,000 square feet of research labs and related areas, providing much-needed space to house state-of-the-art research labs. It would also add a new entrance to The Johns Hopkins Hospital from Monument Street.[20]

Johns Hopkins Regional Physicians

Johns Hopkins Regional Physicians (JHRP), a care model created in 2017, also helped bring more community practices into the Johns Hopkins family.

A Third Pavilion at Green Spring Station

After years of planning and construction, the three-story, 110,000-square-foot Pavilion III at Green Spring Station — the Johns Hopkins Health Care & Surgery Center — opened for its first patients in May 2019.

The light-splashed brick building in the Baltimore suburb of Lutherville is part of an upscale development of shops and restaurants that already housed two Johns Hopkins outpatient care pavilions.

The third Johns Hopkins pavilion at Green Spring Station has six operating rooms, space for oncology treatment and imaging, and a musculoskeletal center that combines orthopaedics and physical medicine and rehabilitation.*

It meant Johns Hopkins could increase the number of patient encounters at Green Spring Station from 550,000 to more than 750,000 per year, said Gill Wylie,

president of Johns Hopkins Medical Management.

Pavilion III opened almost exactly 25 years after Johns Hopkins cut the ribbon on Pavilion I in 1994, and five years after the institution purchased the land, which previously housed tennis facilities. Construction began in 2017.†

Together, the three Johns Hopkins buildings at Green Spring Station offer a full range of medical services, including cardiology, dermatology, ophthalmology, pediatrics, primary care and psychology.‡

In many respects, the opening of Pavilion III marked the culmination of the extraordinary career of Wylie, who was instrumental in developing Pavilion I and ultimately all of the major ambulatory health and surgery facilities for Johns Hopkins from 1994 through his retirement in 2021.§

Gill Wylie

---

* Nitkin, Karen, "A Final Flurry, then Pavilion III Opens," *Dome*, May 24, 2019 https://www.hopkinsmedicine.org/news/articles/a-final-flurry-then-pavilion-iii-opens.
† "A Final Flurry, then Pavilion III Opens."
‡ "Specialists and Medical Services," webpage, Johns Hopkins Health Care & Surgery Center – Green Spring Station, Lutherville, https://www.hopkinsmedicine.org/patient_care/locations/green_spring/medical_professionals.html.
§ Karen Nitkin phone interview with Ronald Peterson, Feb. 24, 2022.

Under the JHRP system, affiliated community practices retain considerable autonomy while benefiting from association with the Johns Hopkins brand.

Physicians and staff members become employees of Johns Hopkins Regional Physicians, but the practices have their own compensation and benefit plans and retain their original names and identities, adding the tagline "a member of Johns Hopkins Regional Physicians."

JHRP has a different model than Johns Hopkins Community Physicians, which is comprised of practices that are wholly part of Johns Hopkins Medicine. With JHRP, participating practices establish a relationship with the corresponding school of medicine department for their specialty.

A team from the department provides clinical oversight, and JHRP practices abide by quality, safety and service standards established through the Johns Hopkins Armstrong Institute for Patient Safety and Quality. Practices also receive support for billing, compliance reporting and other functions.[21]

"The concept was developed because Johns Hopkins leaders wanted a new employment model as an alignment strategy for independent groups," said Kim Sherbrooke, chief operating officer of the Office of Johns Hopkins Physicians (OJHP).[22] The first such collaboration was with Cardiovascular Specialists of Central Maryland, a group of cardiologists affiliated with Howard County General Hospital; the second was with ENTAA Care, a 13-physician otolaryngology and allergy practice in Maryland with six locations: Columbia, Annapolis, Laurel, Glen Burnie, Kent Island and Odenton.

"Patients get convenient, timely care and access to Hopkins expertise," said Sherbrooke. "Affiliated providers enjoy association with a premium brand and expert back-end support with billing, compliance and reporting. And we get a wider referral network and greater opportunities to further research and teaching in the community without incurring financial risk."[23]

Ambulatory Surgery Centers

The growth in ambulatory surgery at Johns Hopkins prompted a structural change.

Johns Hopkins Medicine opened its first ambulatory surgery center (ASC) in 2005, providing outpatient eye surgeries in the Wilmer Eye Institute location at Green Spring Station in Lutherville, Maryland.

By 2018, Johns Hopkins had six ASCs in Maryland — including one in Bel Air that had been acquired by Wilmer, one in Bethesda that was owned by Suburban Hospital, and one in Columbia that was a partnership with the Johns Hopkins Division of Gastroenterology & Hepatology clinic. The others, owned by Johns Hopkins Medicine, are in White Marsh and Columbia.

At the time, each center had its own employment practices, financial systems and other protocols. Lisa Ishii, otolaryngologist and senior vice president of operations for the health system, saw an opportunity to create a better system.

"I said, 'We have six ambulatory surgery centers and three are in one group and three are independent,'" she recalled. "We should probably harmonize and start acting strategically."[24]

Working with Sowers, Rothman and other executive leaders, Ishii led creation of and became president of the Johns Hopkins Surgery Center Series (JHSCS).

The new organization purchased the ASCs owned by Wilmer and Suburban so that all six were jointly owned by the Johns Hopkins Health System and The Johns

The Johns Hopkins Health Care & Surgery Center at White Marsh, shown here, is part of the Johns Hopkins Surgery Center Series, which was created to establish a single organizational governance structure and harmonized policies across the institution's ambulatory surgery centers.

Hopkins University. It created a single organizational governance structure with harmonized policies.[25]

Since that time, a second Johns Hopkins ASC opened in Green Spring Station, and another is planned for the Belward Farm site in Montgomery County.

"We created JHSCS because we recognized that it made a lot more sense for us to manage our ASCs together in a centralized, integrated way," Ishii said. "We also recognized that this approach would help us more effectively expand our surgery center network, which is important because we see ASCs as a significant growth opportunity for us as a health system."

ASCs have become an integral part of the Johns Hopkins strategy to care for patients across the entire continuum of care, Ishii said. "Where health systems and health care in general used to focus mostly on hospitals, we are all now appreciating the benefit of patients having access to our providers in different settings, be they ASCs, ambulatory clinics or even in the home. From the Johns Hopkins perspective, it was appreciating that there are patients and procedures for which an ASC is the most appropriate setting and therefore we should be able to offer that as part of our portfolio of services."[26]

Lisa Ishii

Expansion at Suburban Hospital

Suburban Hospital, in Bethesda, Maryland, broke ground in 2017 for a 300,000-square-foot clinical building featuring technological advances in surgery, conveniently located specialty physician offices and private patient rooms. When the North Building opened in March 2020, it modernized the entire hospital campus and created better flow for patients, families and clinicians.

The North Building, which connects to the original structure by a concourse, nearly doubled the size of the hospital while keeping its bed count at 228 as the entire hospital transitioned to private patient rooms. (Previously, about half were semiprivate.)

The new building has 14 operating rooms, two catheterization labs, a procedure room for endoscopies and bronchoscopies, and a procedure room equipped with imaging equipment. A new hybrid operating room has specialized imaging equipment for use during highly precise vascular and structural heart procedures. Pre-op rooms and recovery bays are also new.

Some of the changes were aimed at enhancing access to lifesaving care, such as creating a dedicated driveway entrance for emergency and trauma vehicles, and placing it as close as possible to operating rooms. Others are less dramatic, such as color-coded patient floors to improve wayfinding and a new prayer and meditation area that's large enough for 12 people.

"Years of work are finally being realized," said hospital president Jacky Schultz. "This is what the community wants. They want to come to the hospital to be in a fresh, bright, private space."

The North Building features a new main entrance with a

Suburban Hospital's entrance and North Building, which opened in March 2020.

soaring, window-filled atrium. Stairwells next to elevators are well-lit and attractive to encourage people to take the stairs when possible. Outside is a sculpture titled "Interplay" by the late artist John Safer, whose works also welcome visitors at Sibley Memorial Hospital and the Wilmer Eye Institute at The Johns Hopkins Hospital.

Other features in the North Building include a medical unit that can be converted to an intensive care unit, a conference room that can become an incident command center, and a fourth floor equipped with a separate ventilation system to isolate patients with infections in the event of a pandemic. The building's flexible design proved valuable immediately, as the COVID-19 pandemic dramatically changed hospital care just as the new building opened (more on COVID in Chapter Five).

Suburban Hospital opened in 1943 and joined the Johns Hopkins Health System in 2009. The North Building marked its first major update since 1979, said Schultz. The work was done by Clark Construction Group, which also built the 1.6-million-square-foot Charlotte R. Bloomberg Children's Center building and the Sheikh Zayed Tower at The Johns Hopkins Hospital in 2011. The architect was Wilmot Sanz.

The addition was made possible in part by $65 million in philanthropic support, including a $10 million investment from the A. James & Alice B. Clark Foundation, continuing a long history of generosity by the Clark Foundation to Suburban Hospital and to Johns Hopkins Medicine and University.[27]

Crisis at All Children's

Less than a year into his tenure, Sowers confronted a crisis.

On Nov. 28, 2018, the *Tampa Bay Times* published a devastating investigative article indicating that the Johns Hopkins All Children's Heart Institute had an unusually high mortality rate.

"The internationally renowned Johns Hopkins had taken over the St. Petersburg [Florida] hospital six years earlier and vowed to transform its heart surgery unit into one of the nation's best," the article said. "Instead, the program got worse and worse."[28]

The mortality rate had tripled since 2015 to become the highest of Florida's 10 pediatric heart programs in a decade. In one 18-month period, at least 11 patients died after operations by the hospital's two principal heart surgeons. Other children suffered life-changing injuries.

"Johns Hopkins' hand-picked administrators disregarded safety concerns the program's staff had raised as early as 2015," said the article. "It wasn't until early 2017 that All Children's stopped performing the most complex procedures. And it wasn't until late that year that it pulled one of its main surgeons from the operating room."[29]

The stunning news brought a decisive response from Johns Hopkins leaders, particularly Sowers. "I am a firm believer that from the day you are made aware of an untoward event, you will be evaluated on what steps you took," he said in 2021. "I just did what I had to do."[30]

The pediatric cardiac program was suspended, and Sowers moved to St.

Petersburg for four months and quickly cleaned house, accepting resignations in early December from the hospital's CEO and other leaders.[31]

"He [Sowers] was very new but he went down there and straightened the ship," said Rothman in 2022. "It was amazing."[32]

Sowers took the role of interim president of the hospital until Thomas D. Kmetz, the former president of Norton Children's Hospital in Louisville, Kentucky, came aboard on Feb. 18, 2019.[33]

Other All Children's leaders took on new responsibilities to advance a culture of safety. George Jallo, medical director of the Institute for Brain Protection Sciences and chief of pediatric neurosurgery, accepted the roles of interim vice dean and physician-in-chief. Melissa Macogay served as vice president and chief nursing officer. Angela Green, senior director of patient safety and quality, became interim chief of patient safety and quality. Joseph Perno, vice chair of the Department of Medicine, became vice president of medical affairs.[34]

Sowers forthrightly and solemnly discussed the situation in a four-minute video to the All Children's community, released Jan. 8, 2019.

"We owe an apology to the families and patients who did not receive the high-quality care that is our promise to all patients. I am deeply sorry for their tragic losses," he said, also extending apologies to the community, to referring providers, and to the All Children's staff.

"To move forward, we must acknowledge our mistakes," he said. "I know personally that many of you courageously spoke out when you had concerns but were ignored or turned away. That behavior is unacceptable and will not be tolerated going forward."

Sowers then outlined steps Johns Hopkins Medicine would take to rectify the problems, including commissioning a monitor to track the hospital's progress on implementing suggested reforms, and consulting with a team of pediatric heart experts on how best to safely restart the Heart Institute.

Meanwhile, the Johns Hopkins Medicine board of trustees hired the law firm Gibson, Dunn & Crutcher LLP for an external review that would identify deficiencies at All Children's and the broader Johns Hopkins Health System.

"You and I must commit to a culture where we are all empowered to speak up and speak out," Sowers concluded in the video. "Johns Hopkins Medicine remains deeply proud of and deeply committed to All Children's Hospital and to you, the dedicated professionals who work tirelessly every day to give hope and healing to our patients and their loved ones."[35]

Gibson Dunn's report, released June 26, 2019, was posted on the hospital's website and included specific recommendations such as empowering physician leaders to ensure their patient safety concerns are heard and addressed; and using data more extensively as a compliance tool and to ensure transparency of clinical outcomes.[36]

On Feb. 15, 2020, James Quintessenza returned to the hospital as chief of cardiovascular surgery and co-director of All Children's Heart Institute. He had been at All Children's for 26 years, including 19 as medical director and chief of pediatric cardiac surgery, leaving in 2016 to take a job as chief of pediatric cardiothoracic surgery at Kentucky Children's Hospital in Lexington.[37]

"This marks the first step in a phased process of restarting the Heart Institute in line with the recommendations of the external experts," the hospital said in a press release announcing Quintessenza's return.[38]

In May 2020, All Children's resumed performing cardiac surgery procedures, gradually working its way up the complexity scale. By April 2021, it had completed more than 100 cardiac operations since the program restarted. Another major stride forward came on Aug. 14, 2020, when the Centers for Medicare and Medicaid Services (CMS) notified All Children's that it had fulfilled the requirements of a systems improvement agreement that it had entered with the hospital in June 2019.

"The purpose of this agreement was to identify gaps and implement sustainable systemic changes to ensure the hospital meets all CMS Conditions of Participation," explained the hospital in a statement. "After a yearlong process, we are pleased to share that CMS completed their review and found that we fully met the Conditions of Participation and remain in good standing to serve pediatric patients."[39]

On July 26, 2021, All Children's welcomed a new president: K. Alicia Schulhof, who had most recently been president of Peyton Manning Children's Hospital in Indianapolis. Schulhof had previously spent more than 17 years leading complex health care organizations, including Indiana University Health and HCA Healthcare.

"Her stellar track record of health care leadership and her keen strategic vision

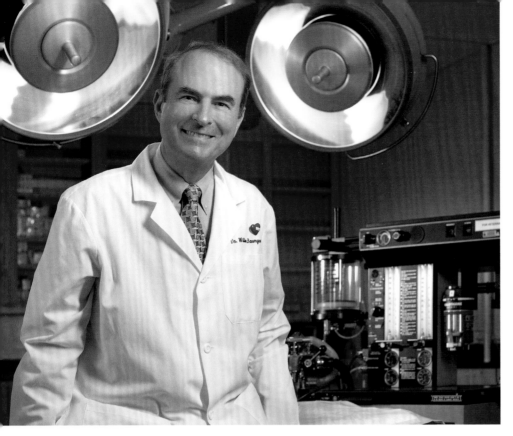

William Baumgartner

make her an extraordinary choice," said Lawrence Repar, chair of the All Children's board of trustees. "Above all, her patient-centric approach ensures that Johns Hopkins All Children's Hospital will continue our legacy of delivering hope and healing to our patients and their families."[40]

Further affirmation that the Florida hospital had emerged from its darkest chapter came with the 2021–2022 rankings from *U.S. News & World Report*, which put All Children's in a tie for best pediatric hospital in Florida.[41]

William Baumgartner Retires

After 36 years of exceptional service to Johns Hopkins Medicine, pioneering cardiac surgeon William Baumgartner retired at the end of 2018.

"One of the most rewarding elements of a career at Johns Hopkins is working with colleagues whose achievements and commitment exemplify excellence," Rothman said. "Bill Baumgartner, both as a cardiac surgeon and through his many clinical leadership roles, has been such a colleague."

Baumgartner arrived at The Johns Hopkins Hospital in 1982 as an assistant professor of cardiac surgery. He re-established Johns Hopkins' heart transplant program and made the hospital a nationally acclaimed center for the surgical treatment of heart failure.

In 1987, Baumgartner and cardiac surgeon Bruce Reitz led a Johns Hopkins medical team in the nation's first successful "domino-donor" three-way transplant surgery.

Baumgartner became a full professor of surgery, served as director of the heart and heart-lung transplant program and was named chief of cardiac surgery. In 1999, he was appointed vice dean of clinical affairs and president of the Clinical Practice Association, which represents 1,700 full-time faculty members.

In 2000, he was named the Vincent L. Gott Professor of Cardiac Surgery, and in 2011, he became senior vice president of Johns Hopkins Medicine's Office of Johns Hopkins Physicians.

A prolific researcher, Baumgartner served as director of the Cardiac Surgery

Research Lab for more than 25 years, sustaining continuous National Institutes of Health funding and authoring more than 375 publications. He mentored more than 45 residents who have become leaders in the field.[42]

Penis and Scrotum Transplant

In March 2018, surgeons at The Johns Hopkins Hospital performed the world's first total penis and scrotum transplant.

The patient, who chose to remain anonymous, had suffered a devastating injury several years earlier from an improvised explosive device while serving in Afghanistan. The 14-hour procedure repaired his abdominal wall, gave him a new scrotum and attached a donor penis.

The first penis transplant had been performed in China in 2006, but the patient's body rejected the new organ.

The Johns Hopkins surgery was the fourth successful penis transplant in the world. The first was in South Africa in 2014, followed by a surgery at Massachusetts General Hospital in 2016 and then another in South Africa in 2017. However, the surgery at Johns Hopkins, which included the entire penis and scrotum as well as the lower abdominal wall, was by far the most extensive transplant.

"Our transplant is different [from previous ones] because it is a much larger piece of tissue," said Richard Redett, Johns Hopkins plastic and reconstructive surgeon and clinical director of the genitourinary transplant program. "An improvised explosive device typically hits the pelvic region. We were able to design a transplant that would include all the tissue to replace the entire defect."

The penile transplant team had been planning for the procedure since 2013, said Redett, who became director of plastic and reconstructive surgery in 2020. They rehearsed on cadavers to understand which arteries, veins and nerves would need connecting and refined their techniques over time. Plastic and reconstructive

Members of the surgical team that performed the historic penis transplant, with **W.P. Andrew Lee,** director of the department of plastic and reconstructive surgery, at center.

Innovation 2023

Johns Hopkins Medicine in 2019 introduced its second five-year strategic plan, called Innovation 2023, which focused on six goals considered essential to the institution's success and stability:

1. Make Johns Hopkins Medicine Easy
2. Support the Well-Being of Our People and Our Communities
3. Push the Boundaries of Science and Education
4. Improve the Quality and Affordability of Health Care
5. Work Like One Organization
6. Aim for Precision in Everything We Do

surgery professor Gerald Brandacher, also on the team, said it also took time to match donor and patient for age, skin tone and certain immunological and viral parameters. All surgeons donated their time to provide the procedure at no cost to the patient.[43]

Johns Hopkins is screening other patients for similar surgery. There are no immediate plans to use the procedure for gender reassignment operations.[44]

Two weeks after the penis surgery, the patient, who also lost both legs in the blast, received bone marrow infusions from the donor. The procedure, pioneered by the same Johns Hopkins team, modulates the immune response that causes patients to reject transplanted organs, so the patient needs only one low-dose maintenance immunosuppression medication per day.

Doctors chose not to give the patient testicles after consulting with bioethicists. They determined that the sperm-generating tissue would confer the potential for having children with genetic material from the donor.

"It's a real mind-boggling injury to suffer, it is not an easy one to accept," said the recipient. "When I first woke up [after the surgery], I felt finally more normal… [with] a level of confidence as well. Confidence… like finally I'm OK now," he said.

The donor's family members also opted for anonymity, but released a statement saying, "We are all very proud that our loved one was able to help a young man that served his country," and noting that their own family includes several veterans.[45]

The surgeons reported the patient's progress in a letter that was printed in the Nov. 7, 2018, edition of the *New England Journal of Medicine*. "It has now been more than one year since the patient received the penile transplant," they wrote.

"He has near-normal erections and the ability to achieve orgasm, as well as substantial improvements in pleasure scores on patient-reported outcome measures. … The patient has returned to school full time and continues to live independently using leg prostheses. He reports an improved self-image and 'feeling whole' again and states that he is very satisfied with the transplant and the implications it carries for his future."[46]

Nickolas Papadopoulos, above. Right: Ken Kinzler and Bert Vogelstein

Bert Vogelstein, Ken Kinzler and Their Triumph for Early Cancer Detection

Bert Vogelstein was born at The Johns Hopkins Hospital in 1949, graduated from the Johns Hopkins University School of Medicine in 1974, and completed his internship and residency in pediatrics at Johns Hopkins.

Now a director of the Ludwig Center, Clayton Professor of Oncology and Pathology, and a Howard Hughes Medical Institute investigator at the school of medicine and Sidney Kimmel Comprehensive Cancer Center, Vogelstein has remained at Johns Hopkins throughout a groundbreaking career that has dramatically changed the way cancer is understood, detected and treated.[47]

Most recently, Vogelstein and Ludwig Center co-director Ken Kinzler led development of CancerSEEK, a noninvasive blood test that screens for eight common cancer types and helps identify the location of the cancer by simultaneously evaluating levels of eight cancer proteins and the presence of cancer gene mutations from circulating DNA in the blood.

A first study showing the effectiveness of the test was published online in *Science* on Jan. 18, 2018.

The test was evaluated on 1,005 patients with nonmetastatic, stages I to III cancers of the ovary, liver, stomach, pancreas, esophagus, colorectum, lung or breast. The median overall sensitivity, or the ability to find cancer, was 70% and ranged from a high of 98% for ovarian cancer to a low of 33% for breast cancer. For the five cancers that have no screening tests — ovarian, liver, stomach, pancreatic and esophageal cancers — sensitivity ranged from 69% to 98%.[48]

"This test represents the next step in changing the focus of cancer research from late-stage disease to early disease, which I believe will be critical to reducing cancer deaths in the long term," said Vogelstein.[49]

The sorrow and frustration of treating pediatric cancer patients during his residency inspired Vogelstein to pursue a career in cancer research. "Without some knowledge of the root cause of a disease, there is little hope that you can do something about it," he later said. "I became convinced that trying to find those causes would be a rewarding way to spend my life."[50]

After two years at the National Cancer Institute, Vogelstein returned to Johns Hopkins in 1978 to direct the cancer center's molecular genetics laboratory.

His first breakthrough was in 1989, when he and colleagues identified p53 gene mutations in colon cancer, a finding that launched a tide of research linking alterations in the gene to other cancers. It is now known as the most common gene mutation in all cancers.

Vogelstein's lab then discovered a series of other mutations linked to p53. The difficulty of these discoveries is akin to finding one typographical error within 20 volumes of an encyclopedia, and then figuring out how it got there.

Vogelstein's prizes and recognitions include the Breakthrough Prize in Life Sciences; Richard Lounsbery Award; Canada Gairdner International Award; Paul Ehrlich and Ludwig Darmstaedter Prize, Paul Ehrlich Foundation; William Allan Award, American Society of Human Genetics; Charles S. Mott Prize, General Motors Cancer Research Foundation; Prince of Asturias Award in Science; and Charles Rodolphe Brupbacher Prize for Cancer Research.[51]

Vogelstein and Kinzler each consistently rank as the most highly cited scientists in clinical medicine.[52]

Kinzler, associate director for laboratory research at the Kimmel Cancer Center, earned his Ph.D. from the Johns Hopkins University School of Medicine in 1988.

He has been recognized for his role in uncovering the genetic alterations linked to the initiation of colon cancer, one of the most common cancers worldwide; development of novel approaches for the molecular analysis of cancer; and more recently, for his role in deciphering the genetic blueprints of many types of cancer.

In the early 1990s, he led scientists in the identification and analysis of the APC gene, the most frequently mutated tumor suppressor gene in colon cancer. In the mid-1990s, he invented SAGE (Serial Analysis of Gene Expression), allowing the first comprehensive analysis of gene expression in human cancer and leading Kinzler to coin the term "transcriptome," referring to genes that are transcribed at a particular point in time.

In the late 1990s and early 2000s, he invented digital polymerase chain reaction (PCR) and other digital genomics approaches for the detection of trace levels of tumor DNA. In the late 2000s, Kinzler and Vogelstein led a team of scientists who decoded the genomes of many cancers, including breast, colon, pancreatic, brain, ovarian and liver.

When Kinzler was elected to the National Academy of Medicine in 2015, Vogelstein described his longtime colleague as "an out-and-out genius, with intuitions and talents that continue to amaze me, even after working so closely with him for two decades."[53]

Thrive Earlier Detection

Vogelstein and Kinzler studied the molecular structure of cancer cells for decades. Based on their insights, their lab developed a blood test that detects cancers early, when treatment can be more effective.

Enter Johns Hopkins Technology Ventures (JHTV), which facilitated the intellectual property protection and subsequent licensing of their game-changing test, called CancerSEEK, to help it move toward the market.

JHTV initially provided custom-built lab space for a startup to develop and test CancerSEEK. It worked with investors on licensing the technology, securing funding in 2019 for what became Thrive Earlier Detection Corp.

In January 2021, diagnostics firm Exact Sciences purchased Thrive in a deal valued at more than $2 billion. Per the terms of its license agreement, Johns Hopkins (including the inventors) was entitled to a share of proceeds from the acquisition. With the backing of an established diagnostic company, CancerSEEK is closer to Vogelstein's goal of making it a lifesaving part of routine preventive medical care.[54]

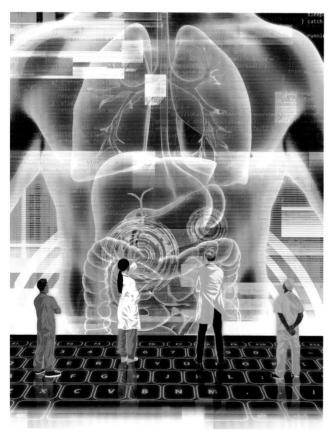

The Felix Project uses an algorithm to detect early-stage pancreatic tumors and abnormalities.

Felix Project

CancerSEEK is one path to early detection; another is using big data and imaging to change the grim statistics of a particularly formidable foe: pancreatic cancer.

Every year, more than 50,000 Americans are diagnosed with pancreatic cancer, and some 93% of patients will die within five years — mostly because their disease wasn't caught in time.[55]

Johns Hopkins radiologist Elliot Fishman is leading an "all-star" research team that is developing a formula to detect early-stage pancreatic tumors and abnormalities. The algorithm — nicknamed Felix — learns to identify healthy and diseased pancreas tissue on patient scans and to distinguish cancer from other abnormalities. Using CT scans, the algorithm can be trained to pick up subtle changes that might indicate an early pancreatic cancer. Felix has a better than 90% accuracy picking up tumors on CT scans, said Fishman.

Eventually, Fishman said, Felix could run on scanners and automatically examine the pancreas whenever the organ appears in one of the millions of scans that are taken annually in the United States.[56]

Funded by the Lustgarten Foundation, the Felix Project is a joint undertaking of the Johns Hopkins radiology, oncology, computer science and pathology departments to use artificial intelligence (AI) to train computers to spot tumors in CT scans when the tumors are still small and localized and can be surgically removed.

It is named after the Felix Felicis potion in Harry Potter books, which makes people successful at everything they do.[57]

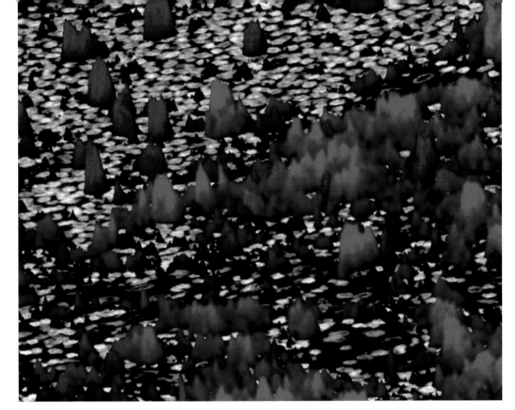

The AstroPath platform gives oncologists valuable information for determining which immunotherapies will be most effective. In the melanoma tissue here, the **PD-L1** expression is shown in red and **PD-1** in blue. Levels of expression are represented by the heights of the peaks.

AstroPath Offers Precision Cancer Immunotherapy

A platform called AstroPath has quite literally taken Johns Hopkins collaborations to new heights.

It has brought together the work of Johns Hopkins astrophysicist Alexander Szalay and Kimmel Cancer Center pathologist Janis Taube to combine astronomic image analysis and mapping with pathology specimens in order to predict which cancers will respond to specific therapies targeting the immune system.

"This platform has the potential to transform how oncologists will deliver cancer immunotherapy," said Drew Pardoll, director of the Bloomberg-Kimmel Institute for Cancer Immunotherapy, in a 2021 news release.[58]

AstroPath's immunofluorescent imaging, using antibodies with fluorescent tags, enables researchers to visualize multiple cellular proteins simultaneously and determine their pattern and strength of expression.

Drew Pardoll

It was built using the image analysis techniques that created the database for the Sloan Digital Sky Survey, a large digital map of the universe architected by astrophysicist Szalay, the Johns Hopkins University Bloomberg Distinguished Professor of Physics and Astronomy and Computer Science.

The sky survey "stitched" together millions of telescopic images of billions of celestial objects, each expressing distinct signatures — just like the different fluorescent tags on the antibodies used to stain the tumor biopsies.

As Pardoll explained: "For the last 40 years, pathology analysis of cancer has examined one marker at a time, which provides limited information. Leveraging new technology, including instrumentation to image up to 12 markers simultaneously, the AstroPath imaging algorithms provide 1,000 times the information content from a single biopsy than is currently available through routine pathology. This facilitates precision cancer immunotherapy — identifying the unique features of each patient's cancer to predict who will respond to a given immunotherapy."[59]

For their work creating AstroPath, Szalay and Taube won a Life Sciences 2021 award at that year's Falling Walls Science Summit, an international event honoring research breakthroughs around the globe.[60]

Janis Taube

Stephanie Reel Retires

When Stephanie Reel retired on July 1, 2020, Johns Hopkins lost a skilled information technology leader and problem solver.

Since 1998, she had been chief information officer and senior vice president for Johns Hopkins Medicine as well as chief information officer and vice provost for information technology for The Johns Hopkins University.

Reel, the longest-serving member of the university's executive leadership team, grew what began as a small team of 75 serving the university and health system to a group of more than 1,300 people who delivered information technology services and solutions —including networking; telecommunications; and clinical, research, and instructional technologies —24 hours a day, seven days a week to Johns Hopkins faculty, staff and students.

Reel, who earned a bachelor's degree in information systems management from the University of Maryland and an M.B.A. from Loyola University in Maryland, joined Johns Hopkins in 1990 as

Stephanie Reel

director of information services for The Johns Hopkins Hospital. She was promoted to senior director of information services for Johns Hopkins Medicine in 1993, and to vice president in 1994.

Among her many achievements, Reel was instrumental in the expanded and enhanced use of electronic patient records at Johns Hopkins Medicine, a project that paved the way for the Johns Hopkins Precision Medicine Application Platform.

Reel also oversaw the development and launch of the Technology Innovation Center, which collaborates with researchers and care providers to strengthen patient care and discovery through the development and commercialization of medical software.

Reel was named a Premier 100 IT Leader by *Computerworld* in 2013, a Hall of Fame Honoree by *CIO* magazine in 2012, and one of the Top 25 Leaders Driving the Healthcare IT Revolution by *InformationWeek Healthcare* in 2011."*

"Serving in these dual roles is unprecedented among academic health systems nationwide, and a testament to Stephanie's vision and leadership ability," wrote Sunil Kumar, provost and senior vice president for academic affairs at The Johns Hopkins University; Daniel G. Ennis, senior vice president for finance and adminis-

tration at The Johns Hopkins University and Robert Kasdin, senior vice president, chief financial officer, and chief operating officer for Johns Hopkins Medicine, in a message to the university and health system communities.†

Starting in October 2021, Dean Zarriello became interim chief information officer for Johns Hopkins Medicine and interim vice provost and chief information officer for The Johns Hopkins University. Zarriello joined Johns Hopkins in 1997 and in 2013 became senior IT director for the Enterprise Technology Services group and chief network officer for The Johns Hopkins University and the Johns Hopkins Health System.‡

*   *Hub* staff report, "Stephanie Reel to Retire as Chief Information Officer of University, Health System," *The Hub*, Dec. 5, 2019, https://hub.jhu.edu/2019/12/05/stephanie-reel-chief-information-officer-retirement/.
†   "Stephanie Reel to Retire as Chief Information Officer of University."
‡   Hub staff report, "Dean Zarriello Named Interim Chief Information Officer for Johns Hopkins University, Medicine," *The Hub*, Oct. 12, 2021, https://hub.jhu.edu/2021/10/12/dean-zarriello-interim-chief-information-officer/#:~:text=Dean%20Zarriello%20named%20interim%20chief,Johns%20Hopkins%20University%2C%20Medicine%20%7C%20Hub.

Center for Psychedelic and Consciousness Research

Johns Hopkins Medicine once again demonstrated its leadership in medical research with the Sept. 4, 2019, opening of the Center for Psychedelic and Consciousness Research, believed to be the first such research center in the country and the largest of its kind in the world.[61]

"Johns Hopkins is deeply committed to exploring innovative treatments for our patients," said Rothman. "Our scientists have shown that psychedelics have real potential as medicine, and this new center will help us explore that potential."[62]

Researchers at the center study if psychedelic substances such as the psilocybin found in "magic mushrooms" can be used to improve brain function and health in both healthy individuals and those affected by conditions such as depression, addiction, anorexia nervosa, Alzheimer's disease and post-traumatic stress disorder.

"Psychedelics are a fascinating class of compounds," said Roland Griffiths, founding director of the center and a professor of psychiatry and of neuroscience. "They produce a unique and profound change of consciousness over the course of just several hours."[63]

Natural psychedelics found in plants and fungi have been ingested for millennia; serious studies into the potential benefits of these hallucinogenic drugs in the 1950s and 1960s showed signs of their therapeutic potential.

But such research slowed to a near halt by 1970. Recreational use of LSD and other psychedelics in the 1960s had created a perception in the popular culture that the drugs were dangerous, addictive, and at least partly to blame for turning a generation of young people into countercultural dropouts.

Griffiths and colleagues in the Behavioral Pharmacology Research Unit led a renaissance in psychedelic research. In 2000, Johns Hopkins became the first institution in the U.S. to win regulatory approval to resume research with psychedelics in healthy volunteers who had no previous experience with them.[64]

With that green light, the group began a run of groundbreaking psilocybin-based experiments, always using rigorous scientific measures and controlled conditions. The group's landmark 2006 study kicked things off by showing that psilocybin can induce mystical experiences just like the spontaneous ones that people have reported for centuries.

"A vast gap exists between what we know of these drugs — mostly from descriptive anthropology — and what we believe we can understand using modern clinical pharmacology techniques," said study leader Griffiths at the time. "That gap is large because, as a reaction to the excesses of the 1960s, human research with hallucinogens has been basically frozen in time these last 40 years."[65]

Next came a 2014 study showing that carefully controlled and monitored use of psilocybin improved success rates for smokers trying to quit. The abstinence rate for study participants was 80% after six months, substantially higher than typical success rates in smoking cessation trials, said study principal investigator Matthew W. Johnson, current director of the center and professor of psychiatry and behavioral sciences at Johns Hopkins.[66]

A study of adults with major depression, published in May 2021 in *JAMA Psychiatry*, suggested that psilocybin may prove effective in a larger population of patients with intractable depression than previously appreciated.[67]

In 2021, Johnson received nearly $4 million from the National Institutes of Health to explore the potential impacts of psilocybin on tobacco addiction. It was the first NIH grant awarded in more than 50 years to directly investigate the therapeutic effects of a classic psychedelic.[68]

Johns Hopkins Medicine is leading the three-year study in collaboration with the University of Alabama at Birmingham and New York University.

Johns Hopkins researchers are studying the brain function and health benefits of psilocybin from mushrooms like these.

The Psilocybin Playlist

A seven-hour and 40-minute playlist, developed by researchers at Johns Hopkins, supported the psychedelic experiences of those who participated in a 2020 study on psilocybin and depression.

Psychologist Bill Richards, a researcher at the Center for Psychedelic and Consciousness Research, chose the music for its ability to guide and support a participant's experience.

The playlist is divided into segments: background music that plays as the participant arrives for their session; music that plays when the drug is starting to take effect, at which point he or she is lying down and wearing eyeshades and headphones; the ascent; the peak; the post-peak; and the "welcome back" music. There are usually two researchers in the room, referred to as guides, who listen to the same music.

The music in each section is deliberately chosen to accompany a particular part of the psychedelic journey. For example, Richards finds that Samuel Barber's iconic "Adagio for Strings" works well when the effects of the psilocybin are steadily intensifying.

"It's going somewhere, it's picking you up and carrying you. It's got some force, some substance," he said.

The playlist is "a nonverbal support system, sort of like the net for a trapeze artist," he said. "If all is going well, you're not even aware that the net is there — you don't even hear the music — but if you start getting anxious, or if you need it, it's immediately there to provide structure."*

In 2021, Johnson and colleagues published a randomized study comparing how well an alternative playlist, including gongs, didgeridoos and singing bowls, compared to the traditional Western classical playlist. They found that the alternative playlist tended to perform a little better than the classical one in terms of helping people quit smoking, suggesting that future work can test different types of music, and more broadly different aspects of the psychedelic session to increase success rates.†

---

*   Shapiro, Marc, "Inside the Johns Hopkins Psilocybin Playlist," *Dome*, Oct. 30, 2020, https://www.hopkinsmedicine.org/news/articles/inside-the-johns-hopkins-psilocybin-playlist.

†   Matthew Johnson email to Karen Nitkin, Jan. 25, 2022.

"The historical importance of this grant is monumental," said Johnson, the principal investigator. "We knew it was only a matter of time before the NIH would fund this work because the data are so compelling, and because this work has demonstrated to be safe. Psilocybin does have very real risks, but these risks are squarely mitigated in controlled settings through screening, preparation, monitoring and follow-up care."[69]

Additional growth was seen in the awarding of two endowed professorships in the area of psychedelic science. In 2020, Griffiths was appointed as the inaugural recipient of the Oliver Lee McCabe, III Professorship in the Neuropsychopharmacology of Consciousness. In 2021, Johnson became the inaugural recipient of the Susan Hill Ward Professorship in Psychedelics and Consciousness.[70]

A Nobel for Gregg Semenza

Oct. 7, 2019, was not a normal Monday for Johns Hopkins genetic medicine professor Gregg Semenza — or for Johns Hopkins Medicine.

It started a little before 4 a.m. with a phone call that most scientists can only imagine: The Nobel Assembly at the Karolinska Institute in Stockholm, Sweden, informed him that he was a co-winner of that year's Nobel Prize in Physiology or Medicine.

"It woke me out of a deep sleep," said Semenza as he settled into his office in the Miller Research Building around 8:30 that morning. In fact, he missed the assembly's first call and went back to sleep until the institute called back several minutes later.

Semenza is the C. Michael Armstrong Professor of Genetic Medicine, Pediatrics, Radiation Oncology and Molecular Radiation Sciences, Biological Chemistry, Medicine and Oncology; and director of the Vascular Program at the Institute for Cell Engineering at the Johns Hopkins University School of Medicine.

The academy recognized him for his groundbreaking discovery of hypoxia-inducible factor 1, or HIF-1, the protein that switches genes on and off in cells in response to low oxygen levels.

Semenza's discovery, along with his additional work clarifying the molecular mechanisms of oxygen regulation in cells, has far-reaching implications in understanding the impact of low oxygen levels in blood disorders, blinding eye diseases, cancer, diabetes, coronary artery disease and other conditions. His research has led to clinical trials for people with kidney cancer and kidney disease.

"I started out asking a really defined question: *How does the body control red blood cell production?*" said Semenza. "And now we know that this is a major mechanism for adaptation to low oxygen."

Semenza shared the award with scientists William G. Kaelin Jr. of the Dana-Farber Cancer Institute and Peter J. Ratcliffe of Oxford University. Kaelin did his residency in internal medicine at The Johns Hopkins Hospital, including a year as chief resident.

Semenza and Kaelin were the 28th and 29th people associated with Johns Hopkins as a faculty member, fellow or graduate to win a Nobel Prize.

Semenza joined three other Nobel Laureates on the university's current faculty: Adam G. Riess, professor in the Department of Physics and Astronomy, who won the prize in physics in 2011; Peter Agre, a 1974 school of medicine graduate, former

Gregg Semenza

About Gregg Semenza

Gregg Semenza, who is director of the Armstrong Oxygen Biology Research Center, director of the Vascular Program in the Institute for Cell Engineering, and a faculty member of the McKusick-Nathans Department of Genetic Medicine and the Johns Hopkins Kimmel Cancer Center, was born in New York City and is the first of five children.

He attended Harvard University for his bachelor's degree. While there, a family friend had a child born with Down syndrome, which inspired his interest in medical genetics.

He earned his M.D. and Ph.D. degrees from the University of Pennsylvania, studying the genetic disorder beta thalassemia. He later went to Duke University to complete his internship and residency in pediatrics. He moved to Johns Hopkins in 1986 for a post-doctoral fellowship in medical genetics; met his wife, Laura Kasch-Semenza; and stayed. They have three children.

Semenza has authored more than 400 research articles and book chapters, which have been cited more than 170,000 times. He serves on the editorial boards of several scientific publications and is a deputy editor of *The Journal of Clinical Investigation.*

He has received the American Heart Association Distinguished Scientist Award, the American Cancer Society Research Professor Award, the Lefoulon-Delalande Grand Prize from the Institut de France, the Stanley J. Korsmeyer Award from the American Society for Clinical Investigation, the Wiley Prize in Biomedical Sciences, the Canada Gairdner International Award and the Albert Lasker Basic Medical Research Award.

He is a founding Fellow of the American College of Medical Genetics, and an elected member of the Academy of the American Association for Cancer Research, Society for Pediatric Research, American Society for Clinical Investigation, Association of American Physicians, National Academy of Medicine and National Academy of Sciences.[*]

---

[*] Nitkin, Karen, "A Day of Celebration for Nobel Winner Gregg Semenza," Dome, Oct. 7, 2019, https://www.hopkinsmedicine.org/news/articles/a-day-of-celebration-for-nobel-winner-gregg-semenza.

professor in the school of medicine, and director of the Malaria Research Institute in the Bloomberg School of Public Health, who won the chemistry prize in 2003; and Carol Greider, former professor and director of molecular biology and genetics in the school of medicine, who won 2009's physiology or medicine prize.

As the new Nobel Laureate telephoned family members and gave news interviews, students and trainees offered congratulations and, in some cases, held back tears. They described Semenza as a mentor who works hard and pushes them to be the best scientists they can be.

Well-wishers included fellow Nobel Laureate Agre, who stopped by the office to congratulate Semenza in person. "I've been looking forward to this day," Agre said. "This is great for Hopkins. All this work was done right here."

Semenza has been at Johns Hopkins since arriving in 1986 for a postdoctoral fellowship in medical genetics. "This is the greatest place to do research, with the greatest colleagues, and the greatest sense of collegiality," he said.

"I always said 'I like my life just the way it is,' I don't foresee any major changes," he told members of his lab as they toasted him with plastic cups of champagne. "My message for everyone training today is: I was once where you are now, and you will one day be where I am. We are very lucky to have this career where we get to follow our interests wherever they may lead."[71]

Ambroise Wonkam Takes the Helm in Genetic Medicine

On Jan. 1, 2022, geneticist Ambroise Wonkam became director of Johns Hopkins Medicine's renowned Department of Genetic Medicine and the McKusick-Nathans Institute of Genetic Medicine. Longtime director David Valle stepped down, while remaining on the Johns Hopkins faculty and continuing his research and clinical work.

Growing up in the central Africa country of Cameroon, Wonkam watched his friends and classmates endure sickle cell disease, a heritable condition of sickle-shaped red blood cells that is marked by intense pain, disability and shortened lifespan. The experience spurred his interest in genetic medicine.

"To make genetic medicine a global good for understanding the variability among all of us, we need to investigate our genome to its fullest," he said. "We have a duty to carry forward this work and make genetic medicine innovations sustainable and equitable for all populations."

Wonkam received his medical degree from the University of Yaoundé in Cameroon, and a doctorate as well as specialist training in medical genetics from the University of Geneva in Switzerland. He also earned a doctorate in human genetics from the University of Cape Town in South Africa.

Ambroise Wonkam

He remained at the University of Cape Town, becoming a professor of medical genetics, then deputy dean of research and director of GeneMAP (Genetic Medicine of African Populations), an effort to increase representation of the genetic diversity of Africa's populations in genetic reference genomes.

Wonkam has authored more than 180 peer-reviewed research publications and leads numerous grants from the National Institutes of Health. His research has included a focus on the psychosocial aspects of sickle cell disease and genetic links to the condition's symptoms. In his research on hearing loss, Wonkam's work revealed that three genes long linked to deafness have no major role among most African populations, making routine testing in these populations unnecessary.

Wonkam, president of the African Society of Human Genetics, is an associate editor of *The American Journal of Human Genetics*, the *American Journal of Medical Genetics* and the *Journal of Community Genetics*. He is an academic editor of *PLOS One* and a member of the Human Genetics editorial board.

In 2021, he won the Gold Scientific Achievement Award from the South African Medical Research Council, and the Alan Pifer Award from the University of Cape Town, in recognition of the excellence of his research. In 2014, he received the Clinical Genetics Society International Award from the British Society for Genetic Medicine.[72]

At Johns Hopkins, Wonkam oversees nearly 40 faculty members who are primary appointees in the Department of Genetic Medicine. With approximately $35 million per year from federal and other funding sources, the department's

David Valle

David Valle, Henry J. Knott Professor at the Johns Hopkins University School of Medicine, developed an interest in science as a young boy, rambling the woods near his upstate New York home.

His 7th grade biology teacher fed his enthusiasm by giving him extra resources, including a microscope that Valle was allowed to take home. That teacher was the first of many "really superb teachers and mentors" who nurtured Valle's love for science, he told the Jackson Laboratory's Search magazine in 2011.*

Valle earned his bachelor of science and medical degrees from Duke University before coming to Johns Hopkins for a pediatric residency in 1969. He joined the faculty in 1975 and quickly established himself in the new-at-the-time field of genetic medicine.

*   Wanner, Mark, "Biology Unlimited," Search magazine, The Jackson Laboratory, October 2011, https://www.jax.org/news-and-insights/2011/october/biology-unlimited.

scientists study genetic and genomic aspects of all disease. The department has trained more than 1,000 medical and graduate students, residents and practitioners.

Each year, faculty and clinical staff in the Department of Genetic Medicine see approximately 4,000 pediatric and adult patients with genetic conditions of the cardiovascular and connective tissue systems and other conditions including inborn errors of metabolism, mitochondrial disorders, malformation syndromes, skeletal dysplasias, intellectual disabilities and neurodegenerative diseases.

Established in 2019, the Johns Hopkins Department of Genetic Medicine is one of 32 departments at the Johns Hopkins University School of Medicine. Previously, genetic medicine was centralized at the McKusick-Nathans Institute of Genetic Medicine, which was formed in 1999 and named for genetics pioneer Victor McKusick and molecular biologist Daniel Nathans.[73]

McKusick, a 1946 graduate of the school of medicine, founded the Division of Medical Genetics in 1957, in essence creating a discipline that would eventually map the human genome and begin the process of understanding the links between genes and disease.[74]

Nathans, a microbiologist, won the 1978 Nobel Prize with fellow Johns Hopkins microbiologist Hamilton Smith for their discovery and first use of restriction enzymes, which made the genetic engineering revolution possible.[75]

He directed the Pediatric Genetics clinic from 1975 to 1987 and the Predoctoral Training Program in Human Genetics from 1989 to 2022.

In 1996, he was founding director of the Johns Hopkins Center for Inherited Disease Research. He became the director of the McKusick-Nathans Institute of Genetic Medicine in 2007. He also was a member of the team that developed the current Johns Hopkins Genes to Society medical curriculum initiated in 2009.

Starting in 1992, he has been a co-organizer of the Johns Hopkins/Jackson Labs McKusick Short Course, an intensive, two-week immersion in human genetics held every summer in Bar Harbor, Maine.

Valle is board-certified by the American Board of Medical Genetics in clinical molecular genetics, clinical biochemical genetics, clinical genetics and pediatrics. He is professor of genetic medicine, pediatrics, ophthalmology, medicine and molecular biology and genetics at the school of medicine, as well as the chief of genetic medicine for The Johns Hopkins Hospital.

His laboratory has discovered the genetic causation for more than 25 diseases, including those responsible for inborn errors of metabolism, inherited retinal degeneration, disorders of cellular organelle biogenesis as well as identification of genetic variations that contribute risk for common disorders such as schizophrenia.

Since 2002, Valle has been a member of the National Academy of Medicine and in 2003 was the recipient of the March of Dimes Colonel Harland Sanders Lifetime Achievement in Genetics. He was the 2014 recipient of the Victor A. McKusick Leadership Award from the American Society of Human Genetics.[76]

He was also the 2016 recipient of the Arno Motulsky-Barton Childs Award for Excellence in Human Genetics Education from the American Society of Human Genetics.[77]

Leadership Changes in Molecular Biology and Genetics

Jeremy Nathans, professor of molecular biology and genetics, neuroscience and ophthalmology, became interim director of the Johns Hopkins Department of Molecular Biology and Genetics in 2020.

Nathans, the son of Johns Hopkins Nobel winner Daniel Nathans, joined the Johns Hopkins faculty in 1988, after earning his undergraduate degree in life sciences and chemistry from MIT and his Ph.D. in biochemistry and M.D. from Stanford University.[78]

His investigations into the mechanisms that allow us to see colors led him to identify the genes that code for color-vision receptors in the light-sensing cones of the retina. This breakthrough finding allowed him to show that variations in these genes cause color blindness. His subsequent investigations defined the molecular mechanisms responsible for the most common inherited forms of retinal degeneration and for vascular disorders of the retina and brain.[79]

He became interim director after Carol Greider, winner of the 2009 Nobel Prize in Physiology or Medicine, stepped down in 2020, returning to her native California to accept a post as a distinguished professor of molecular, cell and developmental biology at the University of California, Santa Cruz.[80]

Greider grew up in Davis, California. She graduated from the University of California, Santa Barbara, with a B.A. in biology in 1983 and earned a Ph.D. in molecular biology in 1987 from the University of California, Berkeley. She worked at Cold Spring Harbor Laboratory on Long Island, N.Y., from 1988 until moving to Johns Hopkins in 1997.[81]

In 2003, Greider became Daniel Nathans Professor and director of the Department of Molecular Biology and Genetics; she had been interim director since 2002.

Greider shared the 2009 Nobel Prize with Elizabeth Blackburn, a professor of biochemistry and biophysics at the University of California, San Francisco, and Jack Szostak of Harvard Medical School, who discovered that telomeres are made up of simple, repeating blocks of DNA building blocks and that they are found in all organisms. Greider, Blackburn and Szostak shared the 2006 Albert Lasker Award for Basic Medical Research for this work.

"What intrigues basic scientists like me is that any time we do a series of experiments, there are going to be three or four new questions that come up when you think you've answered one. Our approach shows that while you can do research that tries to answer specific questions about a disease, you can also just follow your nose," Greider said at the time.[82]

Jeremy Nathans

Carol Greider

Frank Lin

Conversations and Collaborations

It happens often at Johns Hopkins Medicine: A conversation, a passion or an observation becomes the catalyst for an entirely new and often groundbreaking program that takes advantage of the rich opportunities for collaboration across The Johns Hopkins University.

Otolaryngologist Frank Lin created the Cochlear Center for Hearing and Public Health because he recognized that hearing loss was impairing the daily life of his patients.

Neurologist Alexander Pantelyat channeled his love of music to co-found the Johns Hopkins Center for Music and Medicine.

And biomedical engineering professor Jeff Siewerdsen created a Surgery for Engineers program for graduate students interested in transforming the tools that surgeons use in the operating room.

Frank Lin and the Cochlear Center for Hearing and Public Health

Untreated hearing loss contributes to dementia and cognitive decline, among other health problems, and can lead to social isolation and depression.

Lin observed the negative impact hearing loss had on his patients' daily functioning. He decided to address the problem in a big way.

The bipartisan Over-the-Counter Hearing Aid Act, which he helped craft, was signed into law by former President Donald Trump in 2017, making hearing aids more accessible and affordable.[83]

In 2018, Lin, who earned his medical degree and his public health Ph.D. at Johns Hopkins, founded and became director of the Cochlear Center for Hearing and Public Health, the first global research institution focused exclusively on issues related to hearing loss and public health in older adults.

The center is training a generation of researchers and clinicians to study the impact of hearing loss on public health; developing and testing strategies to optimize hearing; and helping to implement effective policies for hearing loss at the local, national and global levels.[84]

The center, based in the Department of Epidemiology at the Johns Hopkins Bloomberg School of Public Health, is affiliated and co-located with both the Johns Hopkins Center on Aging and Health and the Welch Center for Prevention, Epidemiology and Clinical Research.

It was launched with more than $20 million in existing grant funding from the National Institutes of Health, other philanthropic funding, and a $10 million gift from Cochlear Ltd., an Australian company that produces implantable hearing devices.[85]

In the center's first annual report from 2018, Lin wrote:

In discussing our work at the Cochlear Center with others over the past year, whether it's related to our research on hearing and dementia, policy work bringing about over-the-counter hearing aid legislation, or crafting new "frictionless" approaches for obtaining hearing care outside the usual clinical setting, I'm often struck by how commonly I hear a response that is along the lines of, "That makes a lot of sense," or "No one's done that before?"

From there, the person will usually then recall their own personal story of observing a family member or parent struggle with hearing and how it impacted them.

I consider this a good sign. Namely, I believe that often the most important research findings, in hindsight, seem obvious, and the most impactful innovations, in hindsight, also seem obvious.[86]

David Rubenstein

Philanthropist and longtime Johns Hopkins Medicine trustee David M. Rubenstein, founder and co-executive chairman of The Carlyle Group, gave $15 million to the Department of Otolaryngology-Head and Neck Surgery in 2015 to establish the David M. Rubenstein Hearing Center, focusing on restoring functional hearing.[87]

Rubenstein gave an additional $15 million in 2021 to establish the David M. Rubenstein Precision Medicine Center of Excellence, deepening his support for basic science researchers focused on the development of therapeutic approaches to preserve and restore hearing.

"David's initial gift has helped Johns Hopkins researchers make important discoveries in several crucial areas related to hearing and hearing loss," said Rothman in announcing the second gift. "But there is so much more to be done in this area, and once again, David has stepped forward. We are grateful for all that his generosity has made possible so far, and we are even more excited about what this new commitment will allow us to accomplish going forward. In the end, this work will help the millions of people who struggle every day with hearing problems."[88]

Center for Music and Medicine

Pantelyat started playing violin at age 7, and kept making music even while pursuing a medical degree at Temple University and completing his neurology residency at the Hospital of the University of Pennsylvania.

"Violin would have been my vocation, if it were not for medicine," he said.

Pantelyat, whose work as a neurologist focuses on Parkinson's disease, began investigating specific ways to link his clinical work and music even before coming to Johns Hopkins in 2014. His research indicated that musical interventions could ease some of the suffering associated with the neurological disorder.[89]

In January 2015, Pantelyat proposed a collaboration to

Alexander Pantelyat

Peabody Institute Dean Fred Bronstein: a Johns Hopkins Center for Music and Medicine. Bronstein accepted the idea with enthusiasm. Pantelyat soon had strong partners in center co-founder Serap Bastepe-Gray and co-director Sarah Hoover.[90]

Currently, the center brings together clinicians, researchers and musicians across Johns Hopkins Medicine and The Johns Hopkins University to explore links between music and health and possibilities for healing through music.

The center, the first of its kind in the eastern United States, also treats musicians and dancers for injuries and illnesses related to their art, such as carpal tunnel syndrome for instrumentalists, vocal cord problems for singers, or balance disorders for dancers.[91]

While anecdotal evidence linking music and healing is not new, many of the most important questions remain tantalizingly open: What specific benefits can music create? Which illnesses are best suited for musical interventions?

"We are just now scratching the surface of the mechanisms behind all of this," said Pantelyat. "We want to make music and rhythm an integral part of treating illness, and we want to learn more about and implement better treatment for musicians' illnesses worldwide."[92]

One of the center's first projects was a study of how choral singing can affect symptoms of Parkinson's disease, which can include tremors, poor balance, a weakened voice and slurred speech. The study found that group singing improved voice strength and clarity, as well as quality of life. Participants gained confidence and a sense of community.[93]

Setting a template for future studies and their link to long-term community-based initiatives, the center created and supports the ParkinSonics choral group for people with Parkinson's disease and related disorders, which continues to explore the connections between group singing and improved mood and neurologic function.[94]

Subsequent studies showing the benefits of guitar lessons and group drumming for people with Parkinson's disease and their care partners led to the creation of a subsidized guitar lesson program at Peabody Preparatory and a virtual group drumming program.

In 2021, the center hired senior music therapist Kerry Devlin and neurologic music therapist and postdoctoral fellow Kyurim Kang. They became part of a multidisciplinary clinic for atypical parkinsonian disorders at Johns Hopkins Bayview Medical Center, where Devlin is creating a group music therapy program and, together with Pantelyat and collaborators, plans to study music therapy as a tool to improve autobiographical memory in patients with mild cognitive impairment and mild Alzheimer's disease-related dementia.

The center also established the Baltimore Racial Justice Concert Series in 2021 in partnership with Johns Hopkins organizations and medical students to support and organize concerts by Baltimore artists that raise funds for organizations promoting racial and social justice in the city.[95]

Beyond the work of the center, an extensive partnership between the Peabody

Institute and Johns Hopkins Medicine brings the healing power of music and dance to the hospital environment, providing arts experiences to help reduce anxiety, alleviate pain, promote wellness and improve patient and staff experiences.

For example, the Sound Rounds program, which began in spring 2018, provides impromptu concerts by Peabody Conservatory musicians in hallways, waiting rooms and family visiting rooms. (During the COVID-19 pandemic, the program moved to Zoom and primarily served staff.)

"Music has been an integral part of the human experience as long as humanity has been around. It's been intuitively felt to have healing properties, but now we are in a position to study the mechanisms and optimize music-based interventions," said Pantelyat.[96]

Surgineering: A New Approach to Engineering and Surgery

As a Halsted surgery resident at Johns Hopkins, Sandra DiBrito fielded many queries when she taught skills in surgical technique to medical students at the school of medicine.

"The main questions are, 'Am I doing this right? Am I turning my hand the right way?' Medical students are so busy absorbing the technique," she said.

Since the September 2018 launch of the Surgery for Engineers program in the Department of Biomedical Engineering, DiBrito heard a different question: "Why?"

"Surgeons take the tools we use for granted," said DiBrito, who helped develop the program. "Engineers approach everything as a problem. They are about the why."

Program creator Siewerdsen, the John C. Malone Professor and vice chair for

Jeff Siewerdsen, above, established the Surgery for Engineers program to create "surgineers" equipped to improve safety and workflow in operating rooms.

Alan Partin Steps Down as Urology Director

Alan Partin

Mohamad Allaf

After 17 years, urologist Alan Partin announced in November 2021 that he would step down as the Jakurski Family Director of the Brady Urological Institute, urologist-in-chief for Johns Hopkins Medicine and director of the Department of Urology.

Under Partin's leadership, the Brady Institute doubled its research space and enjoyed significant expansion with projects such as the world-renowned Greenberg Bladder Cancer Institute, and new clinical space at the Green Spring Station Pavilion III.

He is known for developing the Partin Tables, which are used for predicting the prognosis for prostate cancer, and for his work developing several innovative tests to identify and track prostate cancer, including the Prostate Health Index.

"Alan is a devoted teacher and mentor who is deeply committed to educating the next generation of clinicians and basic scientists," said Paul Rothman, former dean of the medical faculty and CEO of Johns Hopkins Medicine. "He has trained over 45 residents, including many women and people from underrepresented minority groups. He also has edited numerous urological textbooks and journals, and authored more than 600 scientific articles, publications and presentations."

A summa cum laude graduate of the University of Mississippi, Partin has been at Johns Hopkins for nearly 40 years: He earned his M.D. at the school of medicine and his Ph.D. in pharmacology and molecular sciences. He did his residency at Johns Hopkins and then joined the Brady Institute as an associate professor in 1995.

Partin's research has earned him the British Association of Urological Surgeons' distinguished St. Paul's Medal and the American Urological Association's Gold Cystoscope Award and Distinguished Service Award.[*]

Mohamad Ezzeddine Allaf, a renowned surgeon-scientist known for robotic and minimally invasive procedures, became Partin's successor. Allaf, who worked closely with Partin in the Department of Urology, received his medical education and completed his urology residency at Johns Hopkins.[†]

[*]   "Partin Stepping Down as Director," *Discovery* magazine, winter 2022, https://www.hopkinsmedicine.org/news/publications/Discovery_Winter%202022.pdf.

[†]   "Mohamad Ezzeddine Allaf, M.D.," Johns Hopkins Medicine webpage, https://www.hopkinsmedicine.org/profiles/details/mohamad-allaf#:~:text=Mohamad%20Allaf%20is%20professor%20of,Dr.

clinical and industry collaboration in the Department of Biomedical Engineering, said rapid advances in technologies including imaging, navigation and robotics create new opportunities for surgeons and engineers to work together.

He envisions the "surgineer" as a new kind of engineer — one who will become an increasingly essential member of the clinical workforce. With a foundation in biomedical engineering, expertise in systems and data science, and a genuine understanding of interventional procedures, "the surgineer will be equipped to apply perspectives of systems engineering and data science to improving workflow and patient safety in the OR," he said.

The Surgery for Engineers course takes 20 or so graduate students from the Whiting School of Engineering and the Johns Hopkins University School of Medicine through 13 areas of interventional medicine, including general and specialized surgeries, interventional radiology, and radiation oncology. Over the course of the fall semester, the surgineers scrub in to learn surgical skills, such as the

Trust and Praise from a Beloved Actor

After contending with Parkinson's disease for nearly 30 years, actor Michael J. Fox came to Johns Hopkins in 2018 for removal of a tumor on his spinal cord that was unrelated to his neurological disease. Although benign, the large growth was threatening the 57-year-old actor with paralysis.

Fox had conferred with other neurosurgeons and neurologists, but he says he knew almost immediately after meeting Nicholas Theodore, head of the Department of Neurosurgery's Spine Division, that he had found the right doctor.

After graduating from Georgetown University's medical school in 1991, Theodore spent several years in the Marines, including as head of the neurological surgery division at the Naval Medical Center in San Diego.

Before joining the Johns Hopkins faculty in 2016, he was the director of neurotrauma at the Barrow Neurological Institute in Phoenix. He also was the team neurosurgeon for the NFL's Arizona Cardinals and a consultant for the Diamondbacks baseball team and the Coyotes hockey team.

When the actor pointed out that other experts had shied away from performing the operation, Theodore replied, "Who wants to be the guy who paralyzes Michael J. Fox?"

The Canadian-born actor is a beloved superstar, best known for his roles as Alex P. Keaton in the TV sitcom *Family Ties* and as Marty McFly in the *Back to the Future* movies.

Fox's book offers praise for Theodore and also for Johns Hopkins.

"All the wonderful things I can say about Johns Hopkins — and I'll say them — start with Dr. Theodore," he wrote. "Everybody on his team, from the tip

Nicholas Theodore and Michael J. Fox

of the spear down, was just fantastic. I felt like I was in the best place in the world to be."

The surgery was a success. Although damage done to Fox's spine before the operation could not be reversed, "the big deal was that it wouldn't go any further, and it hasn't," Fox says. "Suffice it to say, I'm in much better shape than I would be if I hadn't had Dr. Theodore."

After the surgery, Fox spent several weeks recovering in The Johns Hopkins Hospital, which he also praised. "I'm grateful for and impressed by the doctors, the therapists and the staff. I [could] feel them wanting a positive outcome so badly for me."*

---

* Grauer, Neil A., "Grateful and Impressed," *Dome*, Dec. 19, 2020, https://www.hopkinsmedicine.org/news/articles/grateful-and-impressed.

---

basics of suturing, cautery and wound closure, and they practice taking biopsies and placing endoscopes.[97]

In the spring semester, eight to 10 students from the fall surgineering course are selected for the program's second offering, which involves shadowing Johns Hopkins surgeons, nurses and staff members inside and outside the operating room. The students concentrate on workflow, the interaction among multiple departments, and the continuous capture and curation of data, all in an effort "to overcome complexity, improve quality and enable new advances in clinical care," said Siewerdsen.

The goal of this new course is not to add additional tools to the surgical arsenal, said Siewerdsen. "Rather, it is to help transform the arsenal into a system of seamless integration: to bury complexity, facilitate mainstream adoption of advanced techniques and help position surgery for the era of data science in medicine."[98]

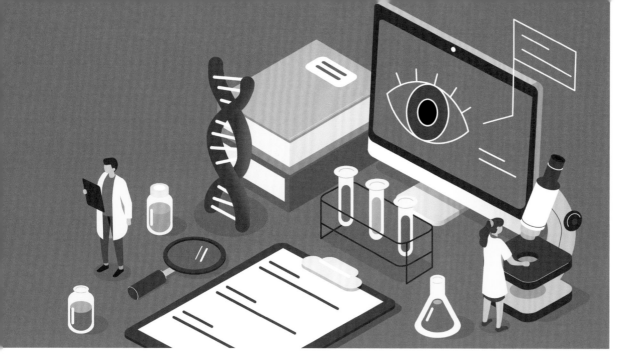

Using Artificial Intelligence to Detect Eye Disease

By 2019, researchers at the Wilmer Eye Institute were pursuing at least a dozen projects that harnessed advances in artificial intelligence to improve clinical care in ophthalmology.

Ophthalmologist Jithin Yohannan, for example, was using deep learning, as well as other machine learning approaches, to predict the likelihood of glaucoma progression and ultimately improve patient outcomes.

And Neil Bressler, Wilmer's James P. Gills Professor of Ophthalmology, was using artificial intelligence to detect age-related macular degeneration (AMD) before a person developed symptoms.[99]

Neil Bressler

The disease, a leading cause of blindness in people over 50, is typically detected in an intermediate stage through careful observation by ophthalmologists, who are looking for areas of debris that have accumulated behind the retina, called drusen.

Bressler and computer scientists at the Johns Hopkins University Applied Physics Lab (APL) pursued a "deep learning" strategy that trained computers to develop an algorithm for identifying drusen on photos of retinas.

The algorithm Bressler's team created accurately identified AMD between 88.4% and 91.6% of the time, which is comparable to the accuracy achieved by human experts. They published the results in *JAMA Ophthalmology* in 2016.[100]

Office for Online Education

Online education offers many advantages: It is convenient for both instructors and students, eliminating travel time and scheduling conflicts. The courses, with options for closed captioning, can help learners overcome language, hearing and learning barriers. And students can learn at their own pace, fast-forwarding through the parts of the lesson they understand, and rewinding to learn more when necessary.

In 2016, the Johns Hopkins University School of Medicine opened the Office for Online Education, with Robert Kearns as director. Kearns had been director of instructional design for the Johns Hopkins University School of Nursing since 2012.

Nancy Hueppchen

In his new role, he would develop, market and evaluate online learning initiatives and distance learning education programs for the school of medicine, working closely with faculty and other leaders across the institution. He also led development of classes and curricula for learners around the world seeking certificates, continuing medical education credits, or new information on particular topics.[101]

Kearns said he and others in his office work closely with faculty to figure out how to add a little dazzle to important but sometimes dry material. "There's a bit of edutainment," said Kearns. "Learners can pick it up, learn a little bit, come back to it later."[102]

Interest in online learning accelerated during the COVID-19 pandemic, when in-person instruction was curtailed by the social distancing requirements put in place to limit the spread of the virus.

Even before the pandemic, only about 25% of Johns Hopkins medical students attended lectures in person, said Nancy Hueppchen, associate professor of gynecology and obstetrics and associate dean for undergraduate medical education.[103] The rest watched lessons that had been recorded and were offered online.

Over the summer of 2020, faculty members and support staff modified their materials to better reflect the realities of online learning. Lectures were updated and shortened when appropriate, and materials were added for independent study.[104]

Center for Medical Humanities and Social Medicine

Can learning about poetry, history or social structures make a student a better health care provider?

Jeremy Greene believes so, as do participants in programs of the Center for Medical Humanities and Social Medicine, which he started in 2017 and directs.

"At this moment in time, there is increasing recognition that we cannot think of health and medicine in American society as something that can be explained by the biological sciences alone," said Greene, also the William H. Welch Professor of Medicine and the History of Medicine.

"Students are looking at the world around them, and they are demanding an approach to medical education that takes into account the broader structural problems facing our society and the tools that physicians will need to be sensitive to the way these problems impact the patients in front of them."[105]

The Center for Medical Humanities and Social Medicine is one example of how educators from throughout the university can work together to forge new and exciting collaborations across schools and disciplines.

The center is a collaboration between the Johns Hopkins University School of Medicine, the Johns Hopkins Bloomberg School of Public Health and the Krieger School of Arts and Sciences.

It gives undergraduate, graduate and medical students opportunities to better study and understand the nuances of health and health care settings. Programs include Medicine, Science, and Humanities; Critical Global Health Studies; Reproduction, Health, and Society; Technology, Data, and Health; Race, Racism, and Health; Social Medicine in Medical Education; and Arts, Humanities, and Health.

The center's inaugural event, in October 2017, was a two-day exploration of the role of text, visual and electronic media in circulation of medical knowledge and practice. Researchers from Johns Hopkins, as well as University of Oxford, Harvard Medical School, Yale School of Medicine and elsewhere attended the Medicine and the Challenge of New Media conference at The Johns Hopkins Hospital and the William H. Welch Medical Library.

Topics included how increased use of electronic medical records has changed

Jeremy Greene

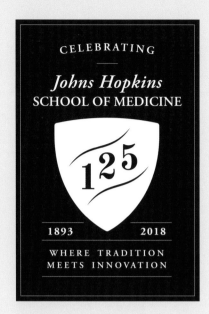

The School of Medicine at 125

When the Johns Hopkins University School of Medicine opened its doors in 1893 to 18 students — including three women — it changed medical education.

While other U.S. medical training institutions in those days required no previous college education for their students, the Johns Hopkins school stipulated that all applicants have a four-year baccalaureate degree. Once accepted, they underwent a rigorous four-year curriculum based on scientific research and clinical observations at patients' bedsides.

In the ensuing years, the Johns Hopkins University School of Medicine has remained a leader in training physicians and scientists.

In 2018, the school celebrated 125 years of medical and scientific excellence and innovation with a year of events around the theme "Where Tradition Meets Innovation." The first event was January 10, when the school of medicine boardroom in the Miller Research Building was officially named for Mary Elizabeth Garrett, who provided the bulk of the funding to launch the medical school and insisted that women be eligible for admission to the school on equal footing with men.*

On June 1, university leaders and elected officials delivered remarks during a symposium at Turner Auditorium that included video highlights recounting the medical school's impact in Baltimore, the state of Maryland, the nation and the world.

"You could say that the mark of a Johns Hopkins faculty member is a reverence for tradition coupled with an almost religious fervor for forward progress in science and medicine," said Paul Rothman, dean of the medical faculty and CEO of Johns Hopkins Medicine, in a March 2018 *Dome* column.†

---

* Smith, Patrick, "Celebrating 125 Years of Medical Education at Johns Hopkins," *Dome*, Jan. 9, 2018, https://www.hopkinsmedicine.org/news/articles/celebrating-125-years-of-medical-education-at-johns-hopkins.

† "Celebrating 125 Years of Medical Education at Johns Hopkins."

---

the way clinical exams are conducted, and how data visualization influences perceptions of public health crises like the Zika outbreak.[106]

XD Bio

In August 2019, the Johns Hopkins University School of Medicine enrolled its first students in the XDBio Cross-Disciplinary Graduate Program in Biomedical Sciences, a training program that bridges basic science and medicine through interdisciplinary research that taps courses and experts across the university.

XDBio students follow a personalized curriculum guided by individual research interests, prior coursework, and future goals.

Students also benefit from the program's unconventional funding structure, which provides grants directly to the student, instead of the usual combination of training grants and research funding that is attached to a particular lab.

"Students in XDBio will be funded for the full five years by the program," explained Peter Espenshade, associate dean for graduate biomedical education. "We expect that this will remove certain constraints and free the student to be more creative and independent."[107]

Peter Espenshade

Johns Hopkins, Enslaved People and a New Look at the Past

The startling news appeared on the front pages of newspapers all over the world.

"Johns Hopkins Reveals that Its Founder Owned Slaves," announced a front-page *New York Times* article on Dec. 9, 2020. "The university and its medical system have long celebrated their founding benefactor as a staunch abolitionist," the *Times* continued. "But newly surfaced documents tell a different story."[108]

The story was especially shocking because it seemed to run counter to what everyone thought they knew about Johns Hopkins, a Quaker who had set aside $7 million in his will to create an interconnected hospital, medical school and university. Shortly before his death, Hopkins wrote that his eponymous hospital must be open to all, "without regard to sex, age or color."[109]

"For most of the last century, our institutions believed Johns Hopkins to be an early and staunch abolitionist whose father, a committed Quaker, had freed the family's enslaved people in 1807," Rothman, Sowers and Johns Hopkins University President Ronald Daniels wrote in a Dec. 9, 2020 letter to the Johns Hopkins community.

"We now have government census records that state Mr. Hopkins was the owner of one enslaved person listed in his household in 1840 and four enslaved people listed in 1850. By the 1860 census, there are no enslaved persons listed in the household."[110]

The communication also explained the history behind the new information, starting with the creation in 2013 of Hopkins Retrospective, "a universitywide initiative to more deeply explore our history."[111]

> *In late spring 2020, we learned for the first time of the possible existence of the 1850 census document indicating that Johns Hopkins was a slaveholder, and we are deeply grateful to Martha S. Jones, [Hopkins history professor], and to Allison Seyler, program manager of Hopkins Retrospective, for their painstaking efforts over the ensuing months to investigate and confirm the connections between this document and our founder, Johns Hopkins.[112]*

Census records show no enslaved person in Hopkins' household from 1860 on, and it is acknowledged that he was a friend of Abraham Lincoln and a fervent supporter of the Union during the Civil War, said Sherita Golden, chief diversity officer for Johns Hopkins Medicine, adding: "This revelation moved us to think more broadly and inclusively about who we want to memorialize on our campus based on their contributions to the history of Johns Hopkins University and Medicine."[113]

Diverse Names and Narratives

In April 2021, Johns Hopkins University and Medicine launched the Diverse Names and Narratives Project, described in the *Hub* as a project to "recognize and more visibly celebrate the names and stories of remarkable people who are part of the institution's history, with a specific focus on individuals from historically

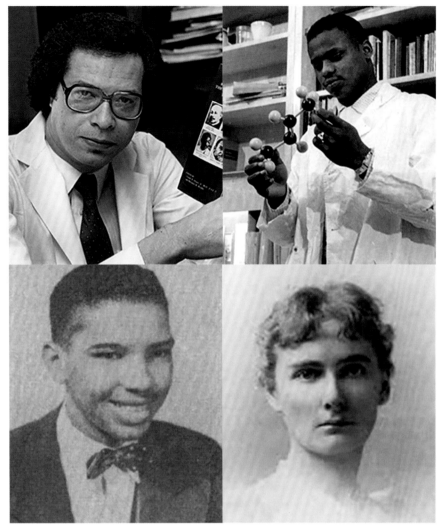

Clockwise from top left: Levi Watkins, Ernest Bates, Florence Bascom and Frederick Scott are being honored for their achievements and contributions.

Alexander Bates, right, accepts an honorary degree on behalf of his grandfather, Ernest Bates, during his 2021 Johns Hopkins University graduation. Ernest Bates in 1958 was the first Black student in the university's School of Arts and Sciences.

marginalized and underrepresented groups."[114]

One of the group's first assignments was to recommend new names for three prominent buildings: The Johns Hopkins Outpatient Center clinics, the Undergraduate Teaching Laboratories and the two undergraduate residential towers known as Charles Commons.

In October 2021, a task force of students, faculty, staff, alumni and community members made their recommendations, which were quickly approved by the boards of Johns Hopkins Medicine, The Johns Hopkins University and the Johns Hopkins Health System:

- The Johns Hopkins Outpatient Center clinics in East Baltimore, known colloquially as JHOC, would be named for Levi Watkins, a pioneer in cardiac surgery and civil rights at Johns Hopkins Medicine, who died in 2015. While the building in which the JHOC sits was named for Robert M. Heyssel when it opened in 1992, the center had not been not named; it serves 300,000 patients annually and is home to medical students and residents, faculty and frontline caregivers.
- The Undergraduate Teaching Laboratories, a 105,000-square-foot facility with a distinct concave, glazed glass façade, which opened at the Homewood campus in 2013, would be named for Florence Bascom, a barrier-shattering geologist who earned a Ph.D. from Johns Hopkins in 1893, becoming the second woman in the U.S. to receive a Ph.D. in geology.
- The two towers of the Charles Commons residence hall in Charles Village, which opened in 2006, would be named for Frederick Scott and Ernest Bates, who were among the university's first Black undergraduate students. Scott enrolled in the School of Engineering in 1945 and was the first Black student to earn a B.A. from Johns Hopkins; he went on to a career as an engineer and editor before his death in 2017. Bates, an acclaimed neurosurgeon and entrepreneur, became the first Black student at JHU's School of Arts and Sciences when he enrolled in 1954. He received an honorary doctorate from JHU in 2021.[115]
- Separate dedication ceremonies were planned to give each building its new name.[116]

The Sheikh Khalifa Stroke Institute

The Sheikh Khalifa Stroke Institute opened Feb. 1, 2018, with locations in Baltimore and in Abu Dhabi, United Arab Emirates (UAE).

The new institute builds on Johns Hopkins' leadership in the treatment of stroke and other neurological conditions, leveraging advances in engineering, gaming technology, artificial intelligence and precision medicine to better diagnose, treat and restore function to people who have had strokes.

It houses three centers of excellence:

- The center of excellence in stroke detection and diagnosis focuses on analyzing the genetic underpinnings of stroke and developing new interventions that might include digital neurological exams or automated bedside diagnosis, or using imaging in new ways to better understand stroke.
- The center of excellence in stroke treatment, recovery and rehabilitation develops new approaches to get blood vessels to regrow, delivering drugs and medications past the blood-brain barrier, and stimulating the brain to coax it to relearn movement post-stroke.
- The Kata Center for Neuro Restorative Technology and Design develops innovative gaming to be integrated into digital therapeutics.

The institute's founding director, Justin C. McArthur, also directs the Johns Hopkins Department of Neurology and is nationally and internationally recognized for his work in studying the natural history, development and treatment of HIV infection, multiple sclerosis and other neurological infections and immune-mediated neurological disorders.[117]

Pablo Celnik, who serves as co-director, is also director of the Johns Hopkins Department of Physical Medicine and Rehabilitation.

The institute was funded with a $50 million gift from the UAE, believed to be the largest ever for a stroke-specific initiative.

"We are grateful for the UAE's gift, which enables us to leverage our considerable strengths in neurology, physical medicine and rehabilitation — in combination with our expertise in biomedical engineering and patient safety — to develop new tools for stroke diagnosis, treatment and recovery," said Rothman. "These efforts will improve the health of millions of people in Baltimore, the UAE and around the world, with the added benefit of bringing down health care costs."

"This new institute will not only generate better outcomes for stroke patients in the UAE and the U.S. but will also strengthen opportunities for collaboration between UAE and U.S. scientists and researchers," said Yousef Al Otaiba, ambassador of the United Arab Emirates to the United States. "We are grateful to Johns Hopkins for their continued leadership in patient-centered medicine and are proud of our longstanding partnership."[118]

Preeti Raghavan, center, is inducted as Sheikh Khalifa Stroke Institute Professor of Stroke Treatment, Recovery and Rehabilitation. Pablo Celnik stands to her left, and Justin McArthur to her right.

New Opportunities for Retired Faculty

The Academy at Johns Hopkins, launched in October 2018, gives Johns Hopkins faculty opportunities to remain involved after they retire — to the benefit of students, trainees and the greater academic world.

The program connects retired clinicians, researchers and educators in the schools of medicine, nursing and public health with one another while encouraging ongoing scholarship. It also provides opportunities to precept, mentor and coach trainees and junior faculty; perform archival research or history of medicine projects; offer critical reviews of proposals, grants and manuscripts; serve on committees and advisory panels for the individual schools; and engage in community service.*

The academy was the brainchild of Janice Clements, who was vice dean for faculty, and Cynthia Rand, senior associate dean for faculty. It was enthusiastically supported and funded by Dean Paul Rothman.†

A renovation in the William H.

Left to right: Bill Baumgartner, inaugural chair of The Academy at Johns Hopkins; Patricia Davidson, dean of the school of nursing; Paul Rothman, dean of the school of medicine, and Janet DiPietro, vice dean of the school of public health.

Welch Medical Library created an academy suite, with comfortable seating, a kitchen, and a business center with five workstations.

More than 100 faculty members and staff attended the academy's opening ceremony on October 2. "We retirees all still possess years of experience and knowledge and would like to stay active and continue to be produc-

tive," said William Baumgartner, former vice dean for clinical affairs at the school of medicine and inaugural chairman of the

academy, who would retire at the end of that year.‡

In its first year, the academy welcomed 109 members, including 89 from the school of medicine, according to its 2019 annual report.

Together, academy members gave 3,400 hours of their time as teachers, mentors and community volunteers. They also published 94 journal articles, eight book chapters and two books.§

The following year, academy members gave 3,362 hours of their time. They responded to the COVID-19 pandemic in 2020 by staffing hotlines, participating in contact tracing and vaccinations, assembling masks and raising funds for community food shortages. They also continued mentoring and teaching, mostly through Zoom.¶

---

* Smith, Linell, "Creating a Senior Brain Trust," *Dome*, Nov. 15, 2018, https://www.hopkinsmedicine.org/news/articles/creating-a-senior-brain-trust.
† "The Academy at Johns Hopkins: 2019 Annual Report," https://www.hopkinsmedicine.org/the-academy/_docs/The_Academy_at_Johns_Hopkins_2019_Annual_Report.pdf.
‡ "Creating a Senior Brain Trust."
§ https://www.hopkinsmedicine.org/the-academy/_docs/The_Academy_at_Johns_Hopkins_2019_Annual_Report.pdf
¶ "The Academy at Johns Hopkins: 2019 Annual Report."

Skip Viragh Cancer Center Opens

On June 8, 2018, about 200 people, including family, friends and business associates of Skip Viragh and Johns Hopkins leaders and donors, celebrated Viragh and the opening of the new outpatient cancer building that bears his name.

Albert P. "Skip" Viragh Jr., a Maryland mutual fund investment leader and philanthropist, was treated at Johns Hopkins for pancreatic cancer and died of the disease in 2003 at age 62.

The grand opening was the culmination of Viragh's vision for a single place where cancer patients could have access to the best and most innovative cancer care.[119]

The building provides clinical services to

A New Vice Dean for Faculty

In 2021, Janice Clements stepped down as vice dean for faculty at the school of medicine, a post she had held since 2000. She remained professor of molecular and comparative pathobiology and the Mary Wallace Stanton Professor for Faculty Affairs.

Maria Oliva-Hemker, the Stermer Family Professor of Pediatric Inflammatory Bowel Disease and professor of pediatrics at the Johns Hopkins University School of Medicine, took the post in August 2021.

Oliva-Hemker earned her undergraduate degree in biology at Georgetown University and her medical degree at Johns Hopkins. She completed a residency in pediatrics and a fellowship in pediatric gastroenterology at Johns Hopkins. She joined the Johns Hopkins faculty in 1993.[*]

In 2011, she was promoted to full professor, becoming the second Hispanic woman and the third woman of color to achieve that title in the school of medicine.[†]

Clements earned her Ph.D. in biochemistry from the University of Maryland and completed two postdoctoral fellowships at Johns Hopkins — one in molecular biology and virology and the other in neurology.

She joined the Johns Hopkins faculty as an assistant professor of neurology in 1979 and then the faculty of the Division of Comparative Medicine in 1988. In 1990, she became the 24th woman to achieve the rank of professor at the Johns Hopkins University School of Medicine.[‡]

As a member of the steering committee of the Johns Hopkins Women's Leadership Council, Clements helped to create the new position of associate dean for diversity and cultural competence, and urged changes in the composition of search committees to be more inclusive of women and minorities. In 2008, she teamed with Barbara Fivush to create the Office of Women in Science and Medicine, which supports female faculty through education, advocacy, mentoring and networking.[§]

As director of the Retrovirus Laboratory, Clements led research on lentiviruses and their role in chronic neurological disease. She developed the first molecular and biochemical tools to study lentivirus molecular biology and was the first to characterize the unusual genome of the lentiviruses. She was also the first scientist to report that HIV is a lentivirus. Her recent discoveries include the use of minocycline, a common antibiotic often used against acne, to protect against viral HIV-related cognitive disease.[¶]

Maria Oliva-Hemker

* "Maria M Oliva-Hemker, M.D., Johns Hopkins Medicine webpage, https://www.hopkinsmedicine.org/profiles/details/maria-oliva-hemker.
† "Maria Oliva-Hemker," Wikipedia webpage, https://en.wikipedia.org/wiki/Maria_Oliva-Hemker.
‡ "Janice Clements, Ph.D." Johns Hopkins webpage, Women in Leadership, https://www.hopkinsmedicine.org/about/women-in-leadership/janice-clements.html.
§ "Johns Hopkins Medicine at 25."
¶ "Janice E. Clements, Ph.D.", Johns Hopkins Medicine webpage, https://www.hopkinsmedicine.org/profiles/details/janice-clements.

patients with solid tumors, accounting for more than 180 current patients daily and 60 to 80 new patients each week.[120]

Funding for the outpatient center came entirely from philanthropy, including a $65 million gift to honor Viragh and a $10 million gift from Under Armour, both given as part of Rising to the Challenge: The Campaign for Johns Hopkins.

"Skip was one of the most influential and innovative people I have ever met," said Daniel Laheru, co-director of the Skip Viragh Center for Pancreatic Cancer, who treated Viragh from his cancer diagnosis in 2002 until his death.

"Skip and his family have helped so many people with pancreatic cancer and, with this building, all cancer patients," said Laheru. "His name is a constant reminder to innovate and to never forget our core mission. Through the generosity of the Viragh family, his memory will endure forever as the heart of this building that will be a beacon of hope for patients and families for years to come."[121]

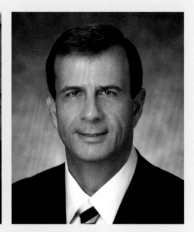

Annemarie Martin-Boyan   **Joanne Pollak**   **G. Daniel Shealer Jr**

The Lawyers Who Lead the Way

The lawyers who serve Johns Hopkins Medicine tend to stay out of the spotlight. They don't generally lead Grand Rounds, interact with patients or publish in prestigious journals.

But their work in the Johns Hopkins Health System Legal Department is essential.

The office provides legal and strategic advice that ensures the health system meets applicable laws and regulations as it navigates mergers and acquisitions, real estate transactions, labor and employment agreements, domestic and international business arrangements, health plan advising, liability claims management, and issues related to compliance, privacy, risk management, audit, corporate governance and more.

Beginning in February 2022, the office has been led by Annemarie Martin-Boyan, senior vice president and general counsel for The Johns Hopkins Health System Corp. and The Johns Hopkins Hospital.

Martin-Boyan oversees all internal and external legal matters for the hospital, health system and its affiliates, and serves as the primary legal adviser to the senior management team, the board of trustees, affiliate hospitals and the boards of those hospitals.

She came to Johns Hopkins with more than 25 years of legal experience, with previous roles that included senior vice president, corporate secretary and general counsel for Penn State Health; associate vice president and deputy general counsel at Rutgers-New Brunswick, in New Jersey; and associate general counsel for the Children's Hospital of Philadelphia.*

Before Martin-Boyan, the post had been filled on an interim basis by Peter Mancino, vice president and deputy general counsel, from September 2021 through February 2022.

G. Daniel Shealer Jr., who retired in September 2021, held the position from December 2011 through September 2021, providing legal support for the Johns Hopkins Health System.

Shealer, who received his bachelor's degree cum laude from Juniata College in Pennsylvania and his law degree from Vanderbilt University in 1984, arrived at Johns Hopkins as deputy general counsel in 1990, after practicing health care law at two Baltimore law firms.

He was appointed to the top legal job when its previous occupant, Joanne Pollak, became chief of staff of the Office of Johns Hopkins Medicine and chair of the Johns Hopkins Medicine Operating Committee.

Pollak had been lead attorney for the health system and Johns Hopkins Medicine since 1994, and was instrumental in the 1996 creation of Johns Hopkins Medicine as an entity combining the school of medicine with the health system.

"The joy and challenge of working at Hopkins is representing some of the greatest medical minds in the country and the world," she said in a 2006 article in University of Maryland School of Law's alumni magazine.†

The article, written to mark her acceptance of the school's Distinguished Graduate Award, described her move to Johns Hopkins from Piper & Marbury (now DLA Piper), where she was a partner and head of the firm's health care group.

"I really learned how to be a lawyer with lots of wonderful role models around me," she said. "Because we didn't have the internet and email, I was allowed to think and learn how to think. I got to sit in on meetings, talk about options, and contribute to decisions that might change over a day or two. My value at Hopkins is incredibly enhanced by the fact that I was taught to do those things."‡

Those skills also served Pollak as chief of staff, a job she held until 2019.¶

---

*   "Annemarie Martin-Boyan, Esq., Named Senior Vice President and General Counsel for Johns Hopkins Health System and The Johns Hopkins Hospital," Johns Hopkins Medicine news release, Dec. 16, 2021, https://www.hopkinsmedicine.org/news/newsroom/news-releases/annemarie-martin-boyan-esq-named-senior-vice-president-and-general-counsel-for-johns-hopkins-health-system-and-the-johns-hopkins-hospital.

†   Kolawowski, Ann Eichler, "Putting Health First, JD, Fall 2006, https://digitalcommons.law.umaryland.edu/cgi/viewcontent.cgi?referer=https://www.google.com/&httpsredir=1&article=1046&context=jd.

‡   "Putting Health First."

Moving Forward with Biophysics

L. Mario Amzel, director of the Department of Biophysics and Biophysical Chemistry since 2006, and interim director for three years before that, stepped down in May 2021. He continued to lead his research laboratory until he died the following August.

Born in Argentina, Amzel earned his doctorate in physical chemistry at the Universidad de Buenos Aires. He came to Johns Hopkins in 1969 as a postdoctoral fellow and remained, becoming a full professor in 1984.

He was a member of the Johns Hopkins team that produced the first high-resolution images of antibody-antigen recognition, the body's reaction to fighting off disease and illness, and his research also involved the cancer-related protein PI3K, which is part of the body's hormone signaling system.

His structural studies encompassed other proteins of great importance in human health and disease, including lipoxygenase, cytosolic quinone reductases, and monooxygenases. In addition, his research led to major discoveries in how proteins regulate sodium channels in cardiac cells, a key factor in enabling the heart to beat.

Amzel mentored and inspired generations of young scientists, and supported Johns Hopkins University programs that provided science opportunities to Baltimore City students, including paid research internships with graduate students and professors.[122]

The director role was filled in May 2021 by Cynthia Wolberger, who had joined the Johns Hopkins faculty in 1991.

Wolberger earned her undergraduate degree from Cornell University and her Ph.D. at Harvard University. She completed postdoctoral work at the University of California, San Francisco, and the Johns Hopkins University School of Medicine.

She studies how DNA packaging proteins—which coil DNA into neat, compact bundles in the cell—turn genes on or off. These proteins, also known as histones, are called to action by the addition of chemical tags, like ubiquitin protein or acetyl chemical groups.

Wolberger is a fellow of the American Association for the Advancement of Science and the Biophysical Society. She has been recognized with the Dorothy Crowfoot Hodgkin Award by The Protein Society for her work in determining the structure of proteins involved in transcriptional regulation and the American Association for Cancer Research Award for Outstanding Achievement in Chemistry in Cancer Research. She is a member of the National Academy of Sciences and the American Academy of Arts and Sciences, and was elected to the National Academy of Medicine in 2021.[123]

She is an oversight committee member and one of the most active users of the Beckman Center for CryoEM at Johns Hopkins, a cryo-electron microscopy facility on the first floor of the Rangos Building, created in 2016 with grants from the Arnold and Mabel Beckman Foundation, the Howard Hughes Medical Institute, the Johns Hopkins University School of Medicine, and private, anonymous donors.[124]

L. Mario Amzel

Cynthia Wolberger

CryoEM is used to determine atomic structures of biological samples. Inventors of the technology, which both simplifies and improves the imaging of biomolecules, were honored with the 2017 Nobel Prize in Chemistry.

In 2016, the Arnold and Mabel Beckman Foundation gave five grants of $2.5 million each to establish five such centers, including one at Johns Hopkins Medicine, with James Berger as its principal investigator.[125] Another grant from the Beckman Foundation, in December 2021, is bringing additional technologies to the center.[126]

Namandjé Bumpus

Namandjé Bumpus Takes the Helm in Pharmacology and Molecular Sciences

Namandjé Bumpus made history in 2020 when she became director of the Department of Pharmacology and Molecular Sciences for the Johns Hopkins University School of Medicine.

The promotion, announced May 13, made Bumpus the first Black woman to lead a department at the Johns Hopkins University School of Medicine and the only Black woman chairing a U.S. pharmacology department at the time.

"Being an African-American woman in science, I had not only the glass ceiling, but the solitude of often being first," she said. "One of the reasons I push and work so hard for these leadership roles is I feel like I need to be there as an advocate, and I need to be there as an example."

Bumpus researches genetic differences that influence how people metabolize antiretroviral drugs used to treat HIV. Her work, she said, aims to correct a longstanding limitation of scientific research.

"Most studies are done on European-American males in their 20s," she said. "There are so many people missing in that picture. We don't know how women are going to respond, or people of other ancestral origins. Our work is finding that many African Americans may respond differently to certain drugs, and now we're looking at how transgender women may respond."

Bumpus earned a bachelor's degree in biology from Occidental College in Los Angeles and a Ph.D. in pharmacology from the University of Michigan. She completed a two-year postdoctoral fellowship in molecular and experimental medicine at the Scripps Research Institute in La Jolla, California, in 2010, and joined Johns Hopkins in 2010 as an assistant professor in the Division of Clinical Pharmacology. She became an associate professor in 2015; from 2015 to 2017, she was the school's first associate dean for institutional and student equity, a role with particular resonance for her.

"That was a graduate student mentoring job with a focus on diversity," she said. "I wound up being a person that a lot of graduate students came to talk to. Luckily, the school gave me the opportunity to do it formally. I had that role for two years and met one-on-one with 450 graduate students from across the university to give them a safe space to talk about their lives and careers."

She became associate dean for basic research in 2018. In addition to her role as department director, Bumpus holds the E.K. Marshall and Thomas H. Maren Professorship in Pharmacology.[127]

On Aug. 1, 2022, Bumpus, on a leave of absence from Johns Hopkins, began a new assignment as the chief scientist of the U.S. Food and Drug Administration. "Dr. Bumpus will be greatly missed, but we are very proud that she will still be part of Johns Hopkins as she takes on this prominent position," wrote Rothman and Executive Vice Dean Landon King in a June 30, 2022 note to colleagues.[128]

Andrew Ewald Succeeds Peter Devreotes as Cell Biology Director

On Dec. 31, 2020, Peter Devreotes stepped down after 20 years as director of the Department of Cell Biology, one of the highest-ranking cell biology departments in the nation. Devreotes remained on the faculty, and cancer researcher Andrew Ewald became director of the department.

Ewald, known for his discoveries of how breast cancer cells leave the primary tumor during metastasis, joined Johns Hopkins in 2008 as an assistant professor in the Department of Cell Biology.

He earned his undergraduate degree in physics from Haverford College and his doctorate in biochemistry and molecular physics from the California Institute of Technology. He completed postdoctoral work at the University of California, San Francisco.

A member of the American Association for Cancer Research, Society for Developmental Biology and the American Society for Cell Biology, Ewald has chaired conferences on cell and cancer biology, given invited lectures across the U.S. and abroad, served as editor of the *Journal of Cell Science* and works closely with metastatic breast cancer patient advocacy organizations.

As of early 2022, he had authored more than 93 peer-reviewed journal articles and book chapters and trained more than 30 students and postdoctoral fellows.[129]

He also co-directs the Cancer Invasion and Metastasis Program at the Johns Hopkins Kimmel Cancer Center, as well as the Johns Hopkins-Allegheny Health Network Cancer Research Fund, a program that provides funding for basic, translational and clinical cancer research. He has secondary faculty appointments in biomedical engineering and oncology.

Devreotes, the Isaac Morris and Lucille Elizabeth Hay Professor at Johns Hopkins, studies how cells sense their surroundings and move directionally in processes like embryogenesis, wound healing and immune response.

He received his B.S. in physics from the University of Wisconsin and completed a Ph.D. in biophysics from The Johns Hopkins University. After a postdoctoral fellowship in biochemistry at the University of Chicago, he joined Johns Hopkins as an assistant professor in 1980, becoming an associate professor in 1985 and full professor in 1987.[130]

He is a member of the National Academy of Sciences, author of more than 218 peer-reviewed journal articles and 30 book chapters, and recipient of the 2019 E.B. Wilson Medal from the American Society for Cell Biology. He has trained more than 70 students and postdoctoral fellows.[131]

Andrew Ewald

Peter Devreotes

ENDNOTES

1 Paul Rothman interview with Karen Nitkin, Jan. 5, 2022.

2 Hoppe, Kim, "Kevin Sowers Named President of Johns Hopkins Health System, Executive Vice President of Johns Hopkins Medicine," *The Hub*, Dec. 11, 2017, https://hub.jhu.edu/2017/12/11/hopkins-health-system-president-kevin-sowers/.

3 Karen Nitkin Zoom interview with Kevin Sowers, Dec. 20, 2021.

4 "Kevin Sowers Named President of Johns Hopkins Health System, Executive Vice President of Johns Hopkins Medicine."

5 Kevin Sowers Named President of Johns Hopkins Health System, Executive Vice President of Johns Hopkins Medicine."

6 Kevin Sowers interview, Dec. 20, 2021.

7 "#25: A Clinical Road Map for the Future of Johns Hopkins Medicine," Johns Hopkins Medicine webpage, Johns Hopkins Medicine at 25, https://www.hopkinsmedicine.org/25-anniversary/events-achievements.html#25.

8 "#25: A Clinical Road Map for the Future of Johns Hopkins Medicine."

9 "Carolyn Carpenter Named National Capital Region President for the Johns Hopkins Health System," Johns Hopkins Medicine news release, Aug. 26, 2020, https://www.hopkinsmedicine.org/news/newsroom/news-releases/carolyn-carpenter-named-national-capital-region-president-for-the-johns-hopkins-health-system.

10 "A Conversation with Kevin Sowers and Carolyn Carpenter," YouTube on Johns Hopkins Medicine channel, https://www.youtube.com/watch?v=5Z8h7QbVWKQ.

11 "Johns Hopkins Community Physicians Opens Practice in McLean," Johns Hopkins Medicine news release, Feb. 2, 2022, https://www.hopkinsmedicine.org/news/newsroom/news-releases/johns-hopkins-community-physicians-opens-practice-in-mclean.

12 Wilen, Holden, "Johns Hopkins Plans First Phase at Belward Farm," *Baltimore Business Journal*, Sept. 24, 2021, https://www.bizjournals.com/baltimore/news/2021/09/24/johns-hopkins-plans-first-phase-at-belward-farm.html.

13 "Johns Hopkins Wins Approval for Development of Belward Farm," Headlines@Hopkins news release, Dec. 2, 1996, https://pages.jh.edu/news_info/news/univ96/dec96/belward.html.

14 Karem, Brian, "High Court Turns Down Belward Farm Case," Montgomery County Sentinel, April 22, 2014, https://www.thesentinel.com/communities/montgomery/news/local/high-court-turns-down-belward-farm-case/article_680d018d-f65e-5d89-b3af-45812f8eb661.html.

15 "Statement from Johns Hopkins Regarding Belward Litigation," Johns Hopkins University news release, Oct. 26, 2010, https://releases.jhu.edu/2012/10/26/belward-rulin/.

16 "Court of Special Appeals Rules for University on Belward," Johns Hopkins University media advisory, Nov. 22, 2013, https://releases.jhu.edu/2013/11/22/belward-appeal/.

17 Wilen, Holden, "Johns Hopkins Plans First Phase at Belward Farm," *Baltimore Business Journal*, Sept. 29, 2021, https://www.bizjournals.com/baltimore/news/2021/09/24/johns-hopkins-plans-first-phase-at-belward-farm.html.

18 Email from Carolyn Carpenter to Karen Nitkin, Feb. 8, 2022.

19 Gunts, Ed, "Johns Hopkins Health System Unveils Plans for $400 Million, 12-story Research Tower," *Baltimore Fishbowl*, Nov. 7, 2019, https://baltimorefishbowl.com/stories/johns-hopkins-health-system-unveils-plans-for-400-million-12-story-research-tower/.

20 "Prepare for Construction to Begin in The Johns Hopkins Hospital Starting Nov. 1," Johns Hopkins Medicine intranet communications, https://intranet.insidehopkinsmedicine.org/news-and-communications/leadership-updates/prepare-for-construction-jhh-nov-1.html.

21 DuVernay, Christina, "Johns Hopkins Regional Physicians Formed to Increase Access and Reach," *BestPractice*, June 18 2018, https://www.hopkinsmedicine.org/office-of-johns-hopkins-physicians/best-practice-news/johns-hopkins-regional-physicians-formed-to-increase-access-and-reach.

22 Email from Kim Sherbrooke to Karen Nitkin, Feb. 1, 2022.

23 "Johns Hopkins Regional Physicians Formed to Increase Access and Reach."

24 Karen Nitkin interview with Lisa Ishii, Nov. 4, 2021.

25 "Profile: Johns Hopkins Surgery Center Series", document provided by Lisa Ishii to Karen Nitkin, Nov. 4, 2021.

26 "Profile: Johns Hopkins Surgery Center Series."

27 Nitkin, Karen, "Campus Expansion Transforms the Care Experience," *Dome*, Jan. 29, 2020, https://www.hopkinsmedicine.org/news/articles/campus-expansion-transforms-the-care-experience.

28 McGrory, Kathleen, and Bedi, Neil, "Heartbroken," *Tampa Bay Times*, Nov. 28, 2018, https://projects.tampabay.com/projects/2018/investigations/heartbroken/all-childrens-heart-institute/.

29 "Heartbroken."

30 Karen Nitkin interview with Kevin Sowers, Dec. 20, 2021.

31 McGrory, Kathleen, and Bedi, Neil, "Top All Children's Executives Resign Following Times Report," *Tampa Bay Times*, Dec. 11, 2018, https://www.tampabay.com/investigations/2018/12/11/top-all-childrens-executives-resign-following-times-report/.

32 Karen Nitkin interview with Paul Rothman, Jan. 5, 2022.

33 McGrory, Kathleen, and Bedi, Neil, "All Children's Hires New Interim President," *Tampa Bay Times*, Feb. 8, 2019, https://www.tampabay.com/investigations/2019/02/08/all-childrens-hires-new-interim-president/.

34 McGrory, Kathleen, and Bedi, Neil, "Three More All Children's Officials Resign Following Times Investigation," *Tampa Bay Times*, Jan. 2, 2019, https://www.tampabay.com/investigations/2019/01/02/three-more-all-childrens-officials-resign-following-times-investigation/.

35 "A Message from Kevin Sowers," https://www.facebook.com/watch/?v=302180810642669.

36 "JHACH Investigation Recommendations," report to Special Committee of the Board of the Johns Hopkins Health System from Gibson, Dunn & Crutcher LLP, June 26, 2019, https://www.hopkinsallchildrens.org/getmedia/1b50f2dd-ecef-4c05-8169-585413c66930/JHACH-Gibson-Dunn-Recommendations.pdf;.aspx.

37 "Dr. James Quintessenza to Join Johns Hopkins All Children's Hospital to Lead Heart Institute," Johns Hopkins All Children's Hospital news release, Oct. 15, 2019, https://www.hopkinsallchildrens.org/ACH-News/Press-Release/Dr-James-Quintessenza-to-Join-Johns-Hopkins-All-C.

38 "Dr. James Quintessenza to Join Johns Hopkins All Children's Hospital to Lead Heart Institute."

39 Manning, Margie, "Johns Hopkins All Children's Hospital Hits Key Milestone," https://stpetecatalyst.com/johns-hopkins-all-childrens-hospital-hits-key-milestone/.

40 "K. Alicia Schulhof Appointed President of Johns Hopkins All Children's Hospital," Johns Hopkins Medicine news release, June 24, 2021, https://www.hopkinsmedicine.org/news/newsroom/news-releases/alicia-schulhof-appointed-president-of-johns-hopkins-all-childrens-hospital.

41 "Johns Hopkins All Children's Hospital Ranked #1 Children's Hospital in Florida," Johns Hopkins All Children's Hospital news release, June 15, 2021, https://www.hopkinsallchildrens.org/ACH-News/General-News/Johns-Hopkins-All-Children-s-Hospital-Ties-for-Top.

42 "A Heritage of Cardiac Surgery Excellence and Medical Leadership," *HopkinsPulse*, Spring 2018, https://www.hopkinsmedicine.org/news/articles/a-heritage-of-cardiac-surgery-excellence-and-medical-leadership.

43 Nitkin, Karen, "First-Ever Penis and Scrotum Transplant Makes History at Johns Hopkins," *Dome*, April 23, 2018, https://www.hopkinsmedicine.org/news/articles/first-ever-penis-and-scrotum-transplant-makes-history-at-johns-hopkins.

44 "Human Penile Allotransplantation," Johns Hopkins Medicine webpage, Reconstructive Transplant Program, https://www.hopkinsmedicine.org/transplant/programs/reconstructive-transplant/penis-transplant.html#eligibility.

45 "First-Ever Penis and Scrotum Transplant Makes History at Johns Hopkins."

46 Redett, Richard J., et al., "Total Penis, Scrotum, and Lower Abdominal Wall Transplantation," New England Journal of Medicine, letter to the editor, Nov. 7, 2019, https://www.nejm.org/doi/full/10.1056/NEJMc1907956.

47 "Bert Vogelstein, M.D., Awarded Breakthrough Prize in Life Sciences," Johns Hopkins Medicine news release, https://www.hopkinsmedicine.org/news/stories/bert_vogelstein_awarded_breakthrough_prize_in_life_sciences/.

48 "Single Blood Test Screens for Eight Cancer Types," Johns Hopkins Medicine news release, Jan. 18, 2018, https://www.hopkinsmedicine.org/news/newsroom/news-releases/single-blood-test-screens-for-eight-cancer-types.

49 "Single Blood Test Screens for Eight Cancer Types."

50 ASCO Post Staff, "At the Forefront of Cancer Genetics, Bert Vogelstein, MD, Calls for Focus on Early Detection and Prevention," The ASCO Post, June 3, 2017, https://ascopost.com/issues/june-3-2017-narratives-special-issue/at-the-forefront-of-cancer-genetics-bert-vogelstein-md-calls-for-focus-on-early-detection-and-prevention/.

51 "Bert Vogelstein, M.D., Wins 2015 Dr. Paul Janssen Award for Biomedical Research," Johnson & Johnson news release, June 17, 2015, https://www.prnewswire.com/news-releases/bert-vogelstein-md-wins-2015-dr-paul-janssen-award-for-biomedical-research-300100540.html.

52 "Kenneth W. Kinzler, Ph.D.," Johns Hopkins Medicine webpage, https://www.hopkinsmedicine.org/profiles/details/kenneth-kinzler.

53 "Johns Hopkins Cancer Biologist Kenneth Kinzler Elected to National Academy of Medicine," Johns Hopkins Medicine news release, Oct. 19, 2015, https://www.hopkinsmedicine.org/news/media/releases/johns_hopkins_cancer_biologist_kenneth_kinzler_elected_to_national_academy_of_medicine.

54 "Johns Hopkins Medicine at 25."

55 Gara, Catherine, "This is Our Manhattan Project," Hopkins Medicine magazine, Spring/Summer 2019, https://www.hopkinsmedicine.org/news/publications/hopkins_medicine_magazine/features/spring-summer-2019/this-is-our-manhattan-project.

56 Richards, Sarah, "New Algorithm Aims for Early Detection of Pancreatic Cancer," Insight, July 10, 2019, https://www.hopkinsmedicine.org/news/articles/new-algorithm-aims-for-early-detection-of-pancreatic-cancer.

57 "New Algorithm Aims for Early Detection of Pancreatic Cancer."

58 "Astronomy Meets Pathology to Identify Predictive Biomarkers for Cancer Immunotherapy," Johns Hopkins Medicine news release, June 10, 2021, https://www.hopkinsmedicine.org/news/newsroom/news-releases/astronomy-meets-pathology-to-identify-predictive-biomarkers-for-cancer-immunotherapy.

59 "Astronomy Meets Pathology to Identify Predictive Biomarkers for Cancer Immunotherapy."

60 "Johns Hopkins Investigators Win Life Sciences Award for Astropath Cancer Mapping Technology," Johns Hopkins Medicine news release, Dec. 9, 2021, https://www.hopkinsmedicine.org/news/newsroom/news-releases/johns-hopkins-investigators-win-life-sciences-award-for-astropath-cancer-mapping-technology.

61 "First-of-Its Kind Psychedelic Research Center Debuts at Johns Hopkins," BrainWise, Physician Update, Jan. 30, 2020, https://www.hopkinsmedicine.org/news/articles/first-of-its-kind-psychedelic-research-center-debuts-at-johns-hopkins.

62 "Johns Hopkins Center for Psychedelic & Consciousness Research," Johns Hopkins Medicine webpage, https://hopkinspsychedelic.org/index/.

63 "First-of-Its Kind Psychedelic Research Center Debuts at Johns Hopkins."

64 "Psychedelics Research and Psilocybin Therapy," Johns Hopkins Medicine webpage, Psychiatry and Behavioral Sciences, https://www.hopkinsmedicine.org/psychiatry/research/psychedelics-research.html.

65 "Hopkins Scientists Show Hallucinogen in Mushrooms Creates Universal 'Mystical' Experience, Johns Hopkins Medicine news release, June 11, 2006, https://www.hopkinsmedicine.org/Press_releases/2006/07_11_06.html.

66 "'Magic Mushrooms' Help Longtime Smokers Quit," Johns Hopkins Medicine news release, Sept. 11, 2014, https://www.hopkinsmedicine.org/news/media/releases/magic_mushrooms_help_longtime_smokers_quit.

67 "Milestone Study Shows Psychedelic Treatment with Psilocybin Relieves Major Depression," Johns Hopkins Medicine webpage, https://clinicalconnection.hopkinsmedicine.org/news/milestone-study-shows-psychedelic-treatment-with-psilocybin-relieves-major-depression.

68 Martinez, Marisol, "Johns Hopkins Receives First Federal Grant for Psychedelic Treatment Research in 50 Years," The Hub, Oct. 20, 2021, https://hub.jhu.edu/2021/10/20/first-nih-grant-for-psychedelics-in-50-years/.

69 "Johns Hopkins Receives First Federal Grant for Psychedelic Treatment Research in 50 Years." /

70 Matthew Johnson email to Karen Nitkin, Jan. 25, 2022.

71 Nitkin, Karen, "A Day of Celebration for Nobel Winner Gregg Semenza," Dome, Oct. 7, 2019, https://www.hopkinsmedicine.org/news/articles/a-day-of-celebration-for-nobel-winner-gregg-semenza.

72 "World Renowned Geneticist and Sickle Cell Disease Expert Takes Helm of Genetic Medicine Department at Johns Hopkins," Johns Hopkins Medicine news release, Dec. 28, 2021, https://www.hopkinsmedicine.org/news/newsroom/news-releases/world-renowned-geneticist-and-sickle-cell-disease-expert-takes-helm-of-genetic-medicine-department-at-johns-hopkins.

73 "World Renowned Geneticist and Sickle Cell Disease Expert Takes Helm of Genetic Medicine Department at Johns Hopkins."

74 Grauer, Neil A., Leading the Way, Johns Hopkins Medicine in Association with The Johns Hopkins University Press, 2012, pp. 85-87.

75 Leading the Way, p. 93

76 "David Valle Honored with ASHG's Victor A. McKusick Leadership Award," Johns Hopkins Medicine news release, July 15, 2014, https://www.hopkinsmedicine.org/news/media/releases/david_valle_honored_with_ashgs_victor_a_mckusick_leadership_award.

77 "David Valle Honored with ASHG's Victor A. McKusick Leadership Award."

78 "Jeremy Nathans, M.D., Ph.D.," Johns Hopkins Medicine webpage, https://www.hopkinsmedicine.org/profiles/details/jeremy-nathans.

79 "Jeremy Nathans, M.D., Ph.D."

80 https://news.ucsc.edu/2020/05/carol-greider.html.

81 "'Telomere' Expert Carol Greider Shares 2009 Nobel Prize in Physiology or Medicine," Johns Hopkins Medicine news release, Oct. 5, 2009, https://www.hopkinsmedicine.org/news/media/releases/telomere_expert_carol_greider_shares_2009_nobel_prize_in_physiology_or_medicine.

82 "'Telomere' Expert Carol Greider Shares 2009 Nobel Prize in Physiology or Medicine."

83 Jensen, Brennen, "New Cochlear Center Will Test Whether Treating Age-Related Hearing Loss Can Delay Dementia," Hopkins Bloomberg Public Health magazine, 2018, https://magazine.jhsph.edu/2018/new-cochlear-center-will-test-whether-treating-age-related-hearing-loss-can-delay-dementia.

84 "Frank Lin, M.D., Ph.D.," Johns Hopkins Medicine webpage, https://www.hopkinsmedicine.org/profiles/details/frank-lin.

85 "New Cochlear Center Will Test Whether Treating Age-Related Hearing Loss Can Delay Dementia."

86 Cochlear Center for Hearing and Public Health 2018 Annual Report, https://jhucochlearcenter.org/sites/default/files/2019-04/2018.Annual_Report_Web_Page_Final_190411V8.pdf.

87 Hub staff report, "Johns Hopkins to Create Center for Hearing Loss Research, Clinical Care," The Hub, Oct. 13, 2015, https://hub.jhu.edu/2015/10/13/rubenstein-gift-to-department-of-otolaryngology/.

88 Hub staff report, Johns Hopkins Department of Otolaryngology-Head and Neck Surgery Receives $15M Contribution," The Hub, Jan. 7, 2021, https://hub.jhu.edu/2021/01/07/david-rubenstein-gift-otolaryngology-head-neck-surgery/.

89 Byrne, Richard, "A Noteworthy Endeavor," *Hopkins Medicine* magazine, Spring/Summer 2017, https://www.hopkinsmedicine.org/news/publications/hopkins_medicine_magazine/features/spring-summer-2017/a-noteworthy-endeavor.

90 "Sarah Adams Hoover" Johns Hopkins Peabody Institute webpage, https://peabody.jhu.edu/explore-peabody/our-leadership/sarah-hoover/.

91 "Treatments for Musicians and Dancers," Johns Hopkins Medicine website, https://www.hopkinsmedicine.org/center-for-music-and-medicine/care-for-performers/index.html.

92 "A Noteworthy Endeavor."

93 Pitts, Jonathan M., "Emerging Hopkins Center Harmonizing Music and Medicine," *Baltimore Sun*, May 26, 2017, https://www.baltimoresun.com/health/bs-hs-music-and-medicine-20170518-story.html.

94 "Center for Music and Medicine."

95 Email from Alexander Pantelyat to Karen Nitkin, Jan. 10, 2022.

96 "Center for Music and Medicine."

97 Achenbach, Sarah, "Make Way for the Surgineer," *Hopkins Medicine* magazine, Winter 2020, https://www.hopkinsmedicine.org/news/articles/make-way-for-the-surgineer.

98 Siewerdsen, Jeff, "Paging the Surgineer," *Hopkins Medicine* magazine, Fall 2018, https://www.hopkinsmedicine.org/news/publications/hopkins_medicine_magazine/forum/fall-2018/paging-the-surgineer.

99 Wilson, Jessica, "The Power of Artificial Intelligence," *Wilmer* magazine, July 1, 2019, https://www.hopkinsmedicine.org/news/articles/the-power-of-artificial-intelligence.

100 "The Power of Artificial Intelligence."

101 "Who/What March 2016," *Dome*, Feb. 25, 2016, https://www.hopkinsmedicine.org/news/articles/whowhat-march-2016.

102 Nitkin, Karen, "Johns Hopkins Medical Education Goes Digital," *Dome*, Sept. 12, 2018, https://www.hopkinsmedicine.org/news/articles/johns-hopkins-medical-education-goes-digital.

103 Nitkin, Karen, "An Academic Year Like No Other," Dome, Sept. 8, 2020, https://www.hopkinsmedicine.org/news/articles/at-the-johns-hopkins-university-school-of-medicine-an-academic-year-like-no-other.

104 "An Academic Year Like No Other."

105 "The Humanities: Critical to Tomorrow's Doctors," *Breakthrough*, Center for Innovative Medicine, Winter 2021, https://www.hopkinscim.org/breakthrough/winter-2021/the-humanities-critical-to-tomorrows-doctors/.

106 Media Medica, program abstracts, https://mediamedicasite.wordpress.com/program/abstracts/.

107 "The Future of Biomedical Education: Part Three," Biomedical Odyssey blog, Oct. 11, 2019, https://biomedicalodyssey.blogs.hopkinsmedicine.org/2019/10/the-future-of-biomedical-education-part-three/.

108 Schuessler, Jennifer, "Johns Hopkins Reveals that its Founder Owned Slaves," *New York Times*, Dec. 9, 2020, https://www.nytimes.com/2020/12/09/arts/johns-hopkins-slavery-abolitionist.html.

109 "Who Was Johns Hopkins?" Johns Hopkins Medicine webpage, About Johns Hopkins Medicine, https://www.hopkinsmedicine.org/about/history/johns-hopkins.html.

110 "Hopkins Retrospective: 1876-Today," Johns Hopkins University webpage, https://retrospective.jhu.edu/.

111 "Reexamining Hopkins History," Johns Hopkins University webpage, Hopkins Retrospective: 1876-Today," https://retrospective.jhu.edu/our-initiatives/reexamining-hopkins-history.

112 Daniels, Ronald J.; Rothman, Paul B.; Sowers, Kevin W., "Reexamining the History of our Founder," Letter to Johns Hopkins community, Dec. 9 2020, https://president.jhu.edu/meet-president-daniels/speeches-articles-and-media/reexamining-the-history-of-our-founder/.

113 Email from Sherita Golden to Karen Nitkin, May 13, 2022.

114 Hub staff report, "Johns Hopkins Launches Effort to Honor its Diverse History through Named Buildings, Programs," The Hub, April 30, 2021, https://hub.jhu.edu/2021/04/30/diverse-names-and-narratives-project-launch/.

115 Hub staff report, "New Building, Facility Names will Honor Four Pioneering Figures from Johns Hopkins History," *The Hub*, Oct. 29, 2021, https://hub.jhu.edu/2021/10/29/diverse-names-narratives-watkins-bascom-bates-scott/.

116 "New Building, Facility Names will Honor Four Pioneering Figures from Johns Hopkins History."

117 "Sheikh Khalifa Stroke Institute," Johns Hopkins Medicine webpage, https://www.hopkinsmedicine.org/neurology_neurosurgery/centers_clinics/khalifa-stroke-institute/index.html.

118 "Johns Hopkins and the United Arab Emirates Launch Sheikh Khalifa Stroke Institute," Johns Hopkins Medicine news release, Feb. 1, 2018, https://clinicalconnection.hopkinsmedicine.org/news/johns-hopkins-and-the-united-arab-emirates-launch-sheikh-khalifa-stroke-institute.

119 "Celebrating the Opening of Skip Viragh Outpatient Cancer Building," *Pancreatic Cancer Matters*, Dec. 26, 2018, https://www.hopkinsmedicine.org/news/articles/celebrating-the-opening-of-skip-viragh-outpatient-cancer-building.

120 "Groundbreaking Set for Additional Kimmel Cancer Center Building, Johns Hopkins Medicine news release, Sept. 9, 2015, https://www.hopkinsmedicine.org/news/media/releases/groundbreaking_set_for_additional_kimmel_cancer_center_building.

121 "Celebrating the Opening of Skip Viragh Outpatient Cancer Building."

122 "In Memoriam: Mario Amzel, Former Department Director of Biophysics and Biophysical Chemistry," Alumni News, Sept. 22, 2021, https://www.hopkinsmedicine.org/news/publications/alumni-news/in-memoriam-mario-amzel-former-department-director-of-biophysics-and-biophysical-chemistry.

123 "Dr. Cynthia Wolberger," Johns Hopkins Medicine webpage, Wolberger Lab, https://wolberger.med.jhmi.edu/people/dr-cynthia-wolberger/.

124 "Dr. Cynthia Wolberger."

125 "Arnold and Mabel Beckman Center for Cryo-EM," Arnold and Mabel Beckman Foundation webpage, Programs We Fund, https://www.beckman-foundation.org/programs/beckman-center-cryo-em/.

126 "New Grant from the Beckman Foundation for FIB-SEM," Johns Hopkins Medicine webpage, Beckman Cryo-EM, Dec. 16, 2021, https://cryoem.jhmi.edu/2021/12/16/new-grant-from-the-beckman-foundation-for-fib-sem/.

127 Nitkin, Karen, "Namandjé Bumpus Makes History with New Department Chair Role," *Dome*, June 8, 2020, https://www.hopkinsmedicine.org/news/articles/namandj-bumpus-makes-history-with-new-department-chair-role.

128 "Namandje Bumpus Appointed as FDA Chief Scientist," message from the Office of the Dean to the school of medicine community and JHM leadership, June 30, 2022.

129 Email from Andrew Ewald to Karen Nitkin, Feb. 11, 2022.

130 "Peter N. Devreotes, Ph.D.," Johns Hopkins Medicine webpage, https://www.hopkinsmedicine.org/profiles/details/peter-devreotes.

131 "Johns Hopkins Cell Biologist Wins Professional Association's Highest Honor," Johns Hopkins Medicine news release, Oct. 1, 2019, https://www.hopkinsmedicine.org/news/newsroom/news-releases/johns-hopkins-cell-biologist--wins-professional-associations-highest-honor.

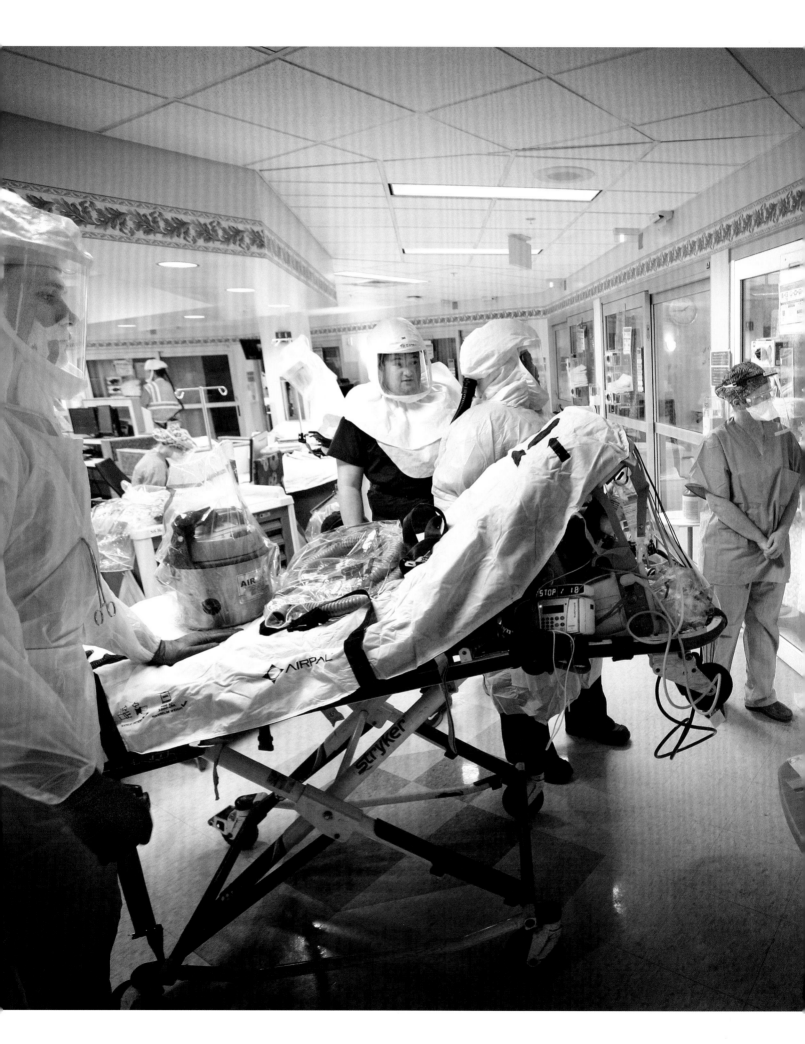

Chapter Five

COVID-19 Changes Everything (2019–)

Members of a critical care transport team wear personal protective equipment in April 2020 as they prepare to enter a patient room at Howard County General Hospital.

T HE FIRST SIGNS OF A PANDEMIC that would change the course of world history came in mid-December 2019.

A cluster of patients in Wuhan, China, reported shortness of breath and fever. On December 31, the unusual pneumonias were reported to the World Health Organization (WHO) in China, and on Jan. 10, 2020, the Centers for Disease Control and Prevention (CDC) had enough information about the novel coronavirus, which it dubbed 2019-nCoV, to publish information on its website.[1]

Meanwhile, infectious disease specialists at Johns Hopkins were already preparing. On Wednesday, Jan. 1, 2020, Johns Hopkins begins its response, activating the alert phase for the Johns Hopkins Medicine Office of Emergency Management, the Office of Critical Preparedness and Response (CEPAR), and the Department of Hospital Epidemiology and Infection Control (HEIC).[2]

"When we first saw the outbreak in Wuhan, we immediately as a team met and realized it would come to the U.S.," said Kevin Sowers, president of the Johns Hopkins Health System and executive vice president of Johns Hopkins Medicine, in December 2021.

"Knowing the science behind pandemics, and viruses like this, we put plans in place — that was back in January 2020. We started ordering supplies and materials, set up the command center."[3]

At the time, little was known about the coronavirus that caused COVID-19, except that it was highly contagious and potentially lethal. Scientists did not know how it was transmitted or why some people got sicker than others. There weren't yet tests, and vaccines were almost a year away. Nobody could know how long the pandemic would linger, how many lives it would claim and disrupt.

Throughout the pandemic, Johns Hopkins Medicine set itself apart as a beacon of light through the seemingly endless confusion, never wavering from its mission to keep staff and patients safe while providing caring, evidence-based treatment and clear-eyed information.

"We all are very proud of how this institution has risen to the challenge," said Dean/CEO Paul Rothman in January 2022, when the health system was coping with a fifth patient surge due to the emergence of a new Omicron variant.

"To this day I feel very good that we've kept our patients and our staff safe the entire time. It's the number one issue," he continued.

"If you look at our patient outcomes, they're top 10 in the country. We've done a great job caring for people. We've used our research arm to really make contributions in terms of understanding the disease and treating the disease.

"We've had a lot of leaders both within medicine and throughout the university who have stepped up and helped lead the country and the world in terms of looking at the science, giving the real guidance that people needed."[4]

Johns Hopkins hospitals began treating patients with COVID-19 in March 2020, the same month the WHO categorized the disease as a global pandemic.[5]

Development of several vaccines at the end of 2020 seemed to spell the beginning of the end, yet the pandemic dragged on. In January 2022, for example, cases and hospitalizations were once again surging, fueled by the highly transmissible Omicron variant.

On May 17, 2022, the U.S. officially passed the sobering milestone of 1 million reported deaths from the disease, according to the Johns Hopkins University and Medicine Coronavirus Resource Center — a website created in January 2020 that quickly become one of the world's most relied-upon sources for COVID-19 information. Worldwide, the death toll exceeded 6.2 million, though both numbers likely understated the actual number of deaths from COVID-19.[6]

At Johns Hopkins hospitals, more than 13,000 patients had been treated for COVID-19 by the end of June 2022, including 1,084 who had died.[7]

The disease and the way people, institutions and governments responded to it seeped into already too-large racial, economic and political cracks in American life, inevitably widening these potent divisions.

Johns Hopkins leaders navigating this bumpy landscape used the best available knowledge to make exquisitely painful decisions, always with the goals of protecting human life, improving health equity and limiting the spread of the virus.

These decisions included, early in the pandemic, cutting back on non-urgent procedures, restricting the number of visitors per patient, and, later, requiring full vaccination and booster shots — with very few exceptions — as a condition for employment.

"We purposely said we'd make our decisions and stick to the science," said Sowers. "What people believe are rights don't necessarily apply in a public health crisis when you're trying to get a pandemic under control."

A key decision, Sowers said, was to collaborate with the University of Maryland Medical System on several key programs and policies, including a field hospital at the Baltimore Convention Center and a vaccination requirement for employees.

"Early on, I went to Dr. Rothman and we agreed that if we tried to do this by ourselves it was not going to work," said Sowers. "I spent time with Dr. Suntha [Mohan Suntha, president and CEO of the University of Maryland Medical System] and we agreed to align. We've become better because of our partnership."[8]

As Rothman and Sowers noted in a July 28, 2020 U.S. News and World Report editorial, "A successful pandemic response requires collaboration among individuals and organizations from the public and private sectors as never before. We've had

to coordinate our efforts across ZIP codes and state and county lines, taking care of all segments of our communities and mindfully protecting our most vulnerable individuals, families and communities.[9]

A History of Preparedness

It's no coincidence that Johns Hopkins became the institution that set the standard for COVID-19 care, research and information.

A century earlier, it had played a similar role during the flu pandemic of 1918, which killed an estimated 21 million people around the world and sparked similar fears, distrusts and divisions in the United States, according to *The Great Influenza: The Epic Story of the Deadliest Plague in History*.

The 2004 book, written by John Barry, describes how Johns Hopkins, with its devotion to rigorous medical education and research, created a first generation of scientists who worked to understand the disease.[10]

More recently, Johns Hopkins Medicine had created a structure for responding to health emergencies with the 2002 launch of the Office of Critical Event Preparedness and Response (CEPAR).

The impetus came in September and October 2001, when U.S. Senate offices and media outlets received letters containing anthrax. Fearing that they had been exposed to the poison, two postal workers sought treatment at Johns Hopkins — one at The Johns Hopkins Hospital and one at Johns Hopkins Bayview Medical Center. The clinicians at The Johns Hopkins Hospital were getting one stream of advice from local and federal agencies, while those at Johns Hopkins Bayview received entirely different directions.

Gabor Kelen, director of emergency medicine, saw the need for a single structure during multisite emergencies. In July 2002, Johns Hopkins Medicine launched CEPAR, with Kelen as chief and James Scheulen as executive director.

Biocontainment Unit

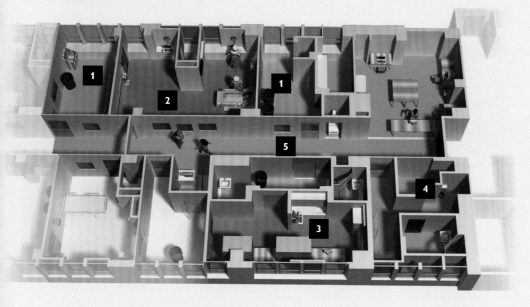

Clinicians learn how to safely put on and take off personal protective equipment.

The illustration shows a portion of the biocontainment unit.

1. Patient rooms have dedicated space for donning (green) and doffing (yellow) personal protective equipment.

2. One of the three private patient rooms.

3. An on-site laboratory for safe and rapid diagnostic testing, with its own donning and doffing rooms.

4. The staff entrance has showering facilities and serves as a clean-in/clean-out anteroom for the entire unit.

5. The special air handling system removes airborne pathogens.

The Johns Hopkins Biocontainment Unit (BCU) opened in the spring of 2015 in response to the Ebola outbreak in West Africa. It quickly became one of the top units in the world for treating people with highly infectious illnesses, while keeping pathogens contained in the unit.

Because of the BCU, Johns Hopkins was better equipped than most healthcare systems to treat patients with COVID-19.

"As one of the 10 federally funded Regional Emerging Special Pathogens Treatment Centers (RESPTCs) we knew that our team would care for the first persons under investigation and the first confirmed cases," said BCU medical director Brian Garibaldi.[*]

The BCU was created with the support of Ronald R. Peterson, then president of the Johns Hopkins Health System and executive vice president of Johns Hopkins Medicine; and Redonda Miller, then senior vice president of

medical affairs for the Johns Hopkins Health System. The medical leads were Garibaldi and Lisa Maragakis, the health system's senior director of infection prevention. Ron Langlotz, director of nursing for the departments of medicine and radiology, was the nursing lead.

The 7,900-square-foot unit at The Johns Hopkins Hospital includes three patient rooms, an on-site laboratory and capability to perform routine surgical procedures, as well as showers and clean-in/clean-out anterooms for health care providers and pass-through autoclaves for infectious medical waste. Its ventilation system is separate from the rest of the hospital.[†]

It stood ready to serve a sprawling region that encompassed Delaware, Maryland, Pennsylvania, Virginia, Washington, D.C. and West Virginia.[‡]

During its first five years of operation, the BCU team trained more than 400 Johns Hopkins staff members in the safe care of patients with serious communicable diseases and participated in regional and national discussions on how to prepare the U.S. health care system for the next infectious disease outbreak.

As the COVID-19 pandemic emerged at the end of 2019, the team pivoted to front-line training sessions to help area hospitals prepare, through measures such as robust screening for respiratory symptoms and recent travel history, and creating personal protective equipment (PPE) ensembles and plans.

The program culminated in a statewide symposium on high-consequence pathogens, hosted by The Johns Hopkins Hospital on Feb. 20, 2020, in partnership with the Maryland Department of Health. About 450 people attended, including representatives from every county in Maryland.[§]

Over the course of four days in March 2020, the BCU and the Hospital Epidemiology and Infection Control (HEIC) teams trained 150 health care workers at The Johns Hopkins Hospital, including clinical educators, clinical support staff, providers and nurses in how to put on (don) and take off (doff) PPE in order to protect against infection. These staff became ambassadors and a resource to reinforce PPE best practices throughout the hospital.[¶]

*   Brian Garibaldi email to Karen Nitkin, April 22, 2022.
†   Callaway, Marian, "If the Worst Happens ..." *Hopkins Medicine* magazine, Summer 2015, https://www.hopkinsmedicine.org/news/publications/hopkins_medicine_magazine/circling_the_dome/springsummer-2015/if-the-worst-happens.
‡   "Johns Hopkins Biocontainment Unit Annual Report: FY 2020," The Johns Hopkins Hospital, https://www.hopkinsmedicine.org/biocontainment-unit/_documents/2020-BCU-Annual-Report-FY20.pdf.
§   "Emerging Infectious Disease Symposium," *Medicine Matters*, Johns Hopkins Department of Medicine blog, Feb. 25, 2020, https://medicine-matters.blogs.hopkinsmedicine.org/2020/02/emerging-infectious-disease-symposium/.
¶   "Johns Hopkins Biocontainment Unit Annual Report: FY 2020."

One of CEPAR's strengths is its flexibility, they both said. Responding to Hurricane Katrina in 2005, CEPAR created the Johns Hopkins Go Team to react to emergencies outside the hospital walls, which became a valuable resource for testing and vaccinating people during COVID-19.

In response to Ebola in 2014, CEPAR was involved in development of a biomedical containment unit (BCU) for patients with highly infectious diseases, which was used extensively during the COVID-19 pandemic.

The COVID-19 pandemic marked the first time CEPAR set up an Incident Command, a key part of the Unified Command structure established at the same time. The Unified Command, chaired by Kelen, was the decision-making body, with Incident Command identifying and resolving issues.

Established on March 2, 2020, the Unified Command's experts and leaders from across the health system and university met around the clock at first, identifying the top problems of the day and setting out to find solutions.

One group might develop infection control protocols, while another led creation of a call center to answer employee questions and a third determined best practices for collecting and analyzing test results.[11]

"CEPAR allows us to tap any Johns Hopkins expert to do whatever needs to be done," said Kelen. "It's like an immune response. When there's no crisis, you don't know it's there. But when you're exposed, it responds to the extent that's needed."[12]

Coronavirus Resource Center

The Coronavirus Resource Center, launched Jan. 23, 2020, instantly became the go-to source for real-time reliable data about the pandemic, followed closely and cited by government leaders, health officials, major media outlets and concerned citizens.

Just days earlier, on January 20, the CDC had confirmed the first U.S. coronavirus case, in a Washington State man who had recently traveled to Wuhan, China.[13]

Lauren Gardner, associate professor and co-director of the Center for Systems Science and Engineering in the Johns Hopkins University Whiting School of Engineering, convened a group of graduate students to create one of the first global maps to track COVID-19 cases and deaths around the world.[14]

It soon became clear that a system of manual data collection and entry could not keep up with the fast-growing pandemic. On January 27, the Johns Hopkins University Applied Physics Lab (APL) joined the effort.

With $250,000 in initial funding from Johns Hopkins University President Ronald J. Daniels, the site evolved on March 3, 2020, to become the Johns Hopkins Coronavirus Resource Center, an interdisciplinary collaboration with experts in disease, public

Lauren Gardner (right), co-creator of the global COVID-19 map, and Beth Blauer, data lead for the Coronavirus Resource Center.

health and engineering from across Johns Hopkins Medicine and University.[15] Subsequent support came from the Stavros Niarchos Foundation and Bloomberg Philanthropies.[16]

By March 2021, the site was using more than 260 sources to track real-time metrics related to cases, testing efforts, test positivity, contact tracing efforts, disparities, hospitalizations, deaths and vaccinations. Information on the site, including news articles, webinars and a free, five-module e-learning program, has been accessed billions of times, helping the public, policymakers and health care professionals worldwide understand and respond to the pandemic.[17]

A hunger for COVID-19 information was also apparent to the marketing and communications team of Johns Hopkins Medicine, which reimagined its website in the early days of the pandemic to provide crucial, timely information.

"We wanted to focus on what are we hearing from the public, what are the most important things that we needed to get answers to and align that with what public health experts are saying," said Aaron Watkins, senior director of internet strategy at Johns Hopkins Medicine. "We communicated and worked on that content on a daily basis."[18]

The site's first coronavirus article, "What is Coronavirus?," was published Jan. 27, 2020; a year later, the Johns Hopkins Medicine coronavirus website had published more than 200 pages of content, attracting 55 million visits.[19]

COVID-19 Arrives

March 12, 2020 — a Thursday — was the day everything changed.

That's when The Johns Hopkins Hospital admitted its first patient with COVID-19, who was treated in the hospital's biocontainment unit (BCU) and discharged March 16, 2020.

The same day, Maryland's governor, Larry Hogan, announced new restrictions to control the spread of the disease, including an immediate ban on all gatherings of more than 250 people and a requirement that all schools close for two weeks, starting March 16. Hospitals were advised to limit visitors, and screen them for flu-like symptoms or possible exposure.[20]

Across the Johns Hopkins Health System, hospitals restricted access and check-in protocols were established. Chairs were removed from waiting rooms and stickers were placed on floors in order to promote the six-foot-apart physical distancing that was recommended to slow spread of the disease.

JHM research labs and offices closed. The school of medicine, along with the rest of The Johns Hopkins University, shifted to remote learning.

The mantra in those days was "flatten the curve," a recognition that hospitals throughout the United States would be overwhelmed if too many people needed treatment at the same time.

As the pandemic ground on, passing the one-year and then the two-year mark, the restrictions would ebb and flow, but the changes brought about by the deadly coronavirus would be felt throughout the organization.

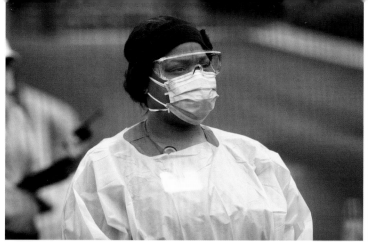

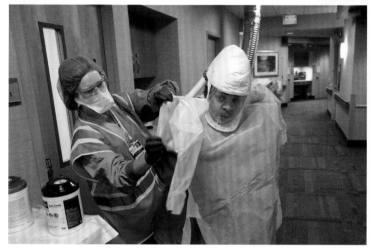

Across Johns Hopkins Medicine, employees rose to the challenges presented by COVID-19.

The First Patient

Though he had expected the day for months, Garibaldi still felt a jolt of adrenaline tinged with fear when he saw the results. The 61-year-old patient had tested positive for the coronavirus that causes COVID-19.

"It was a pretty scary time," recalled Garibaldi, also an associate professor of pulmonary and critical care medicine.

"In addition to concerns about our own safety as providers, and the safety of our families and friends, we really had no idea what to expect clinically from these patients," Garibaldi said in March 2021, in a *Dome* article marking the first year of COVID-19 in the Johns Hopkins Health System. "We didn't know the natural progression of this disease, how to interpret lab data or what treatments we could offer."

The patient had arrived in the emergency department feeling so weak she could barely lift her cell phone.

"I was terrified," she said of learning she had COVID-19. She praised Timothy Niessen, the internal medicine doctor who provided care. "He told me not to be afraid," she recalled.

She was treated in the BCU, never required ventilation and was released four days later. It took a few weeks to regain her strength, and she felt fine ever since, she said in 2021.

"In those early days, I think we were all a little frightened — both patients and their doctors," said Niessen. "There were no defined treatments, and our research programs weren't up and running yet. The people in the room, even in those loud and bulky air purifying respirators, had a huge role in reassuring our patients that whatever comes, we'd confront it together."[21]

Brian Garibaldi

Heba Mostafa (front), director of the molecular virology lab, and Karen Carroll, director of the Division of Medical Microbiology, together developed a COVID-19 diagnostic test.

Leading the Way on COVID Testing

Johns Hopkins was one of the first institutions in the state to provide its own testing for COVID-19.

In the pandemic's earliest days, when screening tests were both scarce and slow, Johns Hopkins clinical microbiologists Karen Carroll and Heba Mostafa created an in-house version of the test.

The breakthrough was achieved with participation from the entire medical microbiology lab as well as "amazing departmental and institutional support," noted Carroll, professor of pathology and director of the Division of Medical Microbiology at Johns Hopkins.

It was important because CDC testing kits, which the Food and Drug Administration (FDA) had approved on Feb. 4, 2020, were scarce, and results, which were reported from state public health laboratories, often took three to five days.

On February 29, the FDA began allowing academic medical centers to develop their own tests, which were fast-tracked for approval.

Johns Hopkins used its test, which analyzed material on a nasal or oral swab, for the first time on March 11, 2020; 81 tests were performed in the first three days and 10 came back positive, said Mostafa, assistant professor of pathology and director of the molecular virology lab.[22]

The need for testing was almost limitless, and Johns Hopkins Medicine rose to the challenge.

The microbiology laboratory validated and implemented 11 different testing platforms, allowing it to rapidly multiply the number of tests it could process, while also limiting supply chain difficulties by switching from one platform to another depending on availability of supplies.

On April 10, 2020, the clinical immunology laboratory, led by immunopathologist Mario Caturegli, launched a serum test to detect the presence of antibodies against the SARS-CoV-2 virus, used to assess COVID-19 prevalence and aid diagnosis.[23]

In November 2020, researchers from the school of medicine and the Johns Hopkins Bloomberg School of Public Health developed a test that used small amounts of saliva to detect antibodies to SARS-CoV-2, and delivered results in hours.[24]

The molecular genetics laboratory led by Christopher Gocke, director of the Division of Molecular Pathology, stood up a high-throughput saliva testing laboratory that was key to bringing university students back to campus that fall.[25]

Point-of-care tests were also hugely important in helping clinicians screen asymptomatic patients and triage those with symptoms.[26] The high volumes of such tests at The Johns Hopkins Hospital were handled by the Point of Care Testing Laboratory directed by pathology professor William Clarke and managed by Jeanne Mumford.[27]

By November 2021, the JHM Department of Pathology systemwide laboratories had performed 1 million tests, an occasion marked with an outdoor celebration in the Peterson Courtyard. Speakers including Rothman, Sowers, Miller, senior lab administrator Al Valentine and Ralph Hruban, pathology director.[28]

Unified Command Center

Perhaps the most frightening thing about COVID-19 in the early days was that the coronavirus that caused it was new. Virtually nothing was known yet about how the virus leapt from person to person and what it did to the human body. Clinicians worried about their own safety as they scrambled to learn which treatments and protocols worked best for which patients.

Lisa Maragakis   **Mary Brown**

The Johns Hopkins Medicine Unified Command Center became an essential resource for providing information and guidance to the entire institution during the rapidly changing crisis.

"This sort of a command center had never been set up before across Johns Hopkins Medicine," said Rob Carter, Johns Hopkins Medicine's senior director of emergency management and co-incident commander along with Maragakis.[29]

Mary Brown, director of emergency management for The Johns Hopkins Hospital, said her team was first alerted to a potential situation on Jan. 1, 2020, when the WHO issued a notice that there had been a cluster of cases of an unidentified pneumonia in mainland China.

The offices of Emergency Management, Critical Event Preparedness and Response (CEPAR) and Hospital Epidemiology and Infection (HEIC) went into "active alert," she said, as they had before for other global health crises, including the emergence of Ebola in 2014 and Zika in 2015.

For COVID-19, the first situational awareness meeting to discuss what was happening and how to prepare took place Jan. 22, 2020.[30]

The JHM Unified Command Center convened March 2, 2022, in The Johns Hopkins Hospital's Chevy Chase Bank Conference Center, led by Kelen,

The Unified Command Center bustled around the clock with high-level decision-making and communication.

Maragakis, senior vice president of operations Lisa Ishii; and Robert Maloney, then senior director of emergency management.[31] The group reported to Rothman and Sowers.

In short order, the Unified Command Center added to its leadership team, bringing in Peter Hill, senior vice president for medical affairs for the health system; Deborah Baker, senior vice president for nursing for the health system; and school of medicine chief of staff Kristy Dickson. When Maloney left, Carter became co-commander of the Incident Command Center.

"Our lives have changed," said Sowers, speaking with Maragakis in an eight-minute video message to employees, filmed in the bustling command center. "We need to take care of each other, support each other in all the ways we can."

He promised that all decisions would be fact-based, and would evolve with new circumstances and information.[32]

Throughout that frightening and frantic spring, the Unified Command Center operated around the clock, with participants wearing masks and maintaining six-foot distances as they assessed the rapidly changing situation, made decisions, and shared information across the health system.

The questions came in a torrent: Which hospitals had capacity for new patients?

What protocols were needed to prevent the spread of the virus? How would testing centers operate?

Every day at 3 p.m., the leaders led a systemwide telephone call, usually with hundreds of people across the health system. The calls, which lasted about an hour, began by addressing any safety-related events or concerns, then moved to an update on COVID numbers and discussion of any changes to COVID guidance. Each hospital, including its ambulatory division, and essential central function, like supply chain, also provided updates.

On June 12, 2020, the command center de-escalated to "watch level." By then, many of the meetings were taking place over Zoom instead of in person; over time, the calls decreased to every two weeks and were just a half-hour long. As of May 2022, the command center had conducted more than 240 meetings and was still in operation, with activity increasing and decreasing as needed.[33]

Rothman noted during the March 31, 2022 Town Hall meeting that the center had provided essential decisions during a pandemic for which "there is no playbook."[34]

Always Enough Supplies

The COVID-19 pandemic disrupted manufacturing and delivery at the very same moment that health care systems around the world desperately needed items such as face masks, protective gowns and disinfectants.

In mid-March, Kelen convened a group called Allocation of Scarce Resources, led by Allen Kachalia, senior vice president of patient safety and quality for Johns Hopkins Medicine and director of the Armstrong Institute for Patient Safety and Quality. The group of about 30 people included Johns Hopkins bioethicists, pharmacists, lawyers, clinicians and others, who created frameworks for equitable distribution of supplies, should they become scarce.

It wasn't until the end of 2020, when vaccines were first approved, that the group's frameworks were pressed into service, Kachalia said.

The group met on Zoom and hashed out the plans, said Kachalia. "We'd get an understanding of the clinical problem and try to leverage the framework for the greatest benefit, with fairness in mind. Then we'd put together a plan and invite people to shape it, thinking through consequences of various decisions."[35]

The plans were endorsed by the state and shared with other Maryland health systems, he said.

Johns Hopkins Medicine never ran out of critical supplies, thanks to its integrated supply chain and consolidated warehouse, said Burton Fuller, health system vice president and chief supply chain officer.

Starting in February 2020, supply chain team members amassed items and created a distribution system for Johns Hopkins hospitals in Maryland and Washington, D.C. They were innovative, even acquiring equipment from unconventional sources such as a car manufacturer that made reusable isolation gowns from air bag material.

They found substitutes when necessary and orchestrated a volunteer assembly line to make tens of thousands of face shields.

With face shields in short supply, volunteers gathered in a Johns Hopkins warehouse to make their own. Supply chain leaders Burton Fuller and Michelle Azotea, right, organized the effort.

Volunteer Mask-Making

In March 2020, dozens of Johns Hopkins volunteers opted to spend hours of their free time in a chilly room of the Johns Hopkins Health System Consolidated Services Center, the off-site warehouse that the health system opened in 2018.

With the garage-style door open for ventilation, medical students, nurses and others sat at long tables, creating face shields and assembling personal protection packs that would help clinicians reuse their surgical masks and N95 filtering respirators.

"I just felt like I needed to do something," said one volunteer, oncology research nurse Lesley Donoho-Cohen, as she cut straps. "The supply of masks is critical to health care provider safety."

Clinicians rely on masks and respirators to protect themselves from infection, in normal times discarding them after a single use.

But masks and other personal protective equipment (PPE) became scarce in January 2020 when the Chinese manufacturers of PPE stopped production so workers could stay home, part of a campaign to halt spread of the novel coronavirus in that country.

"We have a perfect storm of fluctuating supply, surge in demand and high degree of difficulty getting additional products," said Burton Fuller, health system vice president and chief supply chain officer. "Therefore, we are implementing important steps to preserve our PPE."

One day in mid-March, Fuller handed a commercially-made face shield to Michelle Azotea, director of project management and implementation for the Johns Hopkins Health System corporate supply chain. He wanted to know if Johns Hopkins could make 50,000 of them. Quickly.

The idea had come from two women who had made similar mask-protecting shields when volunteering overseas years ago: Polly Trexler, director of operations for Hospital Epidemiology and Infection Control (HEIC) at The Johns Hopkins Hospital, and Melanie Curless, HEIC's assistant director of operations.

On March 14, team members from corporate supply chain gathered at the Consolidated Services Center to put together an assembly line and make the face shields. Over the weekend, they made 1,000 shields, which were then approved by infection control for use.

Kevin Sowers, president of the Johns Hopkins Health System and executive vice president of Johns Hopkins Medicine, described the project and included a signup link in a March 16 blog on his Strengthening Ties blog. "We are experiencing an unprecedented time in our history as a global society, and it will take innovation and, most importantly, teamwork, to navigate these unchartered waters," he wrote.*

---

* Sowers, Kevin, "It Takes a Team: Corporate Supply Chain Staff Come Together to Creatively Address the Need for Face Shields," Strengthening Ties blog, March 16, 2020, https://kevin-sowers.blogs.hopkinsmedicine.org/2020/03/16/it-takes-a-team-corporate-supply-chain-staff-come-together-to-creatively-address-the-need-for-face-shields/.

"We never ran out of any critical supply, knock on wood," said Fuller in 2021, reflecting on the first year of the pandemic. "The pandemic forced us to start acting and functioning like one health system. It will be easier for us to accommodate supply disruptions in the future."[36]

The health system had first moved toward consolidation in 1990, when The Johns Hopkins Hospital and Johns Hopkins Bayview Medical Center combined their purchasing and storage power. The consolidated warehouse opened in Baltimore in 2018.

"Functioning as a single system," said Fuller, "allows us to be more coordinated in our operations, more thoughtful and more efficient."[37]

An Academic Year Like No Other

Johns Hopkins Medicine has about 4,000 medical and graduate students, residents, clinical fellows and postdoctoral research fellows. For each of them, COVID-19 and a renewed focus on racial injustice and disparities created an academic year for the history books, one with challenges galore but opportunities as well.

Most classrooms and labs were shuttered in March 2020, and opened again in August for the 2020–2021 academic year. Still, things were different. Many medical and graduate students took at least some of their classes online. When learning in person or working in labs, they wore personal protective gear and maintained smaller groups than in the past.

Clinical rotations were shorter: six weeks instead of eight for medicine, pediatrics, surgery and women's health, and three weeks instead of four

for neurology, psychiatry and emergency medicine. When the trainees did go on rounds, often only one student, one intern and one resident entered a patient's room at a time, about half the number of pre-COVID times.

COVID-19 accelerated the existing trend toward online learning, said Nancy Hueppchen, associate professor of gynecology and obstetrics and associate dean for undergraduate medical education. For the past decade, she said, attendance has not been required for lectures, though it is for small groups, labs, clinical correlations with real patients and other settings that call for student participation.

Over the summer of 2020, faculty members and support staff updated their materials to better reflect the realities of online learning. Lectures were updated and shortened, when appropriate, and materials were added for independent study.

There's no substitute for dissecting a human cadaver, however. Though gross anatomy lectures were online, dissections were still in person, with eight students at four tables, instead of 18 students at six tables, to accommodate distancing requirements during the pandemic.*

---

* Nitkin, Karen, "At the Johns Hopkins University School of Medicine, an Academic Year Like No Other," *Dome*, Sept. 8, 2020, https://www.hopkinsmedicine.org/news/articles/at-the-johns-hopkins-university-school-of-medicine-an-academic-year-like-no-other.

Building a New Technology Infrastructure

The Johns Hopkins COVID-19 response required a giant leap forward in technology infrastructure.

"When the pandemic arrived, there was just an extraordinary request to change how IT runs," said Peter Greene, JHM chief medical information officer, in a 2022 interview.[38]

Johns Hopkins Medicine clinicians and technology experts worked together to quickly develop and scale up systems to track COVID-19 testing and create self-assessments for employees and patients.

Technology also made it possible for Johns Hopkins Medicine to greatly expand its ability to treat patients through telemedicine, as well as to alleviate the isolation brought about by new visitor restriction policies.

Deanna Hanisch, vice president for health information technology, came to Johns Hopkins in 2011 as an on-site vendor for Epic, and was instrumental in using the medical records system to develop solutions to the challenges posed by the pandemic.

"Epic becomes the mechanism by which we were able to execute on our policies and our mission," she said.[39]

On March 11, 2020, about two dozen representatives from Hospital Epidemiology and Infection Control, Health IT, Johns Hopkins Ambulatory Services and the Department of Pathology convened to form HEAL (HEIC, Epic, Ambulatory, Labs).

Peter Greene

Valeria Fabre **Aaron Milstone** **Manisha Loss**

One of the new group's first challenges was figuring the who, what, where, when and why of COVID-19 testing.

That day, the molecular virology lab at The Johns Hopkins Hospital had performed its first on-site COVID-19 test. At the time, tests were in short supply and were only provided when ordered by doctors.

"We had to work out all these workflows for ordering a test," said Greene. "The criteria around who was eligible changed every week. Locations and volume changed all the time. We brought together these teams that had to work incredibly closely together that had never needed to do that before."[40]

When rapid tests were developed, infectious disease specialists Valeria Fabre and Aaron Milstone were among those working out criteria for when they were preferable to standard tests that could take a day or more for results.

"We had to figure out how doctors would order the test and how we would track the results and report to the state," said Manisha Loss, the medical director of Johns Hopkins Ambulatory Services and a leader of the HEAL group.

Even before the pandemic, Loss, a dermatologist, was involved in finding systems-based ways to improve the patient experience. "My career has been as a liaison between the clinical world and all the people who support us as doctors," she said. "What the pandemic allowed us as an institution to do is bring various groups together to work toward one mission. We learned how to find system solutions and apply them broadly."[41]

Patient Self-Screener

HEAL also tackled the question of how to screen ambulatory patients for COVID-19 symptoms.

Johns Hopkins has been screening patients and visitors at entry points throughout the pandemic, but the impetus to create the Patient Self-Screener tool occurred on May 6, 2020, when Gov. Hogan announced that ambulatory clinics could reopen to nonurgent care, but required patients be screened for COVID-19 symptoms.

The Patient Self-Screener was developed with guidance from Loss; Stephen Sisson, vice president, clinical operations for the Office of Johns Hopkins Physicians; and Kimberly Peairs, vice chair for ambulatory in the Department of Medicine.

The Prodensity App

At the start of the pandemic, each lab could designate no more than three essential personnel — including the principal investigator — to enter, work and maintain experiments, guidance that excluded about half of the research community's 3,687 members.

"It was a logistical nightmare," Geraldine Seydoux, vice dean for basic research at the school of medicine, recalled in 2020. "Keeping track of who was essential, which kept changing, was extremely difficult for us and for the security team members checking our people in. We were literally using printed lists and Post-it Notes. There were bottle-necks of people waiting to get admitted."

Seydoux and George Econo-mas, then executive director of security for Johns Hopkins Medical Institutions, worked with the Johns Hopkins Technology In-novation Center (TIC) to create the Prodensity app, which let us-ers see if there was room in the labs, given current occupancy and physical distancing requirements.* It was soon expanded to cover anyone on-site at The Johns Hop-kins University or Johns Hopkins Medicine.

The check-in systems for both patients and employees remained in place two years into the pandemic and have evolved over time to include a daily employee self-check for symptoms or expo-sure, as well as a mechanism for tracking student and employee vaccinations and boosters.†

* Keating, Michael, "Tracking the Reopening of Labs," *Hopkins Medicine* magazine, Fall 2020, https://www.hopkinsmedicine.org/news/articles/tracking-thereopenngof-labs.

† Hub staff report, "Get to Know Your Prodensity App (and Make Sure You've Got Version 2.1), *The Hub*, July 29, 2021, https://hub.jhu.edu/at-work/2021/07/29/prodensity-app/.

It reduced bottlenecks and limited interactions that could potentially expose staff to the virus. It asked patients a series of questions in English or Spanish about symptoms and possible exposure and gave a green or red digital badge based on the responses, meaning they could either proceed or they were prompted to call the office for further instructions, including possible rescheduling in person or through a telehealth video visit.[42]

Making Rooms for Patients

The BCU became the first unit in the Johns Hopkins Health System to care for persons with confirmed cases of COVID-19, or who were under investigation because of possible exposure.

When the number of cases exceeded BCU capacity, the BCU team, in collaboration with HEIC, helped to set up additional "biomode" units by creating a training program for staff and helping set up the infrastructure for the new units.

In April 2020, engineers, carpenters, painters and other facilities staffers throughout the Johns Hopkins Health System worked at a frenetic pace to create entire units in Johns Hopkins hospitals for treatment of patients with COVID-19.

They added ductwork and HEPA (high-efficiency particulate air) filters to convert regular patient rooms to ones with negative air pressure suitable for patients with the highly infectious coronavirus. They built anterooms where clinicians could safely don or doff their protective gear before treating patients who were diagnosed with COVID-19 or waiting for test results.

They also set up tents for COVID-19 testing, removed chairs from hospital cafeterias to promote physical distancing, and put up signs telling people which entrance doors were locked.

Most Johns Hopkins hospitals had at least a few rooms

Francisco Ruiz Vivas, Todd Heron and Babu Varghese add ductwork and filters to prepare Howard County General Hospital for COVID-19 patients.

that operated at negative pressure or could easily convert to negative pressure for patients with contagious illnesses. The Johns Hopkins Hospital, for example, had created 136 airborne biocontainment rooms in 2001 amid concerns about smallpox or anthrax bioterrorism attacks, and 22 more in 2004 during the SARS pandemic before adding the BCU in 2015.

But as more patients with COVID-19 filled Johns Hopkins hospital beds, facilities employees had to move beyond the usual types of conversions. Ray Littleton, facilities manager for Howard County General Hospital, said his team found a way to use a single HEPA filter for three rooms. "We're taking larger construction HEPA machines and installing them with ductwork in the ceiling," he said.

"I've worked at Hopkins for 27 years," said Cheryl Koch, vice president for operations and facilities at Johns Hopkins Bayview Medical Center in April 2020. "We've certainly had lots of challenges but I would say this one has probably pulled the most people together in the shortest amount of time. Everybody has been working long hours trying to make sure we're ready for the next phase of what's coming."

"Every week we consider the need to convert another unit to COVID capability," said LeighAnn Sidone, chief nursing officer of the 228-bed Suburban Hospital, in April 2020. "We have a plan to stay a unit ahead so we are prepared if the volume rises."[43]

Telemedicine to the Rescue

On a winter day early in 2013, a Johns Hopkins neurologist sitting in a conference room in Green Spring Station in Lutherville, Maryland, provided a 20-minute consultation to a patient with Parkinson's disease who was a two-hour drive away, at his home in Salisbury, Maryland.

The neurologist watched as the patient paced his living room, and the doctor talked to him about depression. At one point, the patient fetched a pill bottle from another room so he could confirm the medication's dosage.

The telemedicine appointment saved the patient nearly a day of travel, plus the hassles of parking, and gave the neurologist a useful glimpse into the patient's home.[44]

Telemedicine had been germinating at Johns Hopkins since the 1970s, with innovative providers creating early telemedicine cart prototypes with TVs powered by car batteries.

In July 2016, under the rubric of the Office of Johns Hopkins Physicians and led by Paul Scheel, director of the Division of Nephrology, and Jennifer Parks, leader of strategic initiatives for OJHP, Johns Hopkins Medicine formed an office to support and integrate telemedicine efforts across the enterprise. Ophthalmologist Ingrid Zimmer-Galler was clinical director of Johns Hopkins Telemedicine, with Rebecca Canino as administrative director.

"Johns Hopkins has a long history of telemedicine innovation, but the projects were isolated in different specialties," said Canino in 2017. "It made sense for us to consolidate and support our remote care initiatives because telemedicine is an important part of our strategic plan, under the Integration pillar."[45]

The office helped departments create the infrastructure they needed to conduct telemedicine interactions while protecting patient privacy. It also helped them navigate some of the payment inconsistencies and regulatory barriers that made telemedicine challenging, particularly in its early days.

In July 2019, Brian Hasselfeld joined Johns Hopkins Medicine as the institution's first medical director for digital health and telemedicine, a job that by the fall of 2019 included leadership of the telemedicine office.

Hasselfeld, a pediatric and internal medicine primary care doctor with Johns Hopkins Community Physicians, earned his medical degree from Tulane University, and did his residency at Brigham and Women's and Boston Children's hospitals. In both New Orleans and Boston, he said, he worked for groups that helped faculty develop digital health care tools and bring them to market.

His work guiding clinicians and shaping the institution's strategy regarding use of telemedicine went into warp speed when the COVID-19 pandemic arrived. Thousands of patients turned to the technology to receive care at a time when in-person visits were discouraged and "physical distancing" was an important risk mitigation tool.

Starting March 16, 2020, the technology was automatically enabled for all Johns Hopkins patients who use MyChart, the portal that connects them to the Epic electronic medical record system.

"Telemedicine is primed to meet the goals of 'physical distancing,' to permit

Brian Hasselfeld

The Child Mobile Treatment Program

Hal Kronsberg

A telemedicine program launched in 2019 has made life easier for young patients with serious, persistent mental health issues, as well as their families and clinicians.

Using telemedicine, psychiatrists can see and assess patients in crisis in the Johns Hopkins Child Mobile Treatment Program through an iPad, so patients do not have to wait to receive care at the program's offices at Johns Hopkins Bayview Medical Center.

The program serves about 90 patients, ages 4 to 24, at any given time. They have

conditions such as schizophrenia or bipolar disorder, and did not have success with outpatient or school-based treatments. To be eligible, children must be at risk for multiple hospitalizations or out-of-home placement, and a special authorization is required.

Bringing a child with mental health issues to a psychiatrist's office is a hassle under the best of circumstances. It's even more difficult for families that rely on public transporta-

tion, that have to tote along siblings because they can't afford child care, and that risk losing a day's pay.

"The use of telemedicine has facilitated psychiatric visits for patients who had previously gone months before they could coordinate a way to come to the clinic," said Hal Kronsberg, psychiatrist for the mobile psychiatry service. "The ability to see my patients more often is always a good thing."*

---

*   Nitkin, Karen, "Telemedicine Connects Young Patients with Psychiatric Care," *Dome*, June 19, 2019, https://www. hopkinsmedicine.org/news/articles/telemedicine-connects-young-patients-with-psychiatric-care.

---

care to the right patient at the right time at the right place," said Hasselfeld in a *Dome* article that month.[46]

Prior to the pandemic, Johns Hopkins Medicine was conducting fewer than 100 telemedicine visits per month at its six hospitals in Maryland, Washington, D.C., and Florida and at more than 40 community physician locations.[47]

Starting in mid-March, when outpatient clinics closed for all but the most urgent cases, the number surged past 5,000 per day. The rapid ramp-up required education materials for clinicians, as well as assistance for patients who were daunted by the technology, said Deanna Hanisch, the vice president for health information technology.

By the summer of 2020, as outpatient offices reopened, telemedicine visits had steadied at about 3,000 per day, or about 20% of ambulatory appointments, said Hasselfeld.

By July 2021, Johns Hopkins Medicine had conducted more than 1 million telemedicine visits for patients across the nation.[48]

In July and August 2020, Johns Hopkins outpatient providers switched telemedicine platforms to Cisco Webex for video visits with internal medicine, primary care, pediatric and family practice patients. The new telemedicine platform gave ambulatory patients one-click access to their virtual appointments, without requiring them to download an app.

"Any type of download obviously represents a technology barrier," said Hasselfeld. "You need internet speed and access in order to download. You need to know how to operate in a new app. All that adds cost, time and complexity for the patient."

The new system, by contrast, uses a browser, so patients and providers merely click a link to join the video appointment with each other.[49]

By December 2021, as in-person care broadly reopened, the monthly number of telemedicine visits was about 35,000, or roughly 600 times pre-pandemic levels.

Going forward, Hasselfeld expected video visits would comprise 20% of all ambulatory patient visits, allowing patients to see providers through more convenient appointment options when clinically appropriate.

Referring to telemedicine as the modern equivalent of a house call, Hasselfeld noted that the technology has led to a medical world that is more accessible, more affordable and often more personal.

"We have a unique opportunity to think about what it means to open up our doors wider," he said.

Although not every ambulatory visit can be virtual, every specialty can offer some benefit through telemedicine, said Canino. "If you come here for surgery, for instance, your pre- and post-op visits, as well as your education and post-discharge follow up, can be done through telehealth, allowing you to recover safely at home."[50]

Using Technology to Help Patients Keep in Touch

The notice sent to all Johns Hopkins Medicine employees on March 13, 2020, outlined a dramatic change in policy:

"To protect our patients, visitors and staff members from the spread of COVID-19, we have enacted Johns Hopkins Medicine-wide visitor guidelines."

Among the new rules, all visitors would have to check in at a front desk and confirm they did not have flu-like symptoms or exposure to someone with COVID-19. Patients in the hospitals could have just one visitor at a time, and patients on appointments or urgent care visits could be accompanied by just one person. No visitors younger than 18 were allowed, unless they were the parents of hospitalized children.

Perhaps hardest of all: No visitors were allowed for patients who had COVID-19 or were under investigation for the disease, except for visitors of patients receiving end-of-life care. "We encourage video meetings between patients and loved ones," said the message.[51]

A collaboration among the Johns Hopkins Medicine patient experience office, information technology (IT) and nursing helped stave off the loneliness, boredom and communication gaps caused by the new policy.

"This just tears at us because it's

Nurse Shelby Cahill helps patient Shawn Hoke navigate the MyChart Bedside tablet, which is equipped with apps to help patients stay connected during hospital stays.

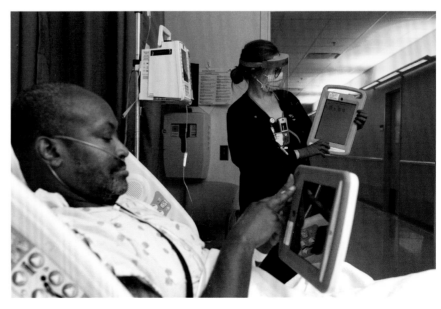

Rickey Jenkins

Amy Kilen

Extraordinary People Extraordinary Moments

Johns Hopkins Medicine highlighted some of the people making a difference during the pandemic in a feature on its website, called Extraordinary People, Extraordinary Moments.

One such person was Rickey Jenkins, a member of the Environmental Services team at Howard County General Hospital, profiled in August 2020.

In addition to disinfecting areas on their routine rounds, Jenkins and his colleagues listened for the "Bat Phone," which rang whenever urgent cleaning services were needed.

The Bat Phone often sent him to the emergency department. Being on the front lines during the pandemic wasn't easy, but Jenkins said support from his family and colleagues helped.

"We have proper personal protective equipment so we can do a good job and keep people safe," he said. "We work together well, because that's what the job requires."*

Another was Amy Kilen, a nurse practitioner with Johns Hopkins Home-Based Medicine (JHOME), who wore a face mask and shield, gown and gloves for her only in-person visit of the day in a photo taken in April 2020. Before COVID-19, she visited several elderly, chronically ill patients in their homes each workday.

Because of the risk of infection, she switched to "televisits" whenever possible. She was reluctant at first to rely on technology to check up on patients, but now looks forward to some aspects of these video visits. "Seeing your older patients on FaceTime is a neat intergenerational connection," she said.*

---

*   "Extraordinary People. Extraordinary Moments."

Carrie Stein   **Lisa Allen**

the opposite of the years we've spent trying to open up visitation in hospitals," said Lisa Allen, chief patient experience officer for Johns Hopkins Medicine, in May 2020. "We totally understand the need to stop the spread of the virus through physical distancing. It's just really painful. We're doing as much as we can to make our communications devices work in a way so families can see patients, even if they [the patients] can't talk, because we know it's comforting."

Carrie Stein, director of clinical informatics, led efforts to make sure patients could easily connect with family members and access information about their condition and care through an app called MyChart Bedside, loaded onto iPads available to patients.

Within weeks of the new restrictions, patients in intensive care units were able to connect with family members through video chats, and had access to amenities including virtual concerts from Peabody Institute musicians; guidance on how to use video, audio and texting apps created in partnership with Patient Education and IT; and consults with doctors that families could join remotely.

The patient experience office worked with marketing and IT to create a COVID-19 virtual resources guide that explains how to set up and use such apps as FaceTime, Google Hangouts, Skype and Zoom. It also included instructions for setting up MyChart, a secure portal that patients and those with their permission can use to access medical records, and CaringBridge, a free blog platform used by patients and patients' contacts to post health updates for friends and family.[52]

Innovations in Treating Patients with COVID-19

Because the coronavirus that caused COVID-19 was new, clinicians had no choice but to "build the airplane as they flew," developing treatments as rapidly as they could. One of the most frightening characteristics of the virus was that it attacked the lungs and respiratory system, depriving patients of oxygen.

Clinicians used high-flow nasal cannulas when they could, but some patients required intubation, a procedure that carried its own set of risks.

Prone Team

Early on, evidence mounted that placing patients on their stomachs during intubation could help clear respiratory secretions by relieving some of the pressure caused by gravity, the heart and diaphragm when lying on the back.

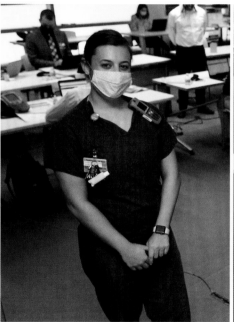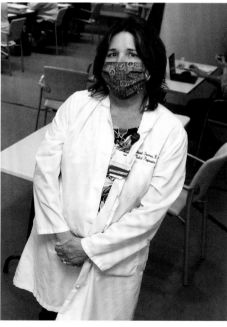

Normally, patients who need to be placed in a prone position are transferred to a medical intensive care unit, but with so many patients being treated for COVID-19, that wasn't always an option.

Nurse Deborah Sherman, program director for clinical informatics and deputy operations chief in The Johns Hopkins Hospital's emergency management command, thought to create a prone team to help manage patients in crisis. Hospital President Redonda Miller loved the idea — and asked Sherman to get it in place quickly.

Within a couple of days, Sherman had assembled a prone team with two outpatient physical therapists and a MICU resource nurse, Nicole Troiani, who conducted training in the hospital's simulation center using mannequins, and created videos and documents showing how to turn over and move patients. Sherman created a document about how to assemble a prone team, used by the other hospitals in the Johns Hopkins Health System and beyond.[53]

Nurses Nicole Troiani (left) and Deborah Sherman, shown in the Unified Command Center, led an effort to place intubated patients with COVID-19 on their stomachs to increase the flow of oxygen to the bottom and back parts of the lungs.

Treating Post-Intensive Care Syndrome

Throughout the spring of 2020, patients with COVID-19 were staying longer than the average three to four days in the intensive care unit (ICU), putting them at greater risk for developing post-intensive care syndrome (PICS) — physical, cognitive and psychological changes that occur after surviving an illness or injury that requires treatment in the ICU.

Megan Hosey, a rehabilitation psychologist at The Johns Hopkins Hospital's medical ICU, was among those working to ensure patients had resources to combat the anxiety and depression created by a toxic swirl of factors: struggling to breathe, being away from family and fear of getting others sick.

Her colleagues in the effort included Dale Needham, a Johns Hopkins critical care physician and researcher who is internationally renowned for advancing PICS research; and Joe Bienvenu, a Johns Hopkins psychiatrist who studies survivors of critical illnesses and intensive care.

"Even in the ICU, we try to help patients see the things that are pleasurable and meaningful in their lives because these [reminders] can protect against anxiety and depression. The sooner you can get back into these activities — or even talking about them — the better," said Hosey.

She also contacted family members to better understand the patient's history and to assure them she was taking care of their loved ones when they were unable to visit.

Another concern was delirium, which can set in when patients are in an unfamiliar environment, under sedation and working to fight infection or recover from an injury. Hosey said the prevalence of delirium in patients with COVID-19 in the ICU was especially high, likely due to the amount of sedation needed to keep patients on ventilators for an extended period of time.

Because delirium can also cause cognitive changes in attention, thinking and memory that last beyond hospitalization, Hosey counseled patients about getting in touch with a mental health provider who understands PICS.

By helping patients manage their health in the ICU as early as possible, Hosey believed patients would be better equipped to engage in the next phase of their recovery. "Very few ICUs in the country provide patients with the amount and intensity of early rehabilitation that we provide at Hopkins," she said in April 2020.[54]

JH PACT: Gerry and Georgene Stephens

Gerry and Georgene Stephens were the first couple to both receive treatment for COVID-19 at The Johns Hopkins Hospital. Their ordeal began in early March 2020, when Gerry, then 69, developed a cough and fever. On March 17, they went to Howard County General Hospital, where they were both tested for COVID-19.

Georgene, 70, was asymptomatic and was sent home. But Gerry, feverish and struggling to breathe, was intubated and taken by ambulance to The Johns Hopkins Hospital, where he was one of the first people with COVID-19 to receive treatment in the hospital's BCU.

The test results, which came back March 18, showed that both Gerry and Georgene had the new coronavirus. Georgene's symptoms emerged later that week. "I started feeling lightheaded and funny," she said, so she called 911 and was taken by ambulance to The Johns Hopkins Hospital, where she was treated and discharged the next day.

Gerry returned home April 8, weak and 35 pounds lighter. As a precaution to prevent transmitting other illnesses between them in their fragile states, the longtime spouses stayed apart for two weeks. They slept in separate bedrooms, used different bathrooms and did not eat together.

"When he came back from the hospital, all I wanted to do was sit on the couch and hold hands," said Georgene. "That's all I wanted to do and I couldn't do that."

She also worried that she wouldn't be able to take care of Gerry.

That's when the Johns Hopkins Post-Acute COVID-19 team (JH PACT) sprang into action. The purpose of JH PACT was to help people like Gerry recover from COVID-19 and combat lingering post-COVID-19 symptoms.

JH PACT was a multidisciplinary program established in April 2020 by Emily Brigham, who was then an adjunct assistant professor in the Division of Pulmonary and Critical Care Medicine, and Ann M. Parker, assistant professor in pulmonary and critical care. It is co-directed by Parker and Alba Azola, assistant professors in the Department of Physical Medicine and Rehabilitation.

"We had long considered starting a program for patients who have been critically ill here at Johns Hopkins," said Parker in 2022. "The COVID-19 pandemic really provided us with an opportunity to expedite that process."[55]

By 2022, PACT had supported hundreds of patients in their recovery from COVID-19 by providing rehabilitation therapy services, including physical, occupational and speech-language therapy, as well as referrals for specialists in psychiatry,

Alba Azola

Ann Parker

Gerry and Georgene Stephens, top right, reunite with members of their care team over Zoom.

Top: Carol Kelehan, Amit Pahwa, Gerry and Georgene Stephens.

Middle: Laura Ridgeway, Adam Kaplin, Allison Madden.

Bottom: Mahendra Damarla, Brian Garibaldi.

psychology, neurology, cardiology, infectious disease, dermatology, nephrology, hepatology, and hematology as needed.

Garibaldi, who treated Gerry, estimated that more than 100 doctors, nurses and other frontline staff members cared for the couple during and after their hospitalizations. That support system included a Johns Hopkins Home Care team that came to the Stephens' home 14 times over 17 days to help Gerry regain his strength. Patients can also access care in Johns Hopkins clinics or through telemedicine.

When Gerry and Georgene were finally able to reunite, Gerry cooked a breakfast of eggs Benedict for his beloved. Because of the pandemic, the couple canceled plans for a 50th anniversary bash that summer. Instead, they savored small pleasures.

"I enjoy eating breakfast together," said Gerry. "I enjoy strolling up and down our street together. We're appreciating the simple things."[56]

Research on Tests and Vaccines

As testing for COVID became more prevalent, clinical microbiologist Heba Mostafa, along with Andrew Pekosz, professor of microbiology and immunology at the Bloomberg School of Public Health, conducted a study to fine-tune the relationship between virus detection and infectiousness.

For the study, published online Oct. 27, 2020, in the journal *Clinical Infectious Diseases*, researchers looked at results from nearly 30,000 COVID-19 tests given between March 11 and May 11, 2020, and found that while SARS-CoV-2 RNA, the virus that causes COVID-19, can be detectable for weeks after the onset of symptoms, it doesn't necessarily denote the presence of infectious virus particles.[57]

"Defining the window of time in which a COVID-19 patient can transmit the virus can help drive more effective isolation practices," said Pekosz, co-senior author of the study.[58]

From the start, the lab sequenced and analyzed the virus in order to identify variants. "Our first genomes revealed significant early diversity that indicated multiple introductions of the virus to our region with the beginning of the pandemic," Mostafa said. The lab was also first, in January 2021, to report a South African variant in Maryland.[59]

The analysis meant the team was able to quickly understand and respond to disease variants, said Mostafa.

In a paper published on Dec. 18, 2021, in *Clinical Infectious Diseases*, Mostafa, Pekosz, associate professor of emergency medicine Eili Klein and others studied the disease's Delta variant and found that it was more transmissible than the original variant, but also that it was not associated with an increase in disease severity in fully vaccinated individuals.[60]

Subsequent research on the Omicron variant indicated that infected patients were often vaccinated but less likely to be admitted or develop severe disease.[61]

The vaccines were remarkably effective at protecting against severe illness, but researchers like Milstone, associate hospital epidemiologist at The Johns Hopkins Hospital, were eager for more nuanced information.

Andrew Pekosz

When Is Remdesivir Most Effective?

A Johns Hopkins-led retrospective study, published in *Clinical Infectious Diseases* in December 2021, found that remdesivir provided clinical improvements in hospitalized patients with COVID-19 who received no or low-flow oxygen, though the antiviral medication was unlikely to help more severely ill patients.*

"We observed that remdesivir is best used as early as possible, before the patient progresses to requiring high levels of oxygen or intubation and mechanical ventilation," said Garibaldi. "Most patients who need that kind of advanced respiratory support are likely past the point where antiviral therapies like remdesivir would be effective."

The study was conducted as part of the COVID-19 Consortium of HCA Healthcare and Academia for Research Generation (CHARGE), a partnership between HCA Healthcare, the Agency for Healthcare Research and Quality, and 10 academic medical centers (including the Johns Hopkins University School of Medicine) across the United States.†

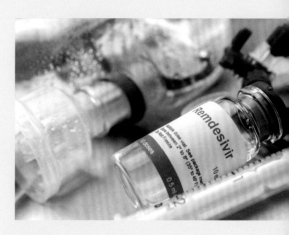

* Garibaldi, Brian T. et al., "Real-World Effectiveness of Remdesivir in Adults Hospitalized with Coronavirus Disease 2019 (COVID-19): A Retrospective, Multicenter Comparative Effectiveness Study," *Clinical Infectious Diseases*, Dec. 15, 2021, https://academic.oup.com/cid/advance-article/doi/10.1093/cid/ciab1035/6463010?guestAccessKey=09167d29-b0cd-4e74-9202-a4a21f951cc6&login=true.
† "Study Supports Use of Remdesivir for COVID-19 Patients on Low-Flow Oxygen or No Oxygen," Johns Hopkins Medicine news release, Jan. 18, 2022, https://www.hopkinsmedicine.org/news/newsroom/news-releases/study-supports-use-of-remdesivir-for-covid-19-patients-on-low-flow-oxygen-or-no-oxygen.

Milstone and colleagues were among the first to study whether a lack of symptoms (such as fever, fatigue or headache) following COVID-19 vaccines indicated a less effective immune response.

The answer was no.

Then they looked at whether the vaccines were more effective in people who had already tested positive for COVID-19. The answer was yes.

Milstone's team began gathering data in June 2020, before vaccines existed. The researchers collected serum samples and survey responses every three to four months from more than 3,000 Johns Hopkins Medicine health care workers who had given informed consent.

A year later, after the workers had received both doses of either the Pfizer/BioNTech or Moderna vaccines, the researchers found that the vaccines were effective even if the recipient didn't experience symptoms following vaccine shots. They published their results, which drew international attention, as an August 2021 research letter in *JAMA Internal Medicine*.

"It wasn't known if a lack of symptoms following vaccination would indicate a less-than-adequate antibody response in people who received either the Pfizer/BioNTech or Moderna vaccines, so we studied an available group of staff from our hospital to see if there were any connections," said Milstone, senior author of the study.[62]

A second study, published in November 2021, explored the durability of immunity after vaccination and found that those who were infected before being vaccinated had more antibodies to protect against disease.[63]

That study, believed to be one of the largest of its kind, looked at a subgroup of nearly 2,000 Johns Hopkins health care workers and found that antibody levels against the COVID-19 virus stayed higher over an extended period of time in people who were infected by the virus and then received two doses of messenger RNA (mRNA) vaccine, compared with those who got immunized but didn't have COVID.

"Health care workers with prior SARS-CoV-2 infection followed by two doses

In March 2020, the Baltimore Convention Center was converted to a 250-bed field hospital (above), led by orthopaedic surgeon James Ficke (right), who had commanded an Army hospital in Mosul, Iraq.

of mRNA vaccine (three independent exposures to spike antigen) developed higher spike antibody measurements than individuals with vaccination alone," they wrote in a research letter in the *Journal of the American Medical Association*.[64]

"This finding adds to our understanding of how immunity against SARS-CoV-2 works, and builds upon an earlier study by our team that showed the mRNA vaccines yielded a robust antibody response, even if a person did not develop significant symptoms following vaccination or did not have a prior SARS-CoV-2 infection," said Milstone at the time.[65]

The study team at Johns Hopkins Medicine was co-led by Diana Zhong and Shaoming Xiao. Along with them and Milstone, the team includes Caturegli, Amanda Debes, Emily Egbert and Elizabeth Colantuoni.[66]

The Baltimore Convention Center Field Hospital

To keep hospitals from being overwhelmed, the state of Maryland in March 2020 established and licensed a 250-bed field hospital at the Baltimore Convention Center, managed jointly by the city's two academic health systems, Johns Hopkins Medicine and the University of Maryland Medical System.

The field hospital's director was James Ficke, director of orthopaedic surgery at The Johns Hopkins Hospital, who drew on his experience as a U.S. Army colonel leading a combat support hospital during one of the most violent years of the war in Iraq.

Chuck Callahan, vice president of population health at the University of Maryland Medical Center, was deputy director. Before joining the University of Maryland, Callahan had been commander of the Fort Belvoir Community Hospital and deputy commander and chief of staff at the National Naval Medical Center and the Walter Reed National Military Medical Center.

The purpose of the facility was to free up beds in area hospitals by caring for

Study Shows Value of Rapid Test

Zishan Siddiqui

Data collected at the Baltimore Convention Center Field Hospital was used for a study comparing a rapid antigen detection test with the polymerase chain reaction (PCR) assay.

The study, published in December 2021 in the American Society for Microbiology journal *Microbiology Spectrum*, involved some 6,000 patients seen at the field hospital during a 10-day period in early 2021. It found that the rapid test was more accurate than expected.

"We found that virus was accurately detected by the rapid antigen test in 87% of patients with COVID-19 symptoms and in 71%

of those who were asymptomatic — rates that surprised us because they were so high," says study lead author Zishan Siddiqui, assistant professor of medicine at the Johns Hopkins University School of Medicine.

"This is a significant finding because the rapid test offers a number of advantages over the PCR test, including time savings, both in sampling and processing; cost savings; and most importantly, ease of distribution and application — basically anywhere — which can help overcome COVID testing disparities in medically underserved communities."*

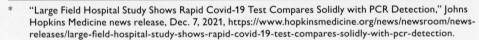

* "Large Field Hospital Study Shows Rapid Covid-19 Test Compares Solidly with PCR Detection," Johns Hopkins Medicine news release, Dec. 7, 2021, https://www.hopkinsmedicine.org/news/newsroom/news-releases/large-field-hospital-study-shows-rapid-covid-19-test-compares-solidly-with-pcr-detection.

patients with COVID-19 who no longer required acute hospital care but weren't well enough yet for discharge.

The beds and supplies were part of a Federal Emergency Management Agency aid package sent to Maryland in response to a request from Gov. Hogan. Troops from the Maryland National Guard unloaded three semitrucks full of supplies into the convention center on March 28.

To oversee the convention center effort, Sowers tapped Ficke and Jennifer Nickoles, the Johns Hopkins Health System's vice president for operations and system integration.

In turn, they appointed Eric Howell, director of Johns Hopkins Bayview's hospital medicine division; and Melinda Kantsiper, associate director of the clinical information management system at Howard County General Hospital, as the chief and deputy chief medical officers for the field hospital.[67]

The field hospital admitted its first patient on April 27, 2020.[68]

In June 2020, the site added COVID-19 testing, sometimes providing more than 1,500 tests a day.[69] And in February 2021, the field hospital began offering antibody therapies to patients with mild to moderate COVID-19 symptoms, to help prevent progression to more serious illness.

On Feb. 8, 2021, it also became Baltimore's first mass vaccination site.

When a mass vaccination site opened at the M&T Bank Stadium on Feb. 25, the field hospital focused on providing vaccinations to historically vulnerable populations in Baltimore City. "We are making progress in reaching the most vulnerable members of our community, but there is more that we can accomplish," said Sowers. "We also strive to provide the information needed for all members of our community to make science-based decisions when it comes to getting vaccinated."[70]

By April 2021, the site had become the nation's longest continuously operating

Infectious diseases specialist Matthew Robinson helped develop a computer program that predicts whether a patient with **COVID-19** faces an imminent bad outcome.

COVID-19 field hospital. In addition to testing up to 1,500 people a day at the height of the winter surge, it was Maryland's largest center for monoclonal antibody infusion therapy, treating more than 1,600 patients between November 2020-April 2021.[71]

The field hospital closed in November 2021, with COVID testing and vaccinations moved to State Center, where it was still managed by Johns Hopkins Medicine and the University of Maryland Medical System.[72]

Precision Medicine for COVID-19

In January 2021, during a terrible winter surge of patients with COVID-19, infectious diseases specialist Matthew Robinson pulled two extra 11-hour overnight shifts caring for several dozen people hospitalized with COVID-19.

Soon after arriving at 9:30 p.m., he logged into a novel computer program he helped develop and, in an instant, real-time clinical data on every coronavirus patient on the ward flashed across the screen.

A colored circle with a number in it appeared next to the name of each patient. Most of the circles were yellow or green. But a handful were bright orange or red, signaling danger.

"Those with orange or red were the ones I needed to pay the most attention to overnight," said Robinson, an assistant professor of medicine in the Johns Hopkins University School of Medicine. "Without that algorithm, I would have struggled to know who [the sickest patients] were."

The algorithm is a computer program known as SCARP, for Severe COVID-19 Adaptive Risk Predictor, developed by Robinson and other scientists at the Johns Hopkins University School of Medicine and the Bloomberg School of Public Health.

"It makes it easier for clinicians to anticipate what will happen to patients and helps them focus on patients who are the sickest," Robinson said.[73]

SCARP was created using COVID-19 data collected through the Johns Hopkins Precision Medicine Center of Excellence for COVID-19, directed by pulmonologist and BCU medical director Garibaldi.

Through the COVID-19 Precision Medicine Analytics Platform Registry (JH-CROWN), the center securely and privately collects information from patients with suspected or confirmed COVID-19, which is used to understand the disease and guide treatment.

Garibaldi was among the first to treat patients coming into the hospital with severe COVID-19. Almost immediately, he began meeting with other frontline clinicians to share observations that could help shape care protocols and clinical trials — at Johns Hopkins and beyond.

"We were only about two or three weeks into the pandemic, when Antony Rosen, vice dean for research for the school of medicine, asked me a simple question: 'What percentage of our patients have had a lab value of this particular amount?' And I replied, 'Gee, Antony, I can tell you anecdotally, but I can't recall all of them.'"

"'We don't have a data repository?' Antony asked. 'We have to create one right now!'"

And so, they did.

In the course of a weekend, the scientists conceived and submitted a plan for what has become the JH-CROWN registry, a collection of data and information about patients having suspected or confirmed cases of COVID-19 infection.

While the main source is Johns Hopkins' electronic medical record system, Epic, the registry also includes data from other sources, such as biospecimen repositories and physiologic device monitoring systems.[74]

By April 2022, more than 50 Johns Hopkins Medicine Institutional Review Board-approved protocols were using the JH-CROWN data set.[75]

That includes SCARP, which looks at factors including a patient's age, body mass index (BMI), lung health and chronic disease, as well as vital signs at the time of admission to calculate the 1-day and 7-day risk of progression to severe disease or death.

"It's a really important tool when you are having discussions with families and patients about what the next week may hold," said Garibaldi. "Having those types of conversations makes it a lot easier to help patients and their families navigate the uncertainties and different treatments options."[76]

A similar tool, the COVID Inpatient Risk Calculator (CIRC), uses assessments at hospital admission to predict the likelihood that a patient with COVID-19 will progress to severe disease or death within seven days of arrival. It is available to health care professionals outside of Johns Hopkins through MDCalc, a free online medical reference that provides point-of-care clinical decision support tools.[77]

Making the most of disparate data sets requires translating data from individual institutions into common models that allow data to be easily shared — work that has been taking place at Johns Hopkins for years through the precision medicine initiatives.

Scott Zeger, a leader of inHealth and a Bloomberg School of Public Health biostatistics professor, calls JH-CROWN "part of a transformation where data provides real-time analytics that help Johns Hopkins doctors, nurses and other health care professionals zero in on precision care for each patient."[78]

The information is used at Johns Hopkins as well as across "mega cohorts" of patients achieved by partnering across institutions to share data. For example, data from the JH-CROWN registry is part of the National COVID Cohort Collaborative, which is collecting and harmonizing data from different institutions across the country into a "data enclave" for use by investigators all over the nation.

"We've cared for more than 12,000 inpatients with COVID-19 at Johns

Scott Zeger

Hopkins, but across the country there have been millions of hospitalized patients," said Garibaldi. "We have so much to learn, in understanding epidemiology and prediction modeling, for example, from these huge cohorts."

He continued, "With COVID-19, there's no one person or group who can possibly tackle all of the different questions that can be answered with this data. There's been an urgency to give access to as many investigators as possible. As a result, we are bringing teams together who might not have known they are working on similar questions. This is an approach to team science that can lead to very rapid discovery — and I'm hopeful it will carry forward for other projects and other data sets."[79]

COVID and Employee Wellness

Johns Hopkins Medicine employees at all levels rose to the challenge of the pandemic, as Rothman noted in November 2021.

"During the 21 months of this pandemic, you have shown incredible dedication to patients, to colleagues and to our cause," he said during a Johns Hopkins Medicine Town Hall. "You've stepped up over and over again to serve the sick and their families. I am so grateful for your dedication and your sacrifice and your bravery. Your actions have brought honor to our institution and our profession."[80]

None of it was easy.

The clinicians on the front lines worked long hours in cumbersome protective gear, sometimes in unfamiliar roles and often worried about their own safety as they tended to frightened and isolated patients.

On top of all that, the measures enacted to slow the spread of the disease meant that many staffers were struggling with new schedules as they changed assignments, worked from home, cared for children who were no longer in school, or coped with family members who were out of work.

For years, Johns Hopkins Medicine had been providing mental health counseling and other emotional support to clinicians. This important work was expanded and intensified during the pandemic.

In 2011, Cheryl Connors, a patient safety specialist with the Armstrong Institute for Patient Safety and Quality, teamed with Albert Wu, a Johns Hopkins professor of health policy and management, and Matt Norvell, a Johns Hopkins Hospital chaplain, to create Resilience in Stressful Events (RISE).

The program was based on Wu's research on second victims, a term he had coined in a *British Medical Journal* article in 2000. The RISE program trains peer responders to provide psychological first aid and emotional support to clinicians who experience distress, remorse and confusion as a result of stressful patient-related events.

As the COVID-19 pandemic approached, members of RISE attended command

Dean/CEO Paul Rothman highlighted the extraordinary work of JHM's faculty, staff and learners during his annual State of JHM address in November 2021.

Wellness carts at Johns
Hopkins hospitals were
stocked with self-care
items, healthy snacks,
and art supplies for
staff, courtesy of the
International Arts + Mind
Lab and the Office of
Well-Being.

center meetings to discuss how to best maintain the well-being and resilience of
clinicians and other staff.[81]

"Leadership astutely recognized that workers were the institution's most
important asset. Distress can be disabling," Wu wrote in an October 2020 editorial
in the *Journal of Patient Safety and Risk Management*.[82]

"I remember thinking it's great that our frontline workers and leaders are taking
care of us, but who is taking care of them?" said Lee Daugherty Biddison, chief
wellness officer for Johns Hopkins Medicine, a year into the pandemic.[83]

RISE leaders collaborated with people representing other resources already
focused on employee wellness. Together, they created MESH (Mental, Emotional
and Spiritual Help), which provided resources for employees across the enterprise
and was led by the Office of Well-Being.

The components of MESH were:

- The Department of Spiritual Care and Chaplaincy, which provides spiritual
  and emotional support to employees as well as patients and their families
- The mySupport employee assistance program, which provides access to free
  counseling and other resources
- RISE, which delivers peer support to health care workers coping with
  stressful patient-related events
- The Office of Well-Being, which promotes work-life balance and includes
  Healthy at Hopkins, which provides wellness support across the institution
- The Department of Psychiatry and Behavioral Sciences, which provides
  timely mental health care to health system employees at The Johns
  Hopkins Hospital and Johns Hopkins Bayview Medical Center.

Recognizing that the pandemic also brought financial strains, Johns
Hopkins established meal and grocery pick-up sites for employees, and created
the Emergency Assistance Support for Employees (EASE) Fund to help with
medical expenses, child care and other costs associated with COVID-19. It
distributed more than $160,000 to more than 200 employees.[84]

Healthy at Hopkins, a key offering of the Office of Well-Being, added more
remote options to its well-being offerings, including meditation sessions through
Zoom, information about housing for clinical staff who needed to isolate from
family, and dance classes run by the Peabody Institute of The Johns Hopkins
University.[85]

When the World's Attention Turns Away

In the spring of 2020, health care providers were hailed as heroes.

The U.S. Air Force Thunderbirds and U.S. Navy Blue Angels flew over The Johns Hopkins Hospital, Johns Hopkins Bayview Medical Center, Sibley Memorial Hospital and Suburban Hospital. Car parades of grateful community members, police officers and firefighters drove through medical campuses. Donations of meals and supplies poured in from local restaurants and the community.

But as the pandemic wore on, the world's attention seemed to turn away. And that was fine with some people who never liked the "hero" label anyway.

Andreas Andreou, the resident chaplain with Nelson 5 since August 2020, said the unit's staff members, as well as others helping to fight the pandemic, shy away from the label of "hero" and refer to their work as a calling.

"If you call them heroes, you're taking that away," he said.*

---

*   Shapiro, Marc, "One Year Later: Are Front-Line Workers Still 'Heroes'?" *Dome*, May 25, 2021, https://www.hopkinsmedicine.org/news/articles/one-year-later-are-front-line-workers-still-heroes.

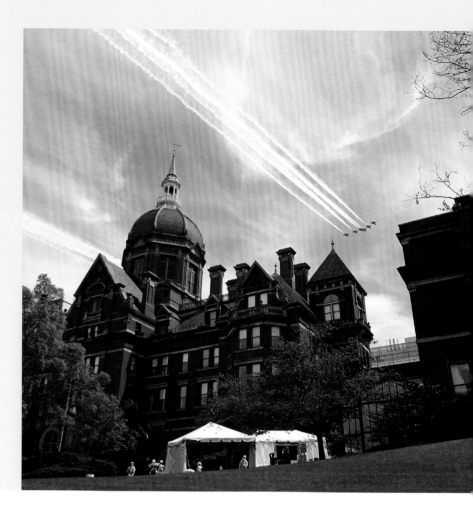

A Carefully Woven Safety Net

"We started by asking them to report on what they were hearing from the field so we could bring together coordinated and comprehensive support," said Deborah Dang, chief nursing director for well-being in the Office of Well-Being.

The result, she said, was "a standardized triage process and an internal referral workflow to connect the unique offerings of each support service."

Some institutions had deemed support staff to be "nonessential workers," and prevented them from coming into the hospital. In contrast, Johns Hopkins Medicine's chaplains and RISE volunteers donned personal protective equipment and rounded on inpatient units to meet with frontline workers and often served as an informal mechanism to connect employees to mental health professionals.

The mySupport team, in turn, could provide counseling or refer Johns Hopkins employees to the psychiatry department's COVID Response Clinic if necessary.

Before the pandemic, RISE had 30 active peer responders, said Wu. During the first surge in the spring of 2020, it used a virtual platform to train 78 more, including clinicians who had been temporarily sidelined by the pandemic.

By March 2022, RISE had helped more than 7,000 clinicians, he said. All six Johns Hopkins hospitals used wellness carts, courtesy of the International Arts + Mind Lab and the Office of Well-Being, that brought healthy treats, art supplies and self-care items around to tired and stressed workers.

"I think this has made it more acceptable for people to understand that sometimes they need help, and it's OK to get help when they need it," Wu said.[86]

COVID-19 Reveals and Highlights Disparities

COVID-19 was new, but health disparities, unfortunately, were not.

As the pandemic progressed during that strange spring of 2020, a bleak but unsurprising trend became apparent: the virus was causing disproportionate harm to people in already disadvantaged communities, particularly people of color.

"The COVID-19 pandemic has brought social and racial injustice and inequity to the forefront of public health," explained the CDC. "It has highlighted that health equity is still not a reality as COVID-19 has unequally affected many racial and ethnic minoritized groups, putting them more at risk of getting sick and dying from COVID-19.[87]

People who were more likely to get sick included those who lived in crowded, multigenerational homes; who traveled by public transportation; who had jobs without the luxuries of paid time off or remote work options; and who lacked adequate access to health care.

"These communities share common social and economic factors, already in place before the pandemic, that increase their risk for COVID-19," said Sherita Golden, vice president and chief diversity officer at Johns Hopkins Medicine, in April 2020.

For those reasons and others, racial tensions were already high when George Floyd was killed.

On May 25, 2020, Floyd, a 46-year-old Black man, pleaded for his life and took his dying breaths with a Minnesota police officer's knee on his neck. The murder, captured on video, sparked marches and protests in America and around the world, aimed at changing a system that did not work the same for everyone.

Black Lives, White Coats

On June 5, 2020, health care workers at all six Johns Hopkins hospitals demonstrated their support for equity and racial justice by kneeling together during a national moment of silence. They joined health care professionals across the country in an event that was organized by White Coats for Black Lives, a national group founded by medical students.[88]

Thomas Elliott, a second-year pediatrics resident at the Johns Hopkins Children's Center and president-elect of the House Staff Diversity Council, read a statement urging Johns Hopkins to continue to work to translate its support for diversity, equity and inclusion into "tangible and perceived impacts for not only the Hopkins community but the Greater Baltimore community as well."

"Today, we kneel in solidarity with White Coats for Black Lives, Black Lives Matter, and the protests and demonstrations around the country that are collectively asking us to dismantle racist policies and cultures in our communities," he said. "When we work together, we can dismantle all forms of oppression and racism within our institution and beyond."[89]

On June 5, 2020, health care workers at all six Johns Hopkins hospitals knelt together during a national moment of silence.

COVID-19 Pandemic Anchor Strategy

Johns Hopkins Medicine and University were already working to address inequities in Baltimore communities that were particularly hard hit by the pandemic.

Through its COVID-19 Pandemic Anchor Strategy, Johns Hopkins leaders worked with Baltimore schools, city and state health departments, and other agencies and organizations to bring COVID-19 information and testing (and later, vaccines) to Baltimore neighborhoods, distribute food, and open crucial lines of communication with community residents and leaders.

"You can give people medicine, but if you're not meeting their basic needs, you're not really treating the whole person," said Golden, a member of the Anchor Strategy group.

In March, as The Johns Hopkins Hospital began treating its first patients with COVID-19, Alicia Wilson, vice president for economic development at The Johns Hopkins University and Johns Hopkins Heath System, was thinking about how to take the hospital's response to the pandemic beyond its walls.

With the support of Sowers and Johns Hopkins University President Ronald J. Daniels, she brought together health system and university leaders to form the COVID-19 Anchor Strategy Work Group.

The group's efforts included:

- **Connection and Education:** Johns Hopkins experts led phone and video conversations and town halls with clergy members, parents, students and older adults on topics such as how to entertain children during the summer, and legal rights in case of eviction. Virtual movie nights, TikTok competitions and trivia games — offered in English and Spanish — provided more opportunities to build connections with young people. Flyers distributed in English and Spanish gave information about physical distancing, mask wearing and other safety measures.

- **Culturally Appropriate Care:** A new program called Juntos connected Latinx patients and their families with bilingual clinicians. "Having someone speak to the patients directly in their primary language is comforting," said Tina Tolson, senior director of operations for Johns Hopkins Medicine Language Services. Juntos was created by Tolson with Kathleen Page, associate professor of medicine; Centro SOL (Center for Salud/Health & Opportunities for Latinos) and the Office of Diversity, Inclusion and Health Equity. It did not replace medical interpretation, but offered additional support, which was especially needed during the pandemic, when visitors were limited, and masks and personal protective equipment made communication more challenging. This team helped to connect patients to helpful community resources in a time of dire need.

Members of the **COVID-19 Pandemic Anchor Strategy Work Group,** left to right: **Nicole McCann, Tina Tolson, Sherita Golden, Panagis Galiatsatos, Alicia Wilson.**

- **COVID-19 Testing and Follow-Up:** Johns Hopkins clinicians traveled to homeless shelters, churches and sober living facilities to provide on-site testing in communities deemed COVID-19 hot spots by the Baltimore City Health Department. They set up temporary testing sites, including one that was under the Jones Falls Expressway, in response to requests from the city or from organizations like Health Care for the Homeless or the Helping Up Mission.

In collaboration with BUILD (Baltimoreans United in Leadership Development), they established a mobile COVID-19 coronavirus testing site at the Sacred Heart of Jesus church in the Highlandtown neighborhood of Baltimore, which serves a mostly Spanish-speaking population.

The teams notified people of their test results and helped them get appropriate care if needed.

One option as a place to live for people with mild or asymptomatic COVID-19 — and for health care workers — was the Lord Baltimore Hotel. Through a partnership between Johns Hopkins, the University of Maryland Medical System and CareFirst, the hotel provided rooms to people who were homeless or living in crowded conditions that made physical distancing difficult.

Anti-Bias Training at the School of Medicine

Jessica Bienstock

Sherita Golden, vice president and chief diversity officer for Johns Hopkins Medicine, remembers walking into a patient's room back when she was an intern. She wore a white coat, with a stethoscope peeking out of her pocket.

"I'm so glad you're here," the patient said. "I have been waiting all day for the TV lady. I can't get the remote control to work."

While Golden stood speechless, the patient's daughter, also in the room, piped up: "Mom, she's your doctor!"

That bystander intervention early in Golden's medical career instantly dispelled tension and resolved the situation. Thinking back on it now, Golden, the Hugh P. McCormick Family Professor of Endocrinology and Metabolism, says she wishes she had thought to deflect the misperception, saying something like, "I'm your doctor, and you wouldn't want me to fix your television because that's not

what I was taught to fix."

The Johns Hopkins University School of Medicine has long addressed issues of race and equity in its curriculum. In 2020, leaders added a three-year program that would teach students and trainees how to address unconscious bias and recognize structural racism in their treatment of patients and their interactions with colleagues.

Faculty, students, trainees, postdocs and fellows were required to complete a virtual training session in unconscious bias, followed by anti-racism and then bystander intervention training, which helps people confront racism as it is happening. For example, patients sometimes say disparaging things about the clinicians caring for them.

Jessica Bienstock, associate dean for graduate medical educa-

tion at the time (and promoted in March 2022 to inaugural director of graduate clinical education and senior associate dean for graduate medical education), said unconscious bias training became a requirement for residency program directors about a year before the pandemic began.

"The damage done to a trainee when a faculty member says nothing is tremendous," she says. "It causes permanent scarring that can influence their confidence and careers. The workshops show faculty members how to step in and step up."

She welcomed the expansion to students and trainees. "This is a defining time in medicine," she says. "It needs to be our goal that the culture of medicine is one that promotes both professional equity and health equity."*

---

* Nitkin, Karen, "New Anti-Bias Training at the School of Medicine," *Dome*, Sept. 8, 2020, https://www.hopkinsmedicine.org/news/articles/new-bias-and-racism-training-at-the-school-of-medicine.

• **Food:** Grab-and-go meals and food pantries were available for employees at The Johns Hopkins Hospital, Johns Hopkins Bayview Medical Center and Sibley Memorial Hospital. Wilson noted that employees may be living in households with reduced incomes because of pandemic shutdowns, and they may have safety concerns about going to a grocery store. Johns Hopkins also worked with the Maryland Food Bank to provide meals to thousands of low-income adults in East Baltimore.[90]

A year into the pandemic, in March 2021, the East Baltimore COVID-19 Food Access Initiative, created by Johns Hopkins and community partners, had delivered 4.6 million meals to Baltimore families.[91]

Sherita Golden

As vice president and chief diversity officer for Johns Hopkins Medicine, Sherita Hill Golden leads the Office of Diversity, Inclusion and Health Equity, collaborating with leaders across Johns Hopkins Medicine to advance diversity and inclusion efforts.

"A diversified workforce is crucial to achieving health equity, enhancing scientific discovery and strengthening ties to our surrounding communities," she said when she began the job in May 2019.*

Golden received her bachelor's degree in biology from the University of Maryland, College Park, her doctorate of medicine from the University of Virginia School of Medicine, and her master of health science degree in clinical epidemiology from the Johns Hopkins Bloomberg School of Public Health.†

Golden, the author of more than 250 articles focused on diabetes, endocrinology and health disparities, is the Hugh P. McCormick Family Professor of Endocrinology and Metabolism, with a joint appointment in the Welch Center for Prevention, Epidemiology and Clinical Research.

In 2015, she became inaugural executive vice-chair of the Department of Medicine at the Johns Hopkins University School of Medicine. In that post, which she held until 2019, she launched a departmentwide civic engagement initiative, resulting in programs that addressed community-related concerns and enhanced employee engagement following Baltimore's civil unrest surrounding the 2015 death of Freddie Gray.‡

Sherita Golden

* "Welcome Letter from Chief Diversity Officer Sherita Hill Golden," Johns Hopkins Medicine webpage, Office of Diversity, Inclusion and Health Equity, https://www.hopkinsmedicine.org/diversity/about-us/letters-from-leadership/welcome-letter-sherita-golden.html.

† "Sherita Hill Golden, M.D.," Johns Hopkins Medicine webpage, Leadership, https://www.hopkinsmedicine.org/about/leadership/biography/sherita-golden.

‡ "Sherita Hill Golden, M.D."

Vaccines and Equity

At 7:30 on the cold and windy morning of Dec. 16, 2020, Kelly Casey got a history-making shot in her left arm.

Seated at a folding table on the concourse of the Thomas B. Turner Building on The Johns Hopkins Hospital's East Baltimore campus, Casey was the first employee at The Johns Hopkins Hospital to be vaccinated against COVID-19.

"I decided to get the vaccine after seeing a lot of pain and suffering in our patients and our co-workers," said Casey, an occupational therapist who worked with patients throughout the hospital, including in COVID-19 units.

Casey received the first of two doses of the COVID-19 vaccine developed by Pfizer Inc. and BioNTech SE, which had received emergency authorization from the FDA on December 11. Another vaccine, developed by Moderna, would receive the same FDA approval on December 18.

The vaccines arrived during a period of record high COVID-19 infections and deaths in Maryland and across the United States.[92]

They didn't immediately end the pandemic, but they marked an important turning point in efforts to control it.

At first, vaccine supplies were extremely limited, and Johns Hopkins Medicine didn't know from week to week how many doses it would receive from the government.

In order to distribute the vaccines equitably, starting with Johns Hopkins

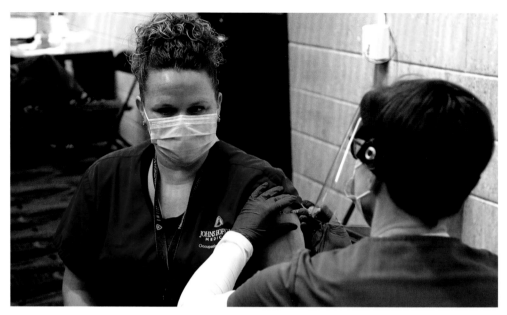

Occupational therapist Kelly Casey was the first staff member at The Johns Hopkins Hospital to receive the COVID-19 vaccine.

employees, Johns Hopkins Medicine launched a Vaccine Equity Task Force, led by Katie O'Conor, an emergency medicine/anesthesiology resident, and Golden.

Other key members of the task force were Deborah Sherman, the nursing informatics director; Carole Martens, director of facilities compliance; and Nicole Iarrobino, who was senior project administrator of patient- and family-centered care.

The group knew that institutionwide emails with information about vaccination opportunities would be missed by many employees who don't work in front of computers, including those in security, food, facilities, patient transport and environmental services.

"The first thing we did was reach out to all the managers of all the departments we thought might be left out and asked them what they wanted," said O'Conor.

In response, task force members went to the departments to provide information and answer questions. They offered posters with QR codes for making vaccine appointments, and held virtual town halls in English and in Spanish.

Golden developed videos and articles with information of particular concern to employees of color, including the demographics of the vaccine's clinical trials, which she noted generally reflected the diversity of the U.S. population.[93]

All JHM employees, including those working remotely, were required to be fully vaccinated or undergo weekly testing by Sept. 1, 2021, and vaccination became a condition or employment or appointment, effective October 1.[94]

By that time, vaccines were widely available and included a single-dose option developed by the Janssen Pharmaceutical Companies of Johnson & Johnson, which had received emergency use FDA approval on Feb. 27, 2021.

In addition to the walk-in vaccination clinic at Turner Auditorium, Johns Hopkins Medicine created pop-up and mobile clinics throughout its hospital system.

Between April and September, 2021, the group organized 11 pop-up vaccine clinics, five vaccine fairs and four mobile clinics in hospital departments, vaccinating 404 Johns Hopkins Hospital employees. Vaccine booster clinics held in the hospital in February and March, 2022, provided a third shot to 146 Johns Hopkins Hospital employees.[95]

Vaccines for the Most Vulnerable

The Vaccine Equity Task Force also led efforts to bring vaccines and information to people most at risk of getting sick or dying of COVID-19.

"For more than a century, Hopkins has gone above and beyond to care for our underserved populations," Rothman said in the November 2021 State of JHM address. "It's no surprise that we have continued this approach during the pandemic, where we know that some of those underserved populations have been the most severely affected by the disease."[96]

On a cool spring Saturday morning in May 2021, O'Conor and nine other Johns Hopkins clinicians and administrators arrived at a city-run syringe exchange site in West Baltimore.

They set up folding chairs and tables, arranged computers and stacked consent forms. They put out single-serve snacks and bottles of water.

That day, they would vaccinate 60 people who were there to swap their used supplies for clean ones. People at the syringe exchange site were already acting to protect their health; O'Conor says she was not surprised that they also lined up for the vaccine.

That may seem like a small number in a city of nearly 576,000, but O'Conor knew that every shot in every arm would bring the pandemic closer to containment, particularly when those arms belong to people most at risk for getting seriously ill or dying of COVID-19.

At the syringe exchange site, participants did not need appointments. Instead, Johns Hopkins administrators registered them on the spot, assuring them that their information would not wind up with law enforcement or immigration officials.

It helped that they were offered a choice between the two-dose Pfizer vaccine

In partnership with local health departments, Johns Hopkins Medicine held pop-up vaccine clinics in neighborhoods with strong Latino populations.

Tapping Community Ties

Centro Sol Co-Director Kathleen Page at a pop-up clinic, talking with a man who was just vaccinated.

Alejandra Flores-Miller walked into the East Baltimore barbershop and asked the owner to turn down the music so she could make an announcement.

Speaking in Spanish, she said that anyone who wanted a COVID-19 vaccine could get one: A Johns Hopkins Medicine clinic at the nearby Sacred Heart of Jesus Church had extra doses.

A few customers stood up, removed the towels draped around their necks, paid and walked over to the church. "Twenty people were in the barbershop, and three went," Flores-Miller later said.

She knew the owner, just as she knew most of the people working in the stores and restaurants of the largely Spanish-speaking neighborhood.

Flores-Miller, research program coordinator in the Division of Infectious Diseases, was part of a team, led by Centro Sol Co-director Kathleen Page, that brought vaccine clinics to the church starting in February 2021.

She had been building relationships in the community since 2008, when she created a Latino Outreach Program for Johns Hopkins to connect residents with resources for preventing and treating HIV.

"The most important thing is that, from the moment you walk in [to the vaccine clinic] to the moment you leave, you can speak Spanish," said Page, an infectious diseases specialist. "If you don't speak the language, it can be intimidating. It's hard to express fears or have questions answered through an interpreter."

Each clinic included a pharmacist, interpreters, administrators to register participants, and clinicians to give the vaccines.

About 200 people received vaccinations at each session, said Page. If extra doses were available, Flores-Miller and other outreach workers tried to find people who could use them.

"We go where we know we'll find members of the Spanish-speaking community, and we start explaining to people, giving them the information that we have so they can come and get vaccinated," said Melissa Cuesta, a senior research program coordinator with the Division of Infectious Diseases.

If they hesitated, she didn't push. "One of the main questions they always ask is, 'Did you get vaccinated?' When we say yes, that gives them more interest."*

---

*   Nitkin, Karen, "Johns Hopkins Taps Community Ties to Bring Vaccines to Latinos in Baltimore," *Dome*, June 4, 2021, https://www.hopkinsmedicine.org/news/articles/johns-hopkins-taps-community-ties-to-bring-vaccines-to-latinos-in-baltimore.

and the single-dose Johnson & Johnson vaccine. "This population, with substance use and unstable housing, doesn't often have access to choices, on their terms," said O'Conor, adding that about 60% chose Pfizer.

"What I love about this is we are adapting the environment to meet the community's needs," said O'Conor.

Her group and other vaccine equity teams at Johns Hopkins Medicine concentrated their efforts to reach people hardest hit by the COVID-19 pandemic, including those who are Black, Latino, Indigenous or elderly.

For example, an initiative that ran from January to April, 2021, in Washington, D.C., vaccinated 1,013 people in 12 public housing complexes, plus 1,532 residents through targeted outreach to additional senior housing properties and surrounding neighborhoods in ZIP codes with low vaccination rates and high rates of COVID-19 cases and mortality.[97]

And clinics at the Sacred Heart of Jesus Church in East Baltimore brought vaccines and information to a primarily Spanish-speaking population.

Sowers personally administered vaccines at several of the clinics.

"In order to achieve high vaccination rates, we need to reach out to every group," said Ben Bigelow, the Johns Hopkins medical student who spearheaded all three vaccine equity efforts across Johns Hopkins Medicine. By March 2022, 431 community vaccination clinics had been held, he said.[98]

COVID Year 2: Vaccines and Variants

As vaccines became widely available in early 2021, it seemed as though the worst of the pandemic would come to an end. COVID-19 cases, hospitalizations and deaths were falling, and restrictions were easing.

Between March and June, 2021, Johns Hopkins Medicine relaxed its guidance for employee travel,[99] went back to letting health care workers replace surgical masks and fit-tested N95 respirators every day (instead of saving and storing them until soiled or damaged),[100] and began allowing more visitors into the hospitals.[101]

Student Debt Initiative

In 2022, the Johns Hopkins University School of Medicine launched a program aimed at reducing and eventually eliminating student debt by replacing education loans with scholarships.

The program, which could help Johns Hopkins create a more diverse class of medical students, was made possible with $50 million each from Carlyle Group co-founder and Johns Hopkins Medicine Board of Trustees chair Bill Conway and his wife, Joanne Conway; board member James Davis; and an anonymous donor. The Johns Hopkins University committed $25 million in endowed matching funds.

The program is focusing first on medical students with the greatest financial need, and will eventually extend to all qualified medical students.

Surveys of students who got into Johns Hopkins but chose another medical school found that financial considerations were the top driver for students from underrepresented racial and ethnic backgrounds.*

* SOM Debt-Free One-Pager, October 2021, email from Kim Hoppe to Karen Nitkin, May 15, 2022.

"Slowly, but surely, parts of our lives that flipped upside down as COVID-19 took hold last year are beginning to tip right-side up again," read a "Dear Colleagues" email to Johns Hopkins Hospital staff, dated May 26, 2021.

The letter noted that Nelson 5, the hospital unit that had cared for COVID-19 patients the longest, was out of biomode for the first time since March 14, 2020, no longer requiring special equipment and ventilation to keep contaminated air from escaping and the negative pressure anterooms where health care workers could safely put on and take off their personal protective equipment.

"This moment feels significant," the note continued. "Even historic. After all, at one point early in the pandemic, 11 units — including six of our ICUs — were in biomode at the same time."[102]

For the first time since the start of the pandemic, Rothman and Sowers held a Town Hall meeting in person, wearing masks on the stage of Hurd Hall and sitting in chairs that had been placed more than six feet apart.

"We are in the 16th month of the COVID pandemic," said Rothman at the June 22, 2021, event, speaking over Zoom and in front of a physically distanced audience. "We have diagnosed over 23,000 people with COVID-19 and have hospitalized about 8,000. We've provided superb care to these patients, really thanks to many of you in this room or watching today and your tireless efforts."[103]

But it wasn't over yet.

The Johns Hopkins Health System, along with the state, nation and world, would experience several more surges of patients with COVID-19 in the next year, brought about by highly infectious Delta and Omicron virus variants.

The bad news was that the surges once again filled hospital beds and taxed clinicians who were already physically and mentally exhausted. Nursing shortages were a real problem, as some frontline workers tested positive for COVID-19 and couldn't come to work and others simply left the profession.

Many nurses relinquished their staff jobs to become traveling nurses, a move that gave them more freedom and money, but was more expensive and less reliable for the health system.

Clinicians at all levels said they were dispirited and surprised that COVID-19 care had taken an ugly, partisan turn, with many people making decisions about masks, vaccines and treatments based on their party affiliation, instead of on science-based public health recommendations.

There was good news, too.

In a bid to lure and keep workers, Johns Hopkins University and Health System, collectively the largest private employer in Maryland, announced in May 2021 that it would adopt a $15 minimum wage across the enterprise.[104]

And a year into the pandemic, COVID-19 was no longer a mystery. Thanks in no small measure to Johns Hopkins researchers and clinicians, much was known about the disease and treatments, and how they varied by individual. Visitor restrictions and mask mandates could be altered as needed, and hospital units could be converted in and out of biomode.

The health system followed a crisis standards of care framework, developed by the Institute of Medicine (now known as the National Academy of Medicine), to

Johns Hopkins Medicine worked to boost vaccine rates throughout the Baltimore and D.C. regions.

adjust expectations and delivery of care at individual hospitals during COVID-19 surges. Hospitals also used a visitor guideline framework to adjust policies as the situation fluctuated between green (least risk) up through yellow, red and purple (highest risk).

On July 7, 2021, all Johns Hopkins Medicine inpatient and outpatient locations in Maryland and the District of Columbia moved to green, low-risk level guidelines, meaning two care partners could accompany most patients.[105]

Then came the Delta variant.

On June 16, 2021, the CDC classified Delta as a "variant of concern," meaning it was likely to be highly transmissible and resistant to established vaccines and treatments.[106]

In short order, Delta became the dominant variant of the SARS-CoV-2 virus in the U.S. Even fully vaccinated people, while protected from the most severe forms of the disease, were susceptible to breakthrough infections.

As hospitalizations climbed again, Johns Hopkins Medicine on July 27 changed the risk level at All Children's Hospital from yellow to red.[107]

On July 30, universal masking requirement were reinstated across all Johns Hopkins Medicine buildings, a change for "fully vaccinated personnel who, since July 1, have had the option to remain unmasked in nonclinical buildings and private areas of clinical buildings where there is no reasonable expectation of encountering a patient, care partner or member of the public," explained a note to employees.[108]

With a rise in breakthrough infections, Johns Hopkins Medicine, following CDC guidance, began offering booster COVID-19 vaccinations starting August 20 to immunocompromised patients.[109]

On October 1, Johns Hopkins Medicine and University began providing boosters to qualifying employees — those who were 65 years and older, had certain underlying medical conditions, or worked or lived in high-risk settings.[110]

On December 3, again following CDC guidance, Johns Hopkins Medicine "strongly encouraged" employees to get the booster if they had received the second dose of the Moderna or Pfizer vaccine more than six months prior, or the single dose of the Johnson & Johnson vaccine more than two months earlier.[111]

Another variant, Omicron, began raising alarm bells in November 2021. It caused a high number of breakthrough cases in vaccinated people, and came just as

Johns Hopkins Medicine went to bat for vaccine access, even making sure the Baltimore Orioles mascot was protected during the Central Intercollegiate Athletic Association (CIAA) Fan Fest at the Baltimore Convention Center in February 2022. Fifty-one people — and one bird — were vaccinated during the event.

people were traveling for holidays and were spending more time indoors because of colder weather.

The surge was so intense that Johns Hopkins Medicine hospitals and outpatient locations moved to purple — the most critical risk level for care partner visitation, starting Jan. 5, 2022.

"Every time health care professionals dared to hope the worst was over, the virus evolved and struck again," observed a Jan. 31, 2022 *Hub* article. It continued:

> *The most recent surge delivered a powerful punch to hospital capacity, not just at Hopkins and in the state but in health systems across the nation.*
>
> *"We've simply never seen this many COVID patients," says Gabor David Kelen, director of the Department of Emergency Medicine and director of the Johns Hopkins Office of Critical Event Preparedness and Response. "During the first surge, when this started, Maryland reached 1,500 COVID patients hospitalized at one time. This time, the state peaked at 3,462. We were snowed in during the second major wave, and now, with Omicron, it's been nearly twice that. Think of a train wreck every day for weeks on end. That's what we are dealing with."*
>
> *The latest surge has, in fact, seen the largest increase of COVID-19-related admissions since the pandemic began. This is compounded by many staff members themselves becoming ill. ...*
>
> *To be sure, this wave is different. Many COVID-19 patients are not as severely ill as those hospitalized earlier in the pandemic, thanks to vaccinations and boosters, and physicians know more now about how to treat COVID-19. Nevertheless, the sheer volume of patients has proved daunting. "Even though there is a smaller proportion of people with COVID who are very sick, it's still a huge number," Kelen says. "We are seeing a fair number of breakthrough infections, less so among the boosted."[112]*

The article's author, Marlene Cimons, asked Garibaldi, who had been on the front lines of the pandemic from the very start, what he thought the future would hold.

"I've been wrong so many times trying to predict what's going to happen that I no longer try," he said.[113]

ENDNOTES

1 "CDC Museum COVID-19 Timeline," Centers for Disease Control and Prevention webpage, David J. Sencer CDC Museum: In Association with the Smithsonian Institution, https://www.cdc.gov/museum/timeline/covid19.html#:~:text=December%2012%2C%202019,shortness%20of%20breath%20and%20fever.

2 Nitkin, Karen, "A Year of COVID-19 Loss, Learning and New Strength," *Dome*, March 8, 2021, https://www.hopkinsmedicine.org/news/articles/a-year-of-covid-19-loss-learning-and-new-strength.

3 Karen Nitkin interview with Kevin Sowers, Dec. 20, 2021.

4 Karen Nitkin interview with Paul Rothman, Jan. 5, 2022.

5 Ghebreyesus, Tedros Adhanom, "WHO Director-General's Opening Remarks at the Media Briefing on COVID-19," World Health Organization news release, March 11, 2020, https://www.who.int/director-general/speeches/detail/who-director-general-s-opening-remarks-at-the-media-briefing-on-covid-19---11-march-2020.

6 "Johns Hopkins University and Medicine Coronavirus Resource Center, Johns Hopkins University and Medicine webpage, https://coronavirus.jhu.edu/.

7 Data supplied to Karen Nitkin by Avinash Gadala, June 28, 2022.

8 Karen Nitkin interview with Kevin Sowers, Dec. 20, 2021.

9 Rothman, Paul B., M.D., and Sowers, Kevin, "Johns Hopkins Medicine: It Takes a Community to Fight a Pandemic," *U.S. News and World Report*, July 28, 2020.

10 Barry, John M., "The Great Influenza," Penguin Random House, Oct. 4, 2005.

11 "A Year of COVID-19 Loss, Learning and New Strength."

12 Johns Hopkins Medicine at 25, Achievement #4: A Unified Response to Medical Crises at Johns Hopkins, https://www.hopkinsmedicine.org/25-anniversary/events-achievements.html#4.13 "CDC Museum COVID-19 Timeline."

14 Surowiec, Justynam " 'We Had to Get This Right' How Johns Hopkins Built the Coronavirus Tracking Global Dashboard: An Oral History," Johns Hopkins Applied Physics Laboratory webpage, Feature Story, https://www.jhuapl.edu/FeatureStory/210426-JHU-COVID-dashboard-oral-history.

15 "Johns Hopkins Medicine at 25 Achievement #4: A Unified Response to Medical Crises at Johns Hopkins."

16 " 'We Had to Get This Right' How Johns Hopkins Built the Coronavirus Tracking Global Dashboard: An Oral History."

17 " 'We Had to Get This Right' How Johns Hopkins Built the Coronavirus Tracking Global Dashboard: An Oral History."

18 Landi, Heather, "How Johns Hopkins Medicine became an information powerhouse during the COVID-19 pandemic, Fierce Healthcare, May 24, 2021, https://www.fiercehealthcare.com/tech/johns-hopkins-medicine-became-information-powerhouse-during-covid-19-pandemic-here-s-how.

19 Nitkin, Karen, "Milestones in a Year of COVID-19 at Johns Hopkins Medicine," *Dome*, March 8, 2021, https://www.hopkinsmedicine.org/news/articles/covid-19-timeline-for-johns-hopkins-medicine.

20 "Governor Hogan Announces Major Actions to Protect Public Health, Limit Spread of COVID-19 Pandemic," The Office of Governor Larry Hogan news release, March 12, 2020, https://governor.maryland.gov/2020/03/12/governor-hogan-announces-major-actions-to-protect-public-health-limit-spread-of-covid-19-pandemic/.

21 "A Year of COVID-19 Loss, Learning and New Strength."

22 Email from Heba Mostafa to Karen Nitkin, March 1, 2022.

23 "COVID-19 Research," Johns Hopkins Medicine website, Johns Hopkins Medicine Pathology, https://pathology.jhu.edu/research/covid-19-research.

24 "Test Could Prove Useful for Large-Scale Screening and Epidemiological Surveys," Johns Hopkins Bloomberg School of Public Health news release, Nov. 13, 2020, https://publichealth.jhu.edu/2020/new-saliva-based-antibody-test-for-sars-cov-2-highly-accurate-in-initial-study#:~:text=A%20new%20saliva%2Dbased%20test,led%20by%20Bloomberg%20School%20researchers.

25 "An Incredible Journey: The Department of Pathology's Response to the COVID-19 Pandemic," *Pathways*, December 2020, https://pathology.jhu.edu/build/assets/department/files/Pathways-December-2020.pdf.

26 Maragakis, Lisa, "JHHS Weekly COVID-19 Testing of Inpatients Launches Week of Dec. 14," Dear Colleagues letter to Johns Hopkins Medicine providers and managers, Dec. 11, 2020, https://intranet.insidehopkinsmedicine.org/heic/_docs/2019-nCoV_jhm_2020_12_11_weekly_testing_inpatients.pdf.

27 "An Incredible Journey: The Department of Pathology's Response to the COVID-19 Pandemic."

28 "1 Millionth Test for Sars-CoV-2," Johns Hopkins Medicine Department of Pathology news release, Nov. 24, 2021, https://pathology.jhu.edu/about/news/2021.

29 Rob Carter phone interview with Karen Nitkin, Feb. 28, 2022.

30 Mary Brown phone interview with Karen Nitkin, March 1, 2022.

31 "A Year of COVID-19 Loss, Learning and New Strength."

32 "COVID-19 President's Message, March 13, 2020," video on JHU webcast, https://webcast.jhu.edu/Mediasite/Play/fd7744be118e4e9e9a343b39dbab37241d.

33 Rob Carter phone interview with Karen Nitkin, Feb. 28, 2022.

34 Johns Hopkins Medicine Town Hall meeting, March 31, 2022.

35 Allen Kachalia phone interview with Karen Nitkin, May 4, 2022.

36 "A Year of COVID-19 Loss, Learning and New Strength."

37 "Johns Hopkins Medicine at 25 Achievement #4: A Unified Response to Medical Crises at Johns Hopkins."

38 Peter Greene telephone interview with Karen Nitkin, Feb. 28, 2022.

39 Deanna Hanisch telephone interview with Karen Nitkin, March 2, 2022.

40 Peter Greene telephone interview with Karen Nitkin, Feb. 28, 2022.

41 Manisha Loss Zoom interview with Karen Nitkin, March 1, 2022.

42 Keating, Michael, "COVID-19 Patient Self-Screener Tool Helps Keep Staff and Patients Safe," Johns Hopkins Medicine website, Coronavirus (COVID-19) Information and Updates, Aug. 27, 2020, https://www.hopkinsmedicine.org/coronavirus/articles/screener-tool.html.

43 Nitkin, Karen, "'All Hands on Deck' as Johns Hopkins Facilities Teams Prepare Hospitals for COVID-19, *Dome*, April 27, 2020, https://www.hopkinsmedicine.org/coronavirus/articles/all-hands-on-deck.html.

44 Nitkin, Karen, "Take the Long (Distance) View," *Dome*, March 2013, https://www.hopkinsmedicine.org/news/publications/dome/dome_march_2013/take_the_long_distance_view.

45 Nitkin, Karen, "Q&A with Rebecca Canino," *Dome*, Aug. 25, 2017, https://www.hopkinsmedicine.org/news/articles/taking-johns-hopkins-telemedicine-to-the-next-level.

46 Nitkin, Karen, "In Fight Against Coronavirus, Telemedicine Ramps Up at Johns Hopkins," Johns Hopkins Medicine webpage, Coronavirus (COVID-19) Information and Updates, March 19, 2020, https://www.hopkinsmedicine.org/coronavirus/articles/telemedicine.html.

47 "Johns Hopkins Medicine Marks 1 Million Telemedicine Visits During COVID-19 Pandemic," Johns Hopkins Medicine news release, July 1, 2021, https://www.hopkinsmedicine.org/news/newsroom/news-releases/johns-hopkins-medicine-marks-1-million-telemedicine-visits-during-covid-19-pandemic.

48 "Johns Hopkins Medicine Marks 1 Million Telemedicine Visits During COVID-19 Pandemic."

49 Nitkin, Karen, "New Platform Makes Telemedicine Easier for Johns Hopkins Patients and Providers," *Dome*, Aug. 24, 2020, https://www.hopkinsmedicine.org/news/articles/new-platform-makes-telemedicine-easier-for-johns-hopkins-patients-and-providers.

50 Smith, Linell, "Telemedicine Improves Care Across Johns Hopkins Medicine," *Dome*, Dec. 15, 2021, https://www.hopkinsmedicine.org/news/articles/telemedicine-improves-care-across-johns-hopkins-medicine.

51 Maragakis, Lisa; Kelen, Gabe; Maloney, Robert, "COVID-19 Visitation Guidance," Dear Colleagues letter to the Johns Hopkins Medicine community, March 10, 2020, https://intranet.insidehopkinsmedicine.org/heic/_docs/2019-nCoV_jhm_2020_03_10.pdf.

52 Shapiro, Marc, "Johns Hopkins Helps Patients and Families Stay Connected When COVID-19 Prohibits Visits," Johns Hopkins Medicine webpage, Coronavirus (COVID-19) Information and Updates, May 11, 2020, https://www.hopkinsmedicine.org/coronavirus/articles/stay-connected.html.

53 Blum, Karen, "Nurses Lead Creation of 'Prone Team' to Help Manage Patients with Coronavirus Who Are in Crisis," Johns Hopkins Medicine webpage, Coronavirus (COVID-19) Information and Updates, May 6, 2020, https://www.hopkinsmedicine.org/coronavirus/articles/prone-team.html.

54 Eddy, Lisa, "At Johns Hopkins, Recovery for Many COVID-19 Survivors Begins in the ICU," Johns Hopkins Medicine webpage, Coronavirus (COVID-19) Information and Updates, April 30, 2020, https://www.hopkinsmedicine.org/coronavirus/articles/icu-recovery.html.

55 "How did COVID-19 Push Medicine Forward?" YouTube video, Johns Hopkins Medicine channel, https://www.youtube.com/watch?v=ceNNvs6HuY4&list=PL6ShoiJMrpsd-9w7bGmlNY70AQYarnjwP.

56 Nitkin, Karen, "Gerry and Georgene Stephens: Savoring Life Together After COVID-19," Johns Hopkins Medicine webpage, Coronavirus (COVID-19) Information and Updates, June 3, 2020, https://www.hopkinsmedicine.org/coronavirus/patient-stories/gerry-georgene-stephens.html.

57 Gniazdowski, Victoria, et al, "Repeated Coronavirus Disease 2019 Molecular Testing: Correlation of Severe Acute Respiratory Syndrome Coronavirus 2 Culture with Molecular Assays and Cycle Thresholds," *Clinical Infectious Diseases*, Oct. 26, 2020, https://academic.oup.com/cid/article/73/4/e860/5940589?login=true.

58 "Covid Story Tip: Does Prolonged COVID-19 Detection Identify People Who Are Infectious Long Term?" Johns Hopkins Medicine news release, Dec. 1, 2020, https://www.hopkinsmedicine.org/news/newsroom/news-releases/does-prolonged-covid-19-detection-identify-people-who-are-infectious-long-term.

59 Nitkin, Karen, "A Year of COVID-19 Loss, Learning and New Strength," *Dome*, March 8, 2021, https://www.hopkinsmedicine.org/news/articles/a-year-of-covid-19-loss-learning-and-new-strength.

60 Luo, Chun Huai et al., "Infection with the Severe Acute Respiratory Syndrome Coronavirus 2 (SARS-CoV-2) Delta Variant Is Associated With Higher Recovery of Infectious Virus Compared to the Alpha Variant in Both Unvaccinated and Vaccinated Individuals," *Clinical Infectious Diseases*, Dec. 18, 2021, https://academic.oup.com/cid/advance-article/doi/10.1093/cid/ciab986/6469786.

61 Heba Mostafa email to Karen Nitkin, March 2, 2022.

62 "COVID-19 NEWS: Vaccines Effective Even Without Post-Shot Symptoms or Prior Infection," Johns Hopkins Medicine news release, Aug. 25, 2021, https://www.hopkinsmedicine.org/news/newsroom/news-releases/covid-19-news-vaccines-effective-even-without-post-shot-symptoms-or-prior-infection.

63 "In Covid-19 Vaccinated People, Those with Prior Infection Likely to Have More Antibodies," Johns Hopkins Medicine news release, Nov. 1, 2021, https://www.newswise.com/coronavirus/in-covid-19-vaccinated-people-those-with-prior-infection-likely-to-have-more-antibodies/?article_id=759957.

64 Zhong, Diana, et al., Durability of Antibody Levels After Vaccination With mRNA SARS-CoV-2 Vaccine in Individuals with or Without Prior Infection, *JAMA Network* research letter, Nov. 1, 2021, https://jamanetwork.com/journals/jama/fullarticle/2785919.

65 "In Covid-19 Vaccinated People, Those with Prior Infection Likely to Have More Antibodies."

66 "In Covid-19 Vaccinated People, Those with Prior Infection Likely to Have More Antibodies."

67 Smith, Patrick, "State Taps Johns Hopkins, University of Maryland to Open COVID-19 Recovery Unit at Baltimore Convention Center," Johns Hopkins Medicine webpage, Coronavirus (COVID-19) Information and Updates, April 23, 2020, https://www.hopkinsmedicine.org/coronavirus/articles/covid-19-recovery-unit.html.

68 "Baltimore Convention Center Field Hospital: One State's Experience During COVID-19," booklet by TRACIE Healthcare Emergency Preparedness Information Gateway, https://files.asprtracie.hhs.gov/documents/baltimore-convention-center-field-hospital-one-states-experience-during-covid-19.pdf.

69 "Baltimore Convention Center Reaches Milestone of 100,000 People Tested since June 2020," Maryland Department of Health news release, April 1, 2021, https://health.maryland.gov/newsroom/Pages/Baltimore-Convention-Center-reaches-milestone-of-100,000-people-tested-since-June-2020.aspx.

70 "Baltimore Convention Center Field Hospital Launches Pilot Program to Vaccinate Most Vulnerable Populations," Office of Governor Larry Hogan news release, March 1, 2021, https://governor.maryland.gov/2021/03/01/baltimore-convention-center-field-hospital-launches-pilot-program-to-vaccinate-most-vulnerable-populations/.

71 "Baltimore Convention Center Reaches Milestone of 100,000 People Tested since June 2020."

72 "BCCFH – COVID Task Force," University of Maryland Medical System webpage, https://www.umms.org/coronavirus/fighting-covid/expanded-clinical-locations/baltimore-convention-center.

73 Cimons, Marlene, "Data-Driven COVID-19 Care at Johns Hopkins Medicine," *Dome*, June 22, 2021, https://www.hopkinsmedicine.org/news/articles/data-driven-covid-19-care.

74 Smith, Linell and DePasquale, Sue, "COVID Speed: Eight Ways COVID-19 Has Pushed Medicine Forward," *Hopkins Medicine* magazine, Spring/Summer 2021, https://www.hopkinsmedicine.org/news/articles/covid-speed#:~:text=When%20Nobel%20Prize%2Dwinning%20physician,of%20change%20in%20medicine%20this.

75 Smith, Patrick, "Johns Hopkins Research Offers Lessons Learned from Patients with COVID-19," *Dome*, Oct. 21, 2020, https://www.hopkinsmedicine.org/news/articles/johns-hopkins-research-offers-lessons-learned-from-covid-19-patients.

76 "Data-Driven COVID-19 Care at Johns Hopkins Medicine."

77 "The Johns Hopkins Precision Medicine Center of Excellence for COVID-19," Johns Hopkins Medicine webpage, Johns Hopkins inHealth, https://www.hopkinsmedicine.org/inhealth/precision-medicine-centers/covid-19/.

78 "Johns Hopkins Research Offers Lessons Learned from Patients with COVID-19."

79 "COVID Speed."

80 Rothman speaking at JHM Town Hall, Nov. 12, 2021.

81 Nitkin, Karen; Broadhead, Lisa; Smith, Linell; Smith, Patrick, "No Room for Error," *Dome*, Jan. 8, 2016, https://www.hopkinsmedicine.org/news/articles/no-room-for-error.

82 Wu, Albert W., Health Worker Well-Being and Resilience: A Red Ball Issue for the COVID-19 Response," *Journal of Patient Safety and Risk Management*, Oct. 27, 2020, https://journals.sagepub.com/doi/10.1177/2516043520968290.

83 "A Year of COVID-19 Loss, Learning and New Strength."

84 "A Year of COVID-19 Loss, Learning and New Strength."

85 Nitkin, Karen, "Johns Hopkins Tends to the Mental, Emotional and Spiritual Health of Staff," Johns Hopkins Medicine webpage, Coronavirus (COVID-19) Information and Updates, April 29, 2020, https://www.hopkinsmedicine.org/coronavirus/articles/health-of-staff.html.

86 Karen Nitkin telephone phone call with Albert Wu, March 22, 2022.

87 "Health Equity Considerations and Racial and Ethnic Minority Groups," Centers for Disease Control and Prevention website, COVID-19, updated Jan. 25, 2022, https://www.cdc.gov/coronavirus/2019-ncov/community/health-equity/race-ethnicity.html#fn6.

88 Hub staff report, "Johns Hopkins 'White Coats' Gather for Black Lives," *The Hub*, June 5, 2020, https://hub.jhu.edu/2020/06/05/white-coats-for-black-lives/.

89 "Johns Hopkins 'White Coats' Gather for Black Lives."

90 Nitkin, Karen, "With Pandemic Anchor Strategy, Johns Hopkins Takes COVID-19 Care to the Community, *Dome*, July 27, 2020, https://www.hopkinsmedicine.org/news/articles/with-pandemic-anchor-strategy-johns-hopkins-takes-covid-19-care-to-the-community.

91 "A Year of COVID-19 Loss, Learning and New Strength."

92 Smith, Patrick, "COVID-19 Vaccinations Begin Across Johns Hopkins Medicine," *Dome*, Dec. 21, 2020, https://www.hopkinsmedicine.org/news/articles/covid-19-vaccinations-begin-across-johns-hopkins-medicine.

93 Nitkin, Karen, "On Street Corners and in Churches, Johns Hopkins Dispenses COVID-19 Vaccines Where They Are Needed Most," *Dome*, June 4, 2021, https://www.hopkinsmedicine.org/news/articles/on-street-corners-and-in-churches-johns-hopkins-dispenses-covid-19-vaccines-where-they-are-needed-most.

94 "COVID-19 Vaccination Now a Condition of Employment or Appointment," Dear Colleagues letter from Paul B. Rothman and Kevin W. Sowers, to Johns Hopkins Health System employees, credentialed medical staff members and Johns Hopkins University School of Medicine faculty and staff members, residents, and clinical and postdoctoral fellows, Sept. 3, 2021, https://intranet.insidehopkinsmedicine.org/heic/_docs/2019_nCoV_jhm_COVID-19_vaccination_now_condition_of_employment_or_appointment.pdf.

95 Email from Deborah Sherman to Karen Nitkin, March 30, 2022.

96 Rothman, Paul B., The State of Johns Hopkins Medicine 2021, YouTube video, on Johns Hopkins Medicine channel, ://www.youtube.com/watch?v=xwOXIZ_sM8s&t=4s.

97 Nitkin, Karen, "Johns Hopkins COVID-19 Vaccine Program Reaches DC's Most Vulnerable," Johns Hopkins Medicine Greater Washington Area publication, Summer 2021, https://www.hopkinsmedicine.org/news/articles/johns-hopkins-covid-19-vaccine-program-reaches-dcs-most-vulnerable.

98 Email from Ben Bigelow to Karen Nitkin, March 16, 2022.

99 "Updated COVID-19 Restrictions," Dear Colleagues letter from Gabe Kelen, Lisa Maragakis and Rob Carter to the Johns Hopkins Medicine community, May 7, 2021, https://intranet.insidehopkinsmedicine.org/heic/_docs/2019-nCoV_jhm_2021_05_07_updated_restrictions.pdf.

100 "Mask and Respirator Conservation Update," Dear Colleagues letter from Lisa Ishii, Lisa Maragakis and Robert Carter to the Johns Hopkins Medicine community, April 2, 2021, https://intranet.insidehopkinsmedicine.org/heic/_docs/2019-nCoV_jhm_2021_04_02_mask_respirator_conservation_update.pdf.

101 "JHM Update on CDC Guidance on Masking for Those Fully Vaccinated," Dear Colleagues letter from Gabe Kelen, Lisa Maragakis and Rob Carter to the Johns Hopkins Medicine community, March 11, 2021, https://intranet.insidehopkinsmedicine.org/heic/_docs/2019-nCoV_jhm_2021_03_11_masking.pdf.

102 "A New Era in our COVID-19 Care," Dear Colleagues letter from Redonda G. Miller, Charles B. Reuland, Peter M. Hill, Deborah J. Baker, to all Johns Hopkins Hospital staff and medical staff, May 26, 2021, https://intranet.insidehopkinsmedicine.org/heic/_docs/2019-nCoV_2021_05_26_jhh_new_era_covid_care.pdf.

103 "JHM Town Meeting June 22: COVID-19 Update; Rejuvenation; and Looking Ahead," Dear Colleagues letter from Paul B. Rothman and Kevin W. Sowers to Johns Hopkins Medicine staff, faculty, trainees and students, June 16, 2021, https://intranet.insidehopkinsmedicine.org/heic/_docs/2019-nCoV_jhm_2021_06_16_town_meeting_june_22_COVID-19_update_rejuvenation_looking_ahead.pdf.

104 Hub staff report, "Johns Hopkins Commits to $15 Minimum Wage for All Employees," The Hub, May 6, 2021, https://hub.jhu.edu/2021/05/06/johns-hopkins-15-dollar-minimum-wage/.

105 "Care Partner Visitation Risk Level at JHM for July 7," message from Johns Hopkins Medicine to Johns Hopkins Medicine nurses, residents, credentialed medical staff members and managers, July 6, 2021, https://intranet.insidehopkinsmedicine.org/heic/_docs/2019-nCoV_jhm_2021_07_06_care_partner_visitation_risk_level_at_JHM.pdf.

106 Crist, Carolyn, "CDC: Delta Variant from India a 'Variant of Concern,'" WebMD, June 16, 2021, https://www.webmd.com/lung/news/20210616/delta-variant-of-concern.

107 "Care Partner Visitation Risk Level at JHM for July 27," Message from Johns Hopkins Medicine to Johns Hopkins Medicine nurses, residents, credentialed medical staff members and managers, July 26, 2021, https://intranet.insidehopkinsmedicine.org/heic/_docs/2019-nCoV_2021_07_26_jhm_care_partner_visitation_risk_level.pdf.

108 "Important Update to Masking Requirement for JHM Personnel," Message from Johns Hopkins Medicine to managers, July 29, 2021, https://intranet.insidehopkinsmedicine.org/heic/_docs/2019-nCoV_jhm_2021_07_29_masking_update.pdf.

109 "JHM Confirms Third COVID-19 Vaccine Dose for Individuals with Qualifying Medical Conditions," Dear Colleagues letter from Gabe Kelen, Lisa Maragakis and Robert Carter to Johns Hopkins Medicine clinicians and managers, Aug. 20, 2021, https://intranet.insidehopkinsmedicine.org/heic/_docs/2019_nCoV_jhm_2021_08_20_JHM_confirms_third_vaccine_dose_individuals_underlying_medical_conditions.pdf.

110 "Pfizer COVID-19 Vaccine Boosters Available at JHM for Eligible Individuals," message from Johns Hopkins University and Medicine to the Johns Hopkins Medicine and The Johns Hopkins University communities, Oct. 1, 2021, https://intranet.insidehopkinsmedicine.org/heic/_docs/2019-nCoV_jhm_2021_10_01_COVID-19_pfizer_vaccine_boosters_available_for_eligible_individuals.pdf.

111 "JHM Personnel Encouraged to Receive a COVID-19 Booster," Dear Colleagues letter from Gabe Kelen and Lisa Maragakis to Johns Hopkins Medicine staff and faculty members, residents, clinical and research postdoctoral fellows, students and providers, Dec. 3, 2021, https://intranet.insidehopkinsmedicine.org/heic/_docs/2019-nCoV_2021_03_12_jhm_covid_booster_email.pdf.

112 Cimons, Marlene, "Hospitals on the Brink," The Hub, Jan. 31, 2022, https://hub.jhu.edu/2022/01/31/omicron-surge-pushes-hospitals-to-the-brink/.

113 https://hub.jhu.edu/2022/01/31/omicron-surge-pushes-hospitals-to-the-brink/.

Acknowledgments

This book was reported, written, edited and designed during a pandemic.

Interviews were conducted over Zoom or by telephone, instead of in person. Discussions about word choices and page layouts likewise took place in a digital world, instead of in the Baltimore office I once shared with the editors and graphic designers who were my collaborators.

Like so many others during this time of physical distancing, I worked from home. Yet I never felt alone.

This book would not exist without **Paul B. Rothman**, dean of the school of medicine and chief executive officer of Johns Hopkins Medicine, from 2012 to 2022. Rothman wisely saw that it was time for a companion piece to **Neil A. Grauer**'s masterful *Leading the Way: A History of Johns Hopkins Medicine* (2012), which traces the extraordinary story of Johns Hopkins Medicine from its founding through 2011.

Rothman shared his memories and insights with me over several hours of interviews, and read multiple drafts to ensure this book accurately reflects his time at the helm. He made sure the book gives credit and attention to the Johns Hopkins Medicine faculty and staff members who make it a world-class institution for patient care, research and medical education.

Dozens of people agreed to interviews for this book, generously giving me their valuable time and attention, even during the most frantic and exhausting days of the COVID-19 pandemic.

Kevin Sowers, president of the Johns Hopkins Health System and executive vice president of Johns Hopkins Medicine, shared his vision for Johns Hopkins Medicine during interviews and through insightful reviews of early drafts. **Theodore DeWeese**, interim dean and CEO, provided unfailing support, along with his trademark enthusiasm and good humor.

Several other Johns Hopkins Medicine luminaries stepped up to the considerable challenge of reading early drafts of *Leading the Change*. They include **Landon King**, **Antony Rosen**, **Roy Ziegelstein**, **William Baumgartner**, **Robert Kasdin** and **Daniel Shealer**, who all devoted hours to the project and offered insightful suggestions that made the book better.

Ronald R. Peterson, a 44-year Johns Hopkins veteran who retired in 2017 as president of the Johns Hopkins Health System and executive vice president of Johns Hopkins Medicine, was a particularly thoughtful reader of early drafts, providing the perspective on Johns Hopkins Medicine that only he possesses.

I am so grateful for the help and support that my colleagues so generously lavished on me and on this project. In particular, editor **Sue De Pasquale** was with me every step of the way, providing brilliant edits to every page. We sometimes spoke several times a day, and her excellent design sensibility, keen attention to word choices, and even-keeled good nature all made this book stronger, and made working with her a joy.

Maxwell Boam, director of identity management and graphic design for Johns Hopkins Medicine, took on the Herculean task of bringing these pages to life with photos, illustrations, charts and designs that beautifully support the words. Alec Gibson, junior graphic designer in the marketing and communications department, provided able assistance.

Michael Keating, director of communications design, was our project lead, a steady presence, who was always willing to step in as sounding board or problem-solver.

I'd also like to thank my wonderful marketing and communications colleagues, who took on extra work so I could focus on *Leading the Change*. They include **Lisa Eddy**, **Judy Minkove**, **Marc Shapiro**, **Linell Smith** and **Patrick Smith**.

Copy editors **Karen Blum**, **Victoria Banks** and **Justin Kovalsky** brought their eagle eyes for typos and considerable love of language to the task of creating a consistent voice and style throughout. Production manager **Maria Patterson** worked with many vendors to obtain and compare costs and schedules. And indexer **Devon Thomas** made everything easy to find.

I'm afraid to list all the writers, editors, photographers and illustrators who provided the documents and images that I relied on so heavily, because I don't want to leave anyone out. Please check the endnotes and credits to see all the people who contributed their talent and hard work to this book.

And last but very much not least, I wish to thank my wonderful family — husband **David**, children **Veronica** and **Sam**, and parents **Barbara** and **Gary**.

William Osler, the famed physician who co-founded The Johns Hopkins Hospital, was known for many things, including his highly quotable observations about medicine and life. "The whole art of medicine is in observation," he said in 1903, speaking to an Academy of Medicine audience.

I hope with this book we have together observed and put on the printed page the story of a remarkable institution during a remarkable decade.

Credits

Photographers

Keith Weller

Norm Barker

Max Boam

Larry Canner

Jon Christofersen

Mike Ciesielski

General Dynamics

Alec Gibson

Neil Grauer

Chris Hartlove

Will Kirk

Maureen Martin

Nigel Parry

Justin Tsucalas

Illustrators

Davide Bonazzi

Francesco Bongiorni

Taylor Callery

Ben Giles

Carole Henaff

Daniel Hertzberg

Gwenda Kaczor

Phil Wrigglesworth

Index

Page numbers in *italics* indicate images.

ferrets, cloning, 11
Ficke, James, 216, *216,* 217
Finance Department, JHHS, 101
Finch, Spencer, 13
Fine, Derek, 127
Fingerhood, Michael, 20, *20*
Fishman, Elliot, 158
Fivush, Barbara, 111, *111*
Fleming, Gloria, *110*
Fletcher, Devona, *123*
Flores-Miller, Alejandra, 230
Floyd, George, 223
food and beverages: community meals, 123; and COVID-19, 221, 222, 226; and employee well-being, 90–91, 221, 222, 226
Ford, Daniel E., 15, 66, *67,* 115–16
Forward. For All of Us. campaign, 183
Fox, Michael J., 173, *173*
Frank L. Coulson Award for Clinical Excellence, 82
Freischlag, Julie, 89, *89,* 109, *111*
Friedman, Keith, *66*
Fuller, Burton, 201, 202, *202*
Furukawa, David, *91*

G

Galiatsatos, Panagis, 25–26, *26, 102, 225*
Gama, Lucio, 102
Gamble, Robert E., 107
Garcia, Alejandro, 107, *107*
Gardner, Lauren, *195,* 195–96
Garibaldi, Brian, 194, 197, *197, 213,* 214, 215, 218–20, 234
Garrett, Mary Elizabeth, 176
Gather Baltimore, 123
Geisinger Health System, 71
Gender Identity Clinic, 131. *See also* Johns Hopkins Center for Transgender Health
Genes to Society curriculum, 1, 24–25, 39
Genetic Medicine, Department of, 163, 165–66
Gey, George Otto, 115, *115*
Al-Ghoson, Huda, *39*
Giardiello, Francis, 70
Gilbert, Paul, *120*
Gilligan, Linda, 40
Gilman, Daniel Coit, 42–43
Gocke, Christopher, 199
Golden, Sherita, *102, 225, 227*; and anti-bias training, 226; career, 227; and COVID-19, 107, 223, 224, 228; and slave ownership by Hopkins, 177; and Vaccine Equity Task Force, 228
Gonzalez, Reina, 23, *23*
Goodman, Barak, 78
Goodspeed, Elizabeth, 24
Go Team, *See* The Johns Hopkins Go Team
Gould, Neda, 94, *94*
Gragnolati, Brian, 66, *66,* 68
Grand Rounds, 68, 77
Gray, Freddie Carlos, Jr., 99, 107, 227
Green, Angela, 151
Green, Ariel, 27
Greenberg, Erwin, 129, *129*
Greenberg, Stephanie Cooper, 129, *129*
Greenberg Bladder Cancer Institute, *See* Johns Hopkins Greenberg Bladder Cancer Institute

Greene, Jeremy, 175, *175*
Greene, Peter, 17, 203, *203,* 204
Green Spring Station: colonoscopy research, 70; expansions, 55, 142, 148, 149; Pavilion III, 5, 147, 172; site acquisition, 147
Gregory, Maria E., *66*
Greider, Carol, 8, 164, 167, *167*
Griffin, Jack, 33, *33*
Griffiths, Roland, 161
Grossi, Rich, 15, 54, *54*
G. Thomas Miller Coaching Program, 82

H

Hackam, David, *120,* 120–21
Hall, Kez, *99*
Haller, Karen, 14, 43, *43*
Hand, Stephon, *28*
Hanisch, Deanna, 203, 208
Hankin, Michael D., 101, *101*
Hanna, Marie, 134, *134*
Harrison, Delegate Hattie, 31
Harry and Jeanette Weinberg Early Childhood Center, 32
Harvey, Abner McGehee, 84, *84*
Harvey/Nelson buildings, 83–84
Hasselfeld, Brian, *207,* 207–9
HEAL (HEIC, Epic, Ambulatory, Labs), 203–5, 206
health disparities: and Cooper, 112; and COVID-19, 196, 217, 223–31; and medical school, 24, 203; and Weisfeldt, 77. *See also* diversity, equity and inclusion
Health Information Technology for Economic and Clinical Health (HITECH) Act, 17
Health Services Cost Review Commission (HSCRC), 11, 53
Healthy at Hopkins, 90–92, 221
Healthy Baltimore 2020, 100
Healthy Beverage Initiative, 90–91
HEBCAC (Historic East Baltimore Community Action Coalition Inc.), 30
HEIC (Hospital Epidemiology and Infection Control), 191, 194, 199, 202
HeLa cells, 31, 115, 116
Hellmann, David, 80, 81, *81,* 82, 83
Henderson, Macey, *110*
Henderson-Hopkins school, 31–32
Henrietta Lacks Building, 116
Henrietta Lacks High School Day, 116
Henrietta Lacks Memorial Lectures, 115
hepatitis C and transplantation, 99, 129
Hernandez, Raquel, 119, *119*
Heron, Todd, *205*
Herzog, William R., Jr., *66*
Heyssel, Robert M., 178
Hicks, Caitlin, *110*
Higgins, Robert, 89, *89, 109,* 109–10, *110*
Highmark Inc., 68–69
High Value Care Committee, 51, 52
high-value health care initiatives, 49–53
High Value Practice Academic Alliance (HVPAA), 52
High Value Practice Alliance, 51
High Value Practice Research Symposium, 52
Hill, Peter, 133, *133,* 200
Himmelfarb, Cheryl Dennison, 71

Historic East Baltimore Community Action Coalition Inc. (HEBCAC), 30
HITECH (Health Information Technology for Economic and Clinical Health) Act, 17
HIV/AIDS: Latino Outreach Program, 230; organ donation and transplantation, *ii,* 99, 127–28
HIV Organ Policy Equity (HOPE) Act, 127
Hodgson, David, 141, 143
Hogan, Larry, 103, 196, 204
Hogans, Beth, 134, *134*
Hoke, Shawn, *209*
HOLA (Hopkins Organization for Latino Awareness), 27
homelessness, 102, 104, 114, 225
Hoover, Sarah, 170
HOPE (HIV Organ Policy Equity) Act, 127
Hopkins (2008), 78
Hopkins, Johns, 43, 177
Hopkins 24/7 (2000), 78
HopkinsLocal, 100–101
Hopkins Organization for Latino Awareness (HOLA), 27
Hopkins Retrospective, 177
Hosey, Megan, 212
Houston, Thomas, *105*
Howard County General Hospital: acquisition of, 142; admissions payments, 53; and CAPRES, 66; and COVID-19, 205, 206, 210, 213; and employee wellness, 92; leadership at, 86; merger, 9
Howard Hughes Medical Institute, 183
Howell, Eric, 217
Hruban, Ralph, 199
HSCRC (Health Services Cost Review Commission), 11, 53
Hueppchen, Nancy, *174,* 175, 203
Huganir, Richard, 33, *33,* 34
Hughes, Helen, *207*
Human Genome Project, 25
human proteome, 35
Hurdon, Elizabeth, *110*
Hurricane Katrina, 195
Hurricane Sandy, 5, 15
HVPAA (High Value Practice Academic Alliance), 52

I

IAM Lab (International Arts and Mind Lab), 33
Iarrobino, Nicole, 228
ICTR (Johns Hopkins Institute for Clinical and Translational Research), 67, 69–71, 115
Incident Command, 195, 199–200
influenza pandemic of 1918, 193
inHealth, 58–59, 61
Innovation 2023, 16, 155
Inova Health System, 70
Institute for Basic Biomedical Sciences, 32–33
Institute for Clinical and Translational Research, *See* Johns Hopkins Institute for Clinical and Translational Research (ICTR)
International Arts and Mind Lab (IAM Lab), 33
Ioannou, JoAnn, 102
iPICS (Prospective Inception Cohort Study), 120
Ishii, Lisa, 148–49, *149,* 200
Israel, Robert, 13